Visual Servoing

Edited by

Rong-Fong Fung

Visual Servoing
Edited by Rong-Fong Fung

Republished by InTech
Janeza Trdine 9, 51000 Rijeka, Croatia

Technical Editor Teodora Smiljanic

Cover Designer Dino Smrekar

Additional hard copies can be obtained from orders@intechopen.com

Visual Servoing, Edited by Rong-Fong Fung
p. cm.
ISBN 978-953-307-095-7

For EU product safety concerns:
IN TECH d.o.o., Prolaz Marije Krucifikse Kozulić 3, 51000 Rijeka, Croatia,
info@intechopen.com or visit our website at intechopen.com.

Meet the editor

Rong-Fong Fung received the M.S. and. Ph.D degrees in Mechanical Engineering from Chung Yuan University and National Taiwan University, Taiwan, in 1982 and 1993, respectively. He was a lecturer, associate professor and professor in the Department of Mechanical Engineering of Chung Yuan University, from 1985 to 2000. He was a visiting scholar of Department of Electrical Engineering of The Ohio State University in1997. He is currently, a professor of Department of Mechanical & Automation Engineering (from Aug. 2000) and Graduate Institute of Electro-Optical Engineering (from Aug. 2004), Dean of College of Engineer (from Aug. 2010), and Chairman of Institute of Engineering Science and Technology (from Aug. 2010), National Kaohsiung First University of Science and Technology (NKFUST), Taiwan. He was Dean of Academic Affairs (Aug. 2004 to July 2007, Aug. 2009 to April 2010, NKFUST) and Chairman of Graduate Institute of Electro-Optical Engineering (Aug. 2004 to July 2005, NKFUST). Professor Fung received the Outstanding Research Award (2005, 2010, NKFUST), Outstanding Industry-Academia Cooperation Award (2010, NKFUST), Outstanding Engineering Professor Award (2006, the Chinese Mechanical Engineering Association, Taiwan), and Outstanding Research Award, 2004, NSC, Taiwan). His main research interest includes dynamics and control of opto-mechatronic systems, and equipment development of automated optical inspection (AOI) for LED and vehicle lamps.

Preface

In the machine vision field, the theory has been studied for many decades. Fortunately, the computer technique is also developed rapidly simultaneously. Recently, these theories of machine vision have been realized practically in variety applications. The machine vision system consists of the optics, electronics, machinery and the computer information technology systematically. The technique is applied widely industrially including vision servoing trajectory motion control, optical measurement and automatic examination, pattern identification and system monitoring and so on. The advantages of the vision system are the non-contact measurement, versatility, cost effectiveness, and practicality. Therefore, the machine vision technology can be investigated and studied enthusiastically to improve the industrial development and human engineering.

In the field of machine vision, there are many technique books about introducing the fundamental theory of vision. But there is not a book about how to employ the vision theory in the market conditions for students or researchers who want to realize the technique of machine vision. It is meaningful to employ the vision theory to the practical application, and the vision theory can also be employed originally by different field researchers.

I am pleasant that the book consists of 10 chapters by different fields about vision applications. The authors in chapters are excellent in their research fields. It is honored to collect the 10 chapters that depict the multiplicity of the vision theory by the authors. For the readers, you can select some kinds of applications you are interesting in this book, and to study the detailed contents in each chapter. This book collects the main studies about machine vision currently in the world, and has a powerful persuasion in the applications employed in the machine vision. The contents, which demonstrate that the machine vision theory, are realized in different field. For the beginner, it is easy to understand the development in the vision servoing. For engineer, professor and researcher, they can study and learn the chapters, and then employ another application method.

The goal of this book is to introduce the visional application by excellent researchers in the world currently and offer the knowledge that can also be applied to another field widely. This present book provides the diversified applications to visual technique. In the content of this book, there are two main parts that consist of vision servoing control (chapters 1~5) and vision servoing application (chapters 6~10).

The completion of this book came at the expense of all authors' long-time effort. I am indebted to Lélio R. Soares Jr, Victor H. Casanova Alcalde, Nils T Siebel, Dennis Peters, Gerald Sommer, Xinhan Huang, Xiangjin Zeng, Min Wang, Rares Stanciu, Paul Oh, Kun-Yung Chen, Rafael Herrejon Mendoza, Shingo Kagami, Koichi Hashimoto, Yuta Yoshihata, Kei Watanabe, Yasushi Iwatani, Koichi Hashimoto, Mika Karaila, Pascual Campoy, Ivan F. Mondragón, Mariko Nakano-Miyatake and Hector Perez-Meana heartily. Moreover, I would be happy to receive any comments, which would be helpful to improve this book.

Editor

Professor Rong-Fong Fung

Department of Mechanical and Automation Engineering,
National Kaohsiung First University of Science and Technology,
1 University Road, Yenchau, Kaohsiung County 824,
Taiwan

rffung@ccms.nkfust.edu.tw

Contents

1

A Modeling and Simulation Platform for Robot Kinematics aiming Visual Servo Control

Lélio R. Soares Jr. and Victor H. Casanova Alcalde
Electrical Engineering Department, University of Brasilia
Brazil

1. Introduction

A robotic system is a mechanical structure built from rigid links connected by flexible joints. The arrangement of links and joints (robot architecture) depends on the task the robot was designed to perform. The robot links have then different shapes and the joints can be of revolute (rotational motion) or prismatic (translation motion) nature. These robots, as described, perform task on an open-loop control scheme, i.e. there is not feedback from the environment (robot workspace) thus it will not notice changes in the workspace. As an attempt to establish a closed-loop control scheme a computer-based vision systems is introduced to detect workspace changes and also to allow guiding the robot (Hutchinson et al., 1996).

At the University of Brasilia to cope with the study and teaching of robotics an educational robotic workstation was built around the Rhino XR4 robot (Soares & Casanova Alcalde, 2006). To implement a vision-guided robot a video camera was installed and integrated to the robot control system. As an alternative for dealing with the real system and for teaching purposes a simulation platform was devised within the Matlab environment (Soares & Casanova Alcalde, 2006). The platform was called *RobSim* and it is based on assembling elementary units (primitives) which represent the robot links, being the joints represented by the motion they perform. This simulation and developing platform then evolved and now it includes robot visual servo control being presented in this work. Within *RobSim* platform control algorithms can be developed for the vision-guided robot to perform tasks before implementing them on the real system.

Simulation tools for either conventional robotic systems (Legnani, 2005; Corke, 1996) and for vision-based systems (Cervera, 2003) do exist, this work presents a unified environment for both systems. The developed simulation tools were assembled as a laboratory platform, where robotic and vision-based algorithms share similar data structures and block building methodologies. Moreover, this platform was developed mainly for educational purposes; later on it was found it can be used for research and design of robotic systems. The graphical presentation is as simple as possible, but allowing an insight and visualization of parts and motions.

The chapter is organized as follows; initially the *RobSim* basic mounting blocks, the primitives, are defined and described. Then, the *RobSim* developed Matlab functions for initialization, motion, computer display and image acquisition are presented. Following, the modeling and simulation capabilities *RobSim* platform offers are presented together with applications to fixed and mobile robots. Further on, vision-based control schemes are briefly discussed. Finally, implementation of visual-based control schemes applied to a robotic workstation consisting of a Rhino XR4 robot and a computer vision system is considered. Image- and position-based visual servoing schemes are implemented.

2. *RobSim* – a modeling and simulation platform for robotic systems

In order to model and simulate the kinematics of robotic systems a software platform named *RobSim* was developed. Three types of basic elements were defined to assembly a model for a vision-guided robotic system: block, wheel and camera. Being basic elements they will be called *primitives*. They will be sufficient to assembly a simulation model for robotic manipulators and robotic vehicles guided by a computer vision system.

2.1 Block primitive

The block primitive is defined as a regular polyhedron with rectangular faces. The faces meet along an edge and three of these intersect orthogonally at a vertex. A block primitive consists then of six faces, twelve edges and eight vertexes. Figure 1 shows a block primitive with its allocated coordinate frame $\{X_b,Y_b,Z_b\}$. The frame orientation is assigned as follows, the X_b-axis along the block length (L), the Y_b-axis along the block width (*W*) and the Z_b-axis along the block height (*H*). A general graphical reference coordinate frame $\{X_g,Y_g,Z_g\}$ is also shown in Figure 1, it indicates the block viewing angle for displaying purposes.

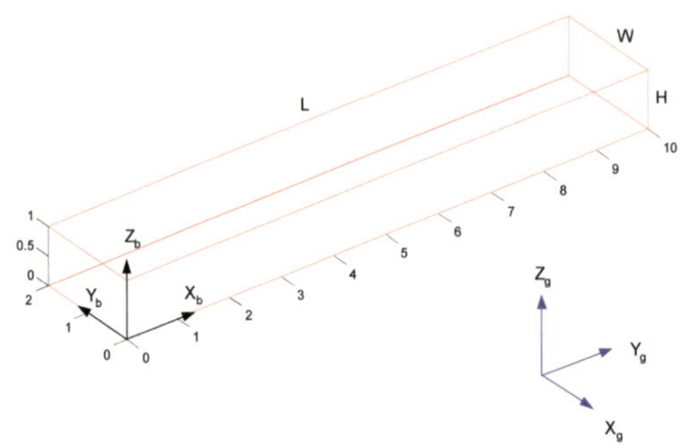

Fig. 1. A Block Primitive

A block primitive will be geometrically defined by nine components: a) eight vectors, each one corresponding to the *3D* coordinates of its vertexes; and b) a character identifying the assigned color to the line edges.

2.2 Wheel primitive

For simulating wheeled mobile robots a wheel primitive is defined. The wheel primitive is defined as two circles of equal radius assembled parallel to each other at a certain distance. The wheel rotation axis passes through the centers of both circles. Figure 2 shows a wheel primitive with its allocated coordinate frame. The wheel primitive coordinate frame

$\{X_b, Y_b, Z_b\}$ is attached to the wheel primitive, being its origin fixed at the middle of the internal line between the circle centers. The Z_b-axis coordinate is fixed along the rotation axis, the X_b-axis along the initial rotation angle (0°).

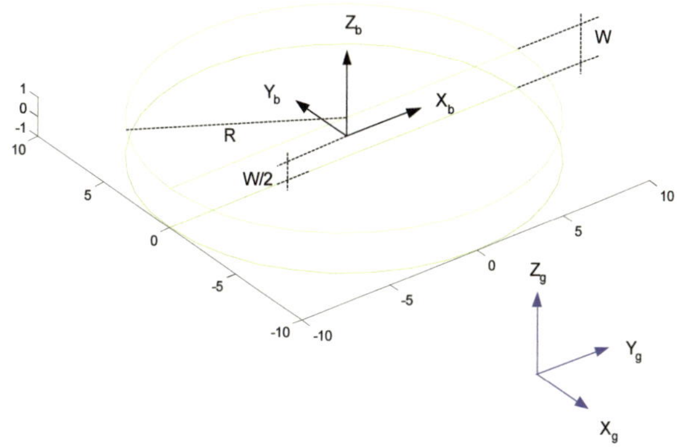

Fig. 2. A Wheel Primitive

A wheel primitive will be geometrically defined by four components: a) the circle radius (R); b) the distance between the circle centers (W); c) the number of points defining both circumferences; and d) the color identifying character.

2.3 Camera primitive

For vision-guided robotic systems, manipulators or mobile robots, video cameras are required. Then, a camera primitive was developed from a modified block primitive. It is a small regular polyhedron with rectangular faces but having a larger opening on one extreme representing the light capturing entrance. Figure 3 shows a camera primitive with its coordinate frame $\{X_b, Y_b, Z_b\}$. The camera primitive coordinate frame is attached to the opposite face, where the image is formed. The coordinate frame center is fixed at the center of this rectangular face, the Z_b-axis along the camera length (L), the X_b-axis along the camera height (H) and the Y_b-axis along the camera width (W). This orientation follows the computer vision convention, so the Z_b-axis coincides with the camera optical axis.
Due to its particular function, a camera primitive will be defined by three groups of components: a) twelve vectors to characterize its vertexes spatial coordinates; b) a color identifying character; and c) the camera intrinsic parameters (subsection 3.4).

3. *RobSim* processing functions

Within the Matlab environment *RobSim* functions for processing the primitives were developed. These functions allow: defining the primitives (initialization functions); moving the primitives (moving functions); and displaying the primitives (displaying functions). An image acquisition function to simulate image capture was also developed.

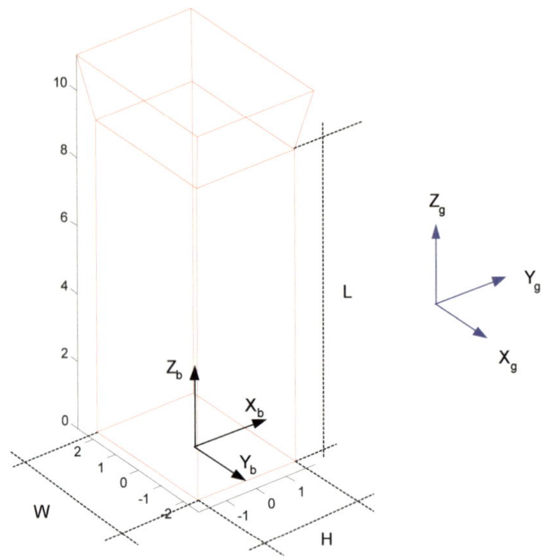

Fig. 3. A Camera Primitive

3.1 Primitives initialization functions

The primitives have to be introduced to the Matlab environment. For that, Matlab structure-type variables (*struct*) are used for initialization of the primitives being the dimensions expressed in centimeters.

Initializing a block primitive - The function to initialize the block primitive *struct* variable has the following syntax:

- *blk=init_block(L,W,H,color)*
  where *L, W, H* and *color* are respectively the length, width, height and line color of the block primitive.

Initializing a wheel primitive - The function to initialize the wheel primitive *struct* variable is

- *circ=init_circ(R,W,n,color)*
  where *R, W, n* and *color* are respectively the radius, width, number of circumference points and line color of the wheel primitive.

Initializing a camera primitive - The function to initialize the camera primitive *struct* variable is

- *cam= init_cam(L,W,H, f,px,py,alpha,u0,v0,color)*
  where *L,W,H,* and *color* are respectively the length, width, height and line color of the camera primitive. The parameters *f, px, py, alpha, u0* and *v0* are the camera intrinsic parameters (Chaumette & Hutchinson, 2006). These camera intrinsic parameters will be further discussed in subsection 3.4.

3.2 Primitives moving functions

Once defined the primitives within Matlab, other functions are necessary for moving the primitives as they simulate the different moving robotic links. For moving the primitives all of its characteristic points have to be moved. A homogeneous transformation (Schilling, 1990) is then applied upon the vectors which define those characteristic points.

Moving a block primitive - Given a struct variable *blk_i* representing an initial block primitive pose (position and orientation), a new variable *blk_o* will represent the final primitive pose as a result of a moving function. For a block primitive the developed moving function is

- *blk_o=move_block(blk_i,***R***,***t***)*

  where **R** and **t** are respectively the rotation matrix (3×3) and the translation vector (3×1) of the homogeneous transformation representing the executed motion.

Moving a wheel primitive - Given *circ_i* representing an initial wheel pose, after applying a moving function the wheel primitive will assume a final pose *circ_o*. For this action the developed moving function is

- *circ_o=move_circ(circ_i,***R***,***t***)*

  where **R** and **t** have the same meaning as the block primitive.

Moving a camera primitive – Similarly, for an initial pose of the camera primitive *cam_i*, a final pose *cam_o* is achieved after a moving function. A developed camera primitive moving function is

- *cam_o=move_cam(cam_i,***R***,***t***)*

  where **R** and **t** have the same meaning as for the block and wheel primitives motion.

3.3 Primitives displaying functions

For displaying primitives specific functions were developed around the Matlab built-in *plot3* function. As the vertexes define the geometry of primitives, for displaying purposes straight lines were drawn to join the vertexes. Thus the displayed primitives look like a *wire-frame* model for solid objects. The graphic displaying functions developed for primitives are

- *plot_block(block)*
- *plot_circ(circ)*
- *plot_cam(cam)*

for the block, wheel and camera primitives respectively. The function argument in the three cases is precisely the *struct* variable that represents the primitive.

3.4 Image acquisition function

A computer-based vision system for robotic systems demands video cameras. A camera coordinate frame is attached to the camera, being ${}^{0}\mathbf{T}_{c}$ the homogeneous transformation matrix relating the camera position (**t**) and orientation (**R**) referred to the base coordinate frame. **R** and **t** constitute the camera extrinsic parameters which together with the intrinsic parameters $\{f, px, py, \alpha, u0, v0\}$ are used to setting up the camera primitive. These intrinsic parameters arise from the perspective projection model (Hutchinson et al., 1996) adopted for the camera and are shown graphically in Figure 4.

An image acquisition function *point_view* was developed to simulate an image point capture and its syntax is

- *pimag=point_view*($\mathbf{p}_{3D}$,**K***i*,${}^{0}\mathbf{T}_{c}$)

  where $\mathbf{p}_{3D}$ is a vector representing the *3D* position of a point in the camera *field-of-view* (FOV), relative to base frame; **K***i* is the matrix of the camera intrinsic parameters; and *pimag* will return the $\mathbf{p}_{imag}$, the 2D position of the image point measured in *pixels*. $\mathbf{K}i$ is arranged as follows

$$\mathbf{K}_i = \begin{bmatrix} -\dfrac{f}{p_x.\cos\alpha} & 0 & u_0 \\ \dfrac{f.\tan\alpha}{p_y} & \dfrac{f}{p_y} & v_0 \\ 0 & 0 & 1 \end{bmatrix} \quad (1)$$

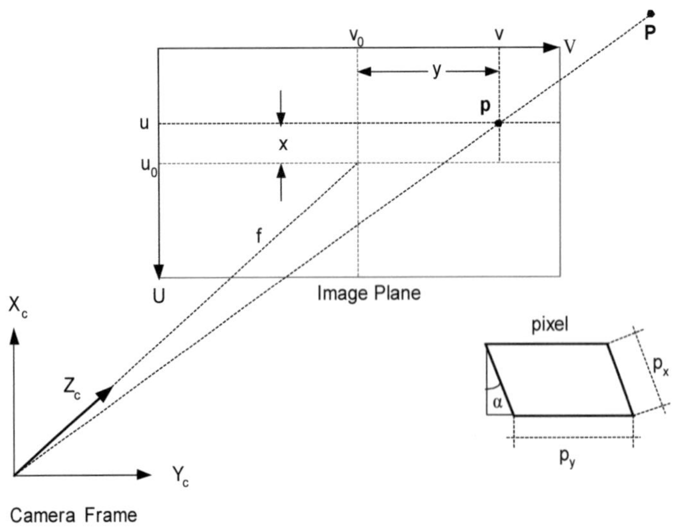

Fig. 4. Perspective Projection Model for the Camera

4. Modeling and simulation of robotic systems kinematics using *RobSim*

A robotic manipulator or vehicle can be considered as a chain of rigid links interconnected by either revolute or prismatic joints. The proposed modeling and simulation tool *RobSim* associates a primitive to a robotic link. By programming the primitive initialization, moving and displaying functions together with Matlab built-in functions it is possible to simulate the kinematical model of any robotic structure. Thus, from these basic structures, the primitives, the kinematics of complex robotic systems can be simulated for analysis and design purposes.

Within *RobSim* the robot joints are not graphically represented or displayed, being their nature (prismatic- or revolution-type) revealed as the motion progresses. For this reason, different colors must be assigned for primitives representing consecutive links.

As primitives are represented by a structure-type variable, the whole set of assembled primitives representing the robot system will be a higher-level structure-type variable.

The kinematical model of a robotic system is determined by applying the Denavit-Hartenberg (DH) algorithm (Schilling, 1990). Transformations between successive links *(k-1)* and (*k*) are characterized by homogenous transformation matrixes like

$$^{(k-1)}\mathbf{T}_k = \begin{bmatrix} \mathbf{R}_{3\times3} & \mathbf{t}_{3\times1} \\ \mathbf{0}_{1\times3} & 1 \end{bmatrix} \tag{2}$$

In which $\mathbf{R}_{3\times3}$ is the rotation matrix representing the relative orientation between frames and $\mathbf{t}_{3\times1}$ is the translation vector representing the relative position between the frames origins. By using DH kinematical parameters $\{\theta, d, a, \alpha\}$, Equation (2) can be written as

$$^{(k-1)}\mathbf{T}_k = \begin{bmatrix} C\theta_k & -C\alpha_k S\theta_k & S\alpha_k S\theta_k & a_k C\theta_k \\ S\theta_k & C\alpha_k C\theta_k & -S\alpha_k C\theta_k & a_k S\theta_k \\ 0 & S\alpha_k & C\alpha_k & d_k \\ 0 & 0 & 0 & 1 \end{bmatrix} \tag{3}$$

In which for rotational joints, θ is the joint variable and C and S represent the cosine and sine functions respectively. To illustrate DH modeling and link-primitive assignment correspondences, Figure 5 shows the coordinate frame assignment for two robotic links. For these links, Figure 6 shows the assembling of primitives

The kinematical model of a particular robot of n joints will be the homogeneous transformation relating the tool-tip coordinate frame (frame n) to the base coordinate frame (frame 0) obtained as

$$^{0}\mathbf{T}_n = {}^{0}\mathbf{T}_1 . {}^{1}\mathbf{T}_2 \cdots {}^{k-1}\mathbf{T}_k \cdots {}^{n-1}\mathbf{T}_n \tag{4}$$

An additional transformation will be necessary for displaying purposes relating the base coordinate frame to the displaying frame ${}^{g}\mathbf{T}_0$ (Fig. 6).

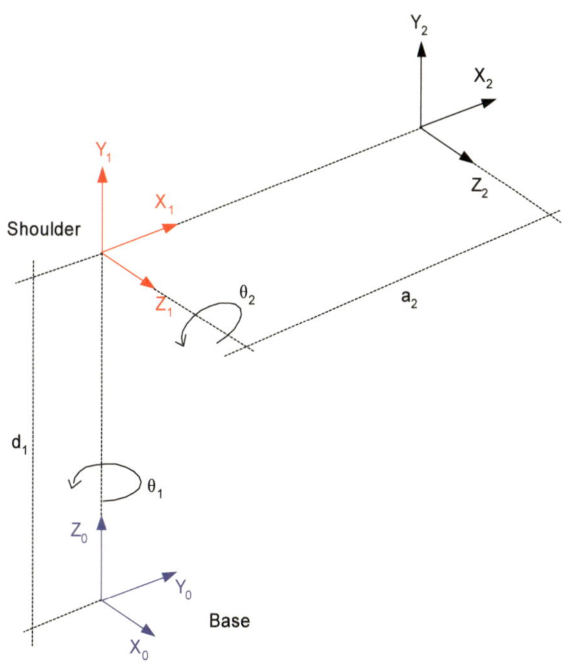

Fig. 5. DH Link Coordinates for two robotic links

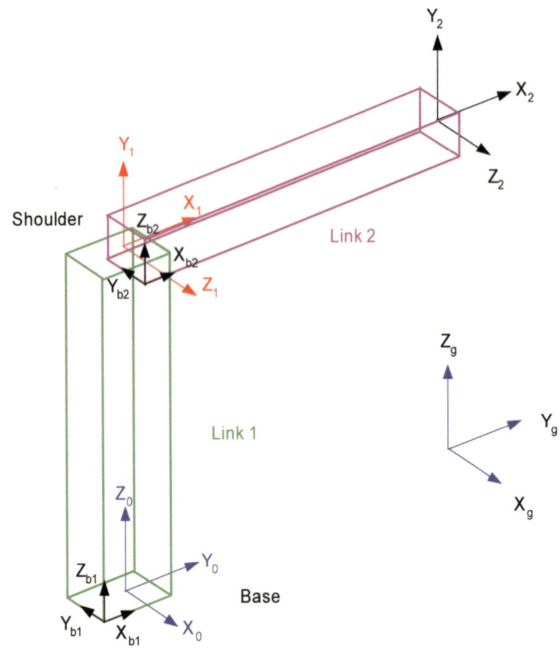

Fig. 6. Assembling Primitives for two robotic links

4.1 *RobSim* modeling and simulation procedure

The different stages to assembly a *RobSim* simulation model for a given vision-guided robotic system are:

1. Allocating link coordinates and determining the kinematical parameters for the robotic system according to the Denavit-Hartenberg (DH) algorithm;
2. Representing the different robot links by the block, wheel or camera primitives as applied;
3. Assembling the chosen primitives through their coordinates as referred to the link coordinates determined by the DH algorithm;
4. Determining the primitives configuration referred to the robot base coordinates;
5. Developing the robotic system initialization as a Matlab *struct* variable, whose variable fields are the individual primitives *struct* representations;
6. Developing the moving and displaying functions for the robotic system from the individual primitives moving and displaying functions;
7. Generating trajectories and executing tasks by controlling the joint variables of the simulation model.

4.2 Simulation of robotic systems

Initially a *RobSim* model for the Rhino XR4 robot will be developed and a simulation test executed. The Rhino XR4, shown in Fig. 7, is an educational desktop robot, classified as a five-axis electric-drive articulated coordinates robot. Around this robot an educational robotic workstation (Soares & Casanova Alcalde, 2006) was built.

Applying the *RobSim* modeling and simulation procedure, link coordinates were allocated and the kinematical parameters for the Rhino XR4 robot obtained, as shown in Figure 8.

Fig. 7. The Rhino XR4 Educational Robot

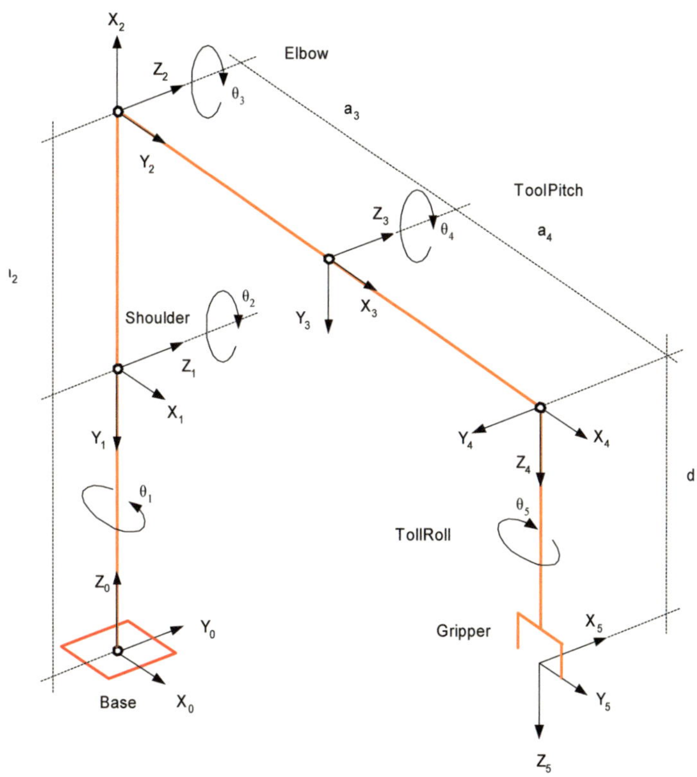

Fig. 8. Kinematical Model for the Rhino XR4 Robot

Only block-type primitives were used to simulate each one of the robot links. For the robot tool three small block primitives were considered to allow simulating the tool opening/closure mechanism. Figure 9 shows the *RobSim* model for the Rhino XR4 at the home position and orientation (initial configuration).

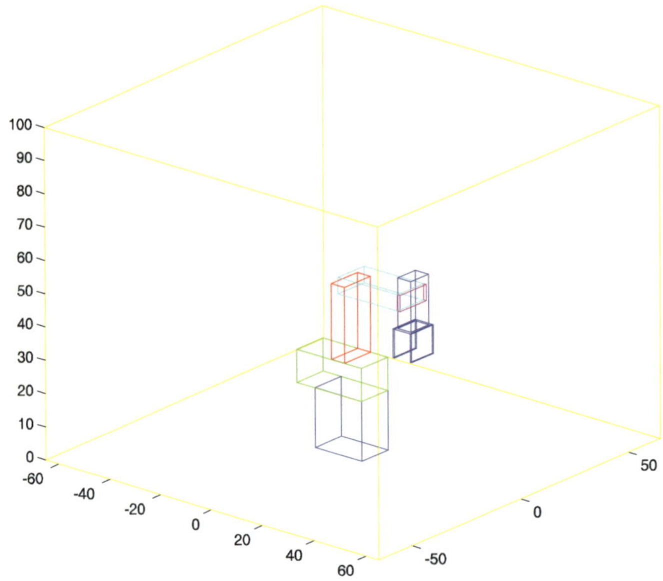

Fig. 9. *RobSim* Model for the Rhino XR4 Robot - Initial Configuration

Figure 10 shows the robot after executing a moving function towards a final configuration.

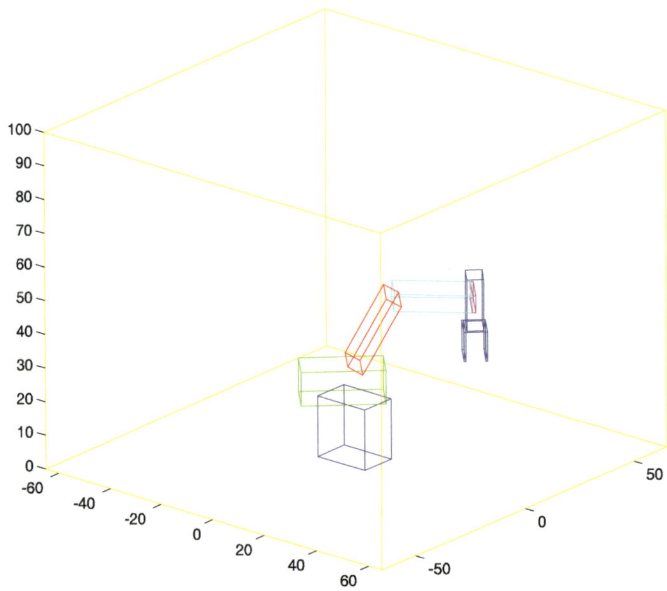

Fig. 10. *RobSim* Model for the Rhino XR4 Robot - Final Configuration

As part of a research project, prototypes of an inspection mobile robot were devised. The *RobSim* platform was particularly suitable to analyze the robots kinematics. The envisaged mobile robot will travel along suspended cables and will execute vision-guided maneuvers in order to overcome obstacles. Figures 11 and 12 show *RobSim* models of two prototypes.

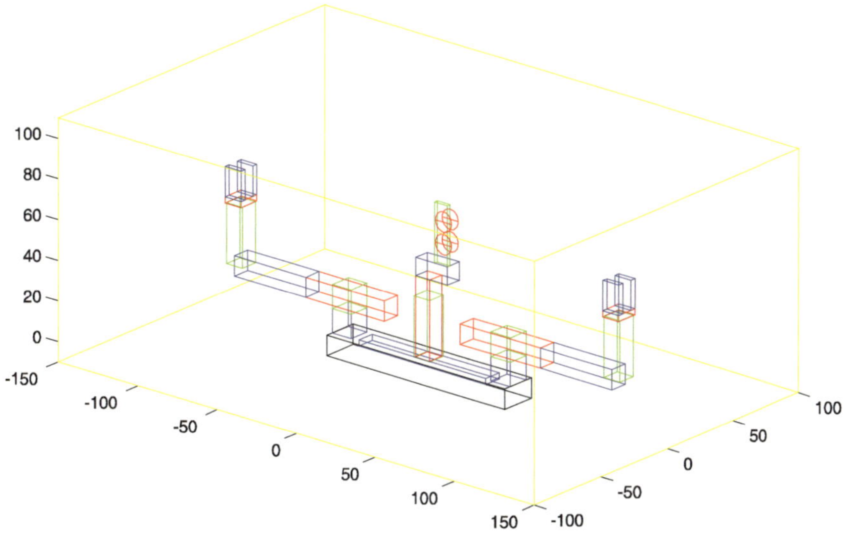

Fig. 11. *RobSim* Model of an inspection mobile robot (Soares & Casanova Alcalde, 2008)

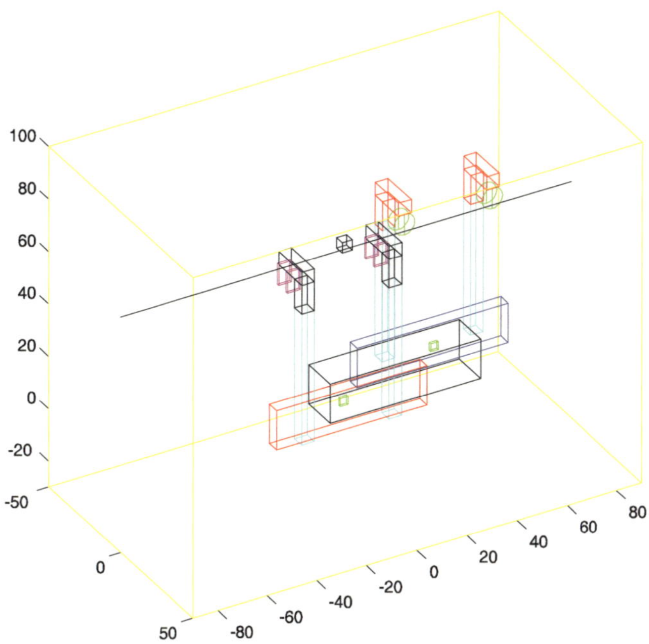

Fig. 12. *RobSim* model of another inspection mobile robot (Soares & Casanova Alcalde, 2008)

5. Visual servo control of robotic systems

Visual servo control of robotic systems uses visual data to implement a feedback control loop to guide the robot in performing a certain task. Therefore the chosen machine vision strategy has to be considered into the robotic system dynamics. The camera for image capture can be mounted on the robot end-effector, or fixed at a certain place to observe the robot workspace. The first approach is called an *eye-in-hand* configuration and the second, an *eye-to-hand* configuration. Other possibilities combining schemes are also possible (Chaumette & Hutchinson, 2007). A variant of the *eye-to-hand* configuration consists on mounting the camera on another robot or on a pan/tilt structure in order to improve the viewing angle. A single camera arrangement for gathering visual data lacks information about depth measurements. Algorithms for position and orientation (pose) estimation could then be introduced or two-cameras can be used to implement a stereo-vision scheme to calculate depth information. This section discusses briefly the main visual-based control schemes. First, a characterization of the control error for a visual servo control strategy is discussed. Then, the position- and the image-based visual servo control schemes are discussed. Some considerations about the system stability are finally pointed out.

5.1 Characterization of the control error for visual servo control schemes

In visual servo control schemes the image coordinates of points of interest are captured. These measurements constitute a set of image measurements represented by $\mathbf{m}(t)$. From these measurements an actual *visual features vector* $\mathbf{s}$ is calculated to represent the actual value of k visual features. It is defined as $\mathbf{s}(\mathbf{m}(t),\mathbf{a})$ (Chaumette & Hutchinson, 2006), where $\mathbf{a}$ is a set of parameters that represent additional knowledge about the system. Vector $\mathbf{a}$ can be an approximation of the camera intrinsic parameters or *3D* models of objects being observed. The desired *visual features vector* is represented by $\mathbf{s}^*$, usually constant, being changes in $\mathbf{s}$ dependent only on camera motion. The objective of the visual servo control is therefore to minimize a *visual features error vector* $\mathbf{e}(t)$ defined by

$$\mathbf{e}(t) = \mathbf{s}(\mathbf{m}(t),\mathbf{a}) - \mathbf{s}^* \tag{5}$$

The visual servo control schemes depend on how the visual features vector $\mathbf{s}$ is determined, as it will be seen in the following subsections. To minimize the visual features error vector $\mathbf{e}(t)$ (Equation 5) a common approach is to implement a velocity controller. Defining the spatial velocity of the camera $\mathbf{V}_c = [\mathbf{v}_c\ \boldsymbol{\Omega}_c]^T$, being $\mathbf{v}_c$ the instantaneous linear velocity of the origin of the camera frame and $\boldsymbol{\Omega}_c$ the instantaneous angular velocity of the camera coordinate frame. A relation is then established between the time derivative of $\mathbf{s}$ and $\mathbf{V}_c$

$$\dot{\mathbf{s}} = \mathbf{L}_s.\mathbf{V}_c \tag{6}$$

Where $\mathbf{L}_s$ is a k×6 matrix related to $\mathbf{s}$ called the image interaction matrix or also a *feature Jacobian*. Assuming a constant $\mathbf{s}^*$ as usual, and using Equations (5) and (6) results in

$$\dot{\mathbf{e}} = \mathbf{L}_s.\mathbf{V}_c \tag{7}$$

A simple strategy could be adopted, for example, an exponential decay of the error ($\dot{\mathbf{e}} = -\lambda.\mathbf{e}$) for a certain λ>0. Then using Equation (7) and the Moore-Penrose pseudo-inverse matrix $\mathbf{L}_s^+$, $\mathbf{V}_c$ the input of the robot velocity controller will be given by

$$\mathbf{V}_c = -\lambda.\mathbf{L}_s^+.\mathbf{e} \tag{8}$$

For a full rank $\mathbf{L}_s$, the pseudo-inverse will be $\mathbf{L}_s^+ = (\mathbf{L}_s^T.\mathbf{L}_s).\mathbf{L}_s^T$ and $\|\mathbf{V}_c\|$ and $\|\dot{\mathbf{e}} - \lambda.\mathbf{L}_s.\mathbf{L}_s^T.\mathbf{e}\|$ will turn to be minimal. For a square matrix $\mathbf{L}_s$, Equation (8) would be $\mathbf{V}_c = -\lambda.\mathbf{L}_s^{-1}.\mathbf{e}$. As in

practice it is impossible to know $\mathbf{L}_s$ and $\mathbf{L}_s^+$, an approximation or estimative, for the pseudo-inverse must be determined, this approximation will be denoted as $\hat{\mathbf{L}}_s^+$.
As mentioned, depending upon the way the visual features vector **s** is established, different visual servoing schemes are possible. Two schemes are considered: a) the image-based visual servo control (IBVS); and b) the position-based visual servo control (PBVS).

5.2 Image-Based Visual Servo control scheme (IBVS)

In this scheme the image features to be determined can be: image-plane coordinates of points of interest, regions of interest of the image, parameters that define straight lines over the image, etc. From these features a visual features vector $\mathbf{s}(\mathbf{m}(t),\mathbf{a})$ is established. Considering the simplest situation, the image measurements vector $\mathbf{m}(t)$ consists of the pixel coordinates of the set of image points of interest. Finally, vector **a** consists of the installed camera intrinsic parameters. In this situation the interaction matrix $\mathbf{L_s}$ can be easily determined. As shown in Figure 4, for a *3D* point ${}^{S_c}\mathbf{P}=[X\ Y\ Z]^T$ referred to S_c, the camera coordinate frame, its projection onto the image plane will be a *2D* point with coordinates ${}^{S_c}\mathbf{p}=[x\ y\ f]^T$, where f is the camera focal length. From geometrical relation (Figure 4) x and y are given by

$$x=\frac{fX}{Z} \qquad y=\frac{fY}{Z} \tag{9}$$

By using the camera intrinsic parameters (f, p_x, p_y, u_0, v_0, α), u and v, **p** coordinates referred to the image plane, are given by

$$u=u_0-\frac{X}{Z}\frac{f}{p_x.\cos\alpha} \qquad v=v_0+\frac{Y}{Z}\frac{f}{p_y}+\frac{X}{Z}\frac{f.\tan\alpha}{p_y} \tag{10}$$

From Equation (10), given X, Y and Z it is possible to calculate u and v. But in the other way round it is not possible to calculate Z, the depth of **P** relative to the camera frame.
Time derivatives of x and y (velocities) in Equation (9) results in

$$\dot{x}=\frac{\dot{X}-x\dot{Z}}{Z} \qquad \dot{y}=\frac{\dot{Y}-y\dot{Z}}{Z} \tag{11}$$

The *3D* velocity of point **P** (S_c coordinates) is related (Hutchinson et al., 1996) to the camera linear and angular velocities, $\mathbf{V}_c$ and $\mathbf{\Omega}_c$ respectively, as

$$\dot{\mathbf{P}}=-\mathbf{V}_c-\mathbf{\Omega}_c\times\mathbf{P} \tag{12}$$

or

$$\dot{X} = -v_x - \omega_y.Z + \omega_z.Y$$
$$\dot{Y} = -v_y - \omega_z.X + \omega_x.Z \qquad (13)$$
$$\dot{Z} = -v_z - \omega_x.Y + \omega_y.X$$

Substituting Equation (13) into Equation (11) and with $\mathbf{p} = [x\ y]^T$ results in

$$\dot{\mathbf{p}} = \mathbf{L}_p.\mathbf{V}_c \qquad (14)$$

where $\mathbf{L}_p$ is given by

$$\mathbf{L}_p = \begin{bmatrix} -\frac{f}{Z} & 0 & \frac{x}{Z} & \frac{x.y}{f} & -\frac{f^2+x^2}{f} & y \\ 0 & -\frac{f}{Z} & \frac{y}{Z} & \frac{f^2+y^2}{f} & -\frac{x.y}{f} & -x \end{bmatrix} \qquad (15)$$

Matrix $\mathbf{L}_p$ then depends on $\mathbf{P}$ coordinates, on $\mathbf{p}$ coordinates and on the camera intrinsic parameters. Any control scheme using this $\mathbf{L}_p$ must estimate Z, the depth of $\mathbf{P}$ relative to the camera frame. Due to $\mathbf{L}_p$ dimension, to control a six axis robot, a minimum of three points will be necessary, so $k \geq 6$. For a visual features vector $\mathbf{s} = (\mathbf{p}_1, \mathbf{p}_2, \mathbf{p}_3)$ three interaction matrixes $\mathbf{L}_{p1}$, $\mathbf{L}_{p2}$ and $\mathbf{L}_{p3}$ must be stacked. To avoid local minimal solutions more than three points are usually considered. For N points, $\mathbf{L}_p$ will be a $2N\times6$ matrix.
The main advantage of the IBVS schemes results form the fact that the visual features error is defined only in the image domain, not being necessary any parameter or variables estimation in the *3D* space. A disadvantage is lack of information about the scene depth.

5.3 Position-Based Visual Servo control scheme (PBVS)

In position-based visual servo control schemes the visual features vector **s** is defined using the camera pose (position and orientation) relative to a reference coordinate frame. Determining the camera pose from a set of measurements in one image requires the camera intrinsic parameters and the *3D* model of the object being observed, this is the classical *3D* localization problem. As the PBVS approach needs *3D* reconstruction it is prone to fail due to calibration errors. The general PBVS will not be treated here, only a particular case implemented with a robotic manipulator and a stereo-vision device whose simulation in the *RobSim* platform is reported in Section 6.
From *2D* image data captured by each of a two cameras arrangement (stereo vision) it is possible to reconstruct the *3D* pose of an object in the cartesian manipulator workspace. Once the specification of a desired pose of an object handled by the robot end-effector is given, it is possible to define an error between the actual object pose and the desired one. Since this error is specified in the *3D* workspace and the robot joints are actuated in order to cancel it, this kind of procedure can be considered a position-based control scheme.

5.4 Some considerations about stability

Vision-based control systems have non-linear and highly coupled dynamics. For stability analysis Lyapunov direct method can be applied. A particular Lyapunov function would be

$$V = \frac{1}{2}\|e(t)\|^2 \qquad (16)$$

In case of IBVS, by using Equations (7) and (8) the time derivative of $V(t)$ is

$$\dot{V} = \mathbf{e}^T.\dot{\mathbf{e}} = -\lambda.\mathbf{e}^T.\mathbf{L}_s.\hat{\mathbf{L}}_s^+.\mathbf{e} \tag{17}$$

A global asymptotic stability is assured if $\dot{V}$ is positive definite or

$$\mathbf{L}_s.\hat{\mathbf{L}}_s^+ > \mathbf{0} \tag{18}$$

If the number of image features k is equal to the camera DOF and a proper control scheme is implemented, then full rank $\mathbf{L}_s$ and $\hat{\mathbf{L}}_s^+$ matrixes will result and the stability condition (Equation 18) will be assured if a well approximated $\hat{\mathbf{L}}_s^+$ is determined (Chaumette & Hutchinson, 2006). But considering a robot with 6 DOF under a IBVS control, where k is usually greater than 6, then the stability condition could never be assured. The resultant $k{\times}k$ matrix in Equation (18) would have at most a rank of 6, then a nontrivial null space will exist and local minima will result.

6. Visual servo control of a robotic manipulator using *RobSim*

The RobSim platform can help designers to analyze a robotic manipulator under a control scheme. To illustrate this approach a visual servo control scheme is applied to a robotic workstation consisting of the Rhino XR4 robot and a computer vision device. Visual servo control uses visual information to control the pose (position and orientation) of the robot end-effector in order to perform a specified task.

6.1 An image-based visual servoing scheme within *RobSim*

For camera simulation within the RobSim platform it is necessary to set up the camera primitive (Section 3), i.e. introduce the camera intrinsic and extrinsic parameters into its initialization, moving and displaying functions. Using the perspective projection model (Hutchinson et al., 1996) two reference frames are of concern: the camera reference frame, Sc, and the sensor reference frame, Ss. The camera reference frame is the one attached to the primitive camera as shown in Figure 3. Given a point P, represented in the Sc frame as ${}^{S_c}\mathbf{P} = [X\ Y\ Z]^T$, its *2D* projection point $\mathbf{p}$ onto the image sensor plane referred to the S_s frame will be, in homogeneous coordinates, ${}^{S_c}\mathbf{p}_h = [u\ v\ 1]^T$, being its pixel coordinates calculated from Figure 4. Executing the *RobSim* image acquisition function $\mathbf{p}_{imag}$=*point_view*($\mathbf{p}_{3D}$,$\mathbf{K}_i$,${}^o\mathbf{T}_c$) (Subsection 3.4) is possible to simulate a (Chaumette & Hutchinson, 2006)point capture as the camera moves. The $\mathbf{p}_{3D}$ vector, a workspace point relative to the base coordinates, is measured in centimeters. The $\mathbf{p}_{imag}$ vector, the *2D* corresponding point onto the image plane, is measured in pixels.

The *RobSim* features for visual servo control will be shown in a vision-guided operation with the Rhino XR4 robot. Figure 13 shows the robot *RobSim* model at its home pose (initial configuration) with a camera attached to its end-effector (gripper), so with the 5 DOF motion capability the robot allows. Resting over the base plane there is a cube (a block primitive) with color marks (asterisks) at its four top vertexes. Figure 13 also shows a window displaying the cube image as captured by the camera, in which the cube is represented by the four top color marks. An additional mark represents the image plane center.

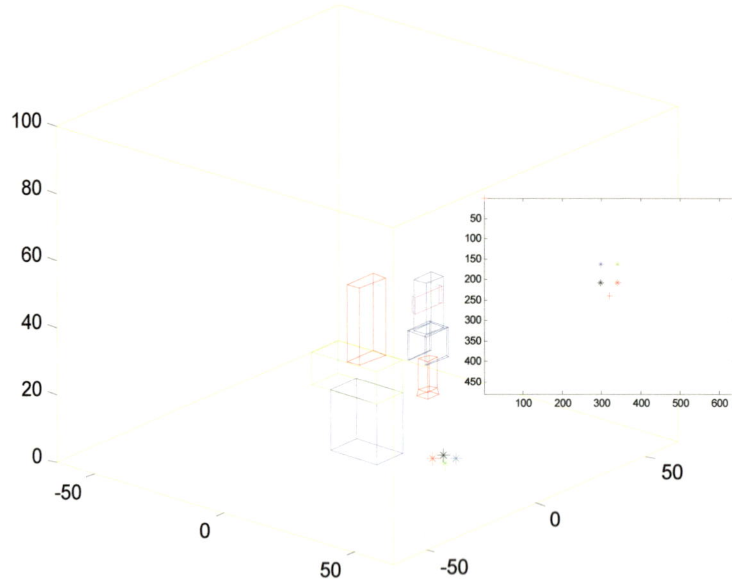

Fig. 13. *RobSim* vision-guided operation - initial configuration

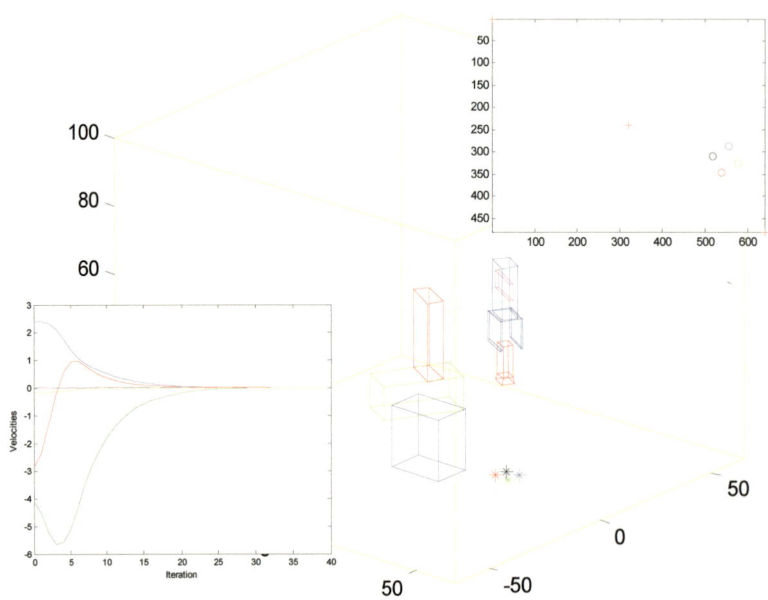

Fig. 14. *RobSim* vision-guided operation - new configuration

In this image-based servoing scheme the visual features error vector is defined as the difference between current and desired cube vertex positions. An exponential decoupled decay for this error was imposed by a velocity control policy. Camera reference velocities were then obtained using the image interaction matrix. In turn, the joint reference velocities for the robot joints controllers were obtained from the robot Jacobian.

Figure 14 shows the robot after executing a moving command towards a new configuration while the cube remains fixed. The window image shows the cube image, represented by the correspondent color marks (now circle marks). Another window shows the time variations of the camera velocity components. Visual information can be then used to guide the robot to describe a trajectory from an initial configuration to a new configuration through individual joint control.

6.2 A position-based visual servoing scheme within *RobSim*

Here, the PBVS architecture was implemented to simulate a vision-guided placing operation with the Rhino XR4 robot and a stereo-vision system with two cameras in the robot workspace. The object to be handled is a cube represented by a block-type primitive. Three marking points are located at three vertexes of the cube in order to visually represent the cube for translation and rotation displacements. Figure 16 shows the initial configuration of the robotic manipulator with the cube being grasped by the end effector, the cube initial pose (green) and the cube final pose (cyan).

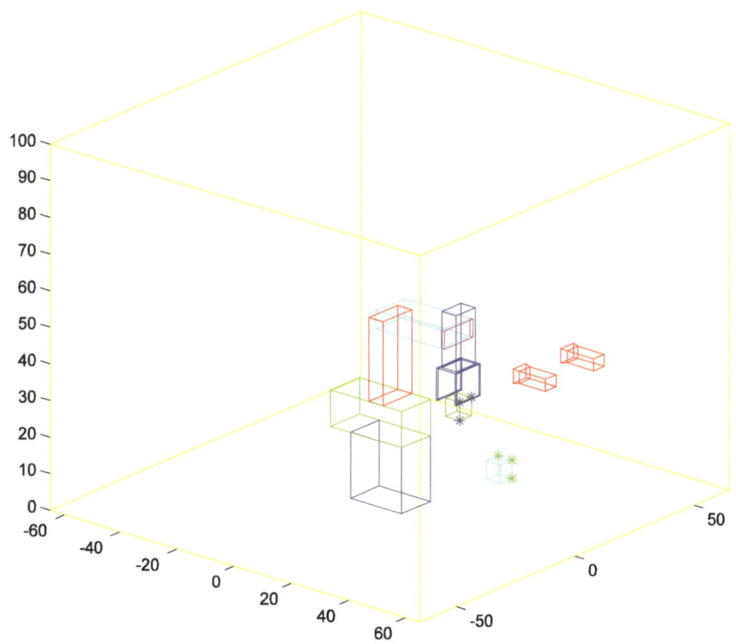

Fig. 15. Vision-guided placing operation - initial configuration

A computer vision algorithm is not required in this case because the object is synthetic and a simple one. Determination of the coordinates of the three vertexes that identifies the cube is performed by the stereo-vision system (Hutchinson, 1996). The coordinates of the three identifying vertexes representing the cube at its initial pose are, $\mathbf{p}_{a1}$ (middle vertex), $\mathbf{p}_{b1}$ and

$\mathbf{p}_{c1}$. The corresponding three coordinates at the final pose are $\mathbf{p}_{a2}$, $\mathbf{p}_{b2}$ and $\mathbf{p}_{c2}$. From these points four *3D* vectors are generated: $\mathbf{P}_{ab1}$ pointing from $\mathbf{p}_{a1}$ to $\mathbf{p}_{b1}$; $\mathbf{P}_{ac1}$ pointing from $\mathbf{p}_{a1}$ to $\mathbf{p}_{c1}$; $\mathbf{P}_{ab2}$ from $\mathbf{p}_{a2}$ to $\mathbf{p}_{b2}$; and finally $\mathbf{P}_{ac2}$ from $\mathbf{p}_{a2}$ to $\mathbf{p}_{c2}$. All these vectors are normalized before use.

To describe the robot joint dynamics a first-order model without dissipation is considered. Once the end-effector velocity vector $\dot{\mathbf{r}}(t)$ (translational and rotational motion) referred to the base frame coordinates is known, the robot inverse kinematics model can be used to determine the joint velocities vector $\dot{\mathbf{q}}(t)$ (Schilling, 1990). These velocities vectors are related by the pseudo-inverse of the robot Jacobian matrix, $\mathbf{J}(\mathbf{q})$ as

$$\dot{\mathbf{q}}(t) = \mathbf{J}^{+}(\mathbf{q}).\dot{\mathbf{r}}(t) \tag{19}$$

The end effector velocity $\dot{\mathbf{r}}(t)$ is known as the *screw* velocity, consisting of a linear velocity along a line and an angular velocity around that line. Its first three elements are the linear velocities $\mathbf{T}_r = [v_x\ v_y\ v_z]^T$ and its last three elements $\mathbf{\Omega}_r = [\omega_x\ \omega_y\ \omega_z]^T$ the angular velocities, being all components referred to the base coordinate frame. Thus, the end effector velocity is

$$\dot{\mathbf{r}}(t) = [\mathbf{T}_r\ \mathbf{\Omega}_r]^T \tag{20}$$

A task function characterizing position and orientation errors of the cube handling task was implemented. By vector analysis, it can be shown that if $\mathbf{P}_r = (\mathbf{P}_{ab1}\times\mathbf{P}_{ac1})\times(\mathbf{P}_{ab2}\times\mathbf{P}_{ac2}) = \mathbf{0}$ (where × denotes vector cross product), the handled cube attains the reference or desired orientation, in the particular cases where $\mathbf{P}_{ab1}$ and $\mathbf{P}_{ab2}$ or $\mathbf{P}_{ac1}$ and $\mathbf{P}_{ac2}$ have the same direction. The angular control velocity is adjusted as $\mathbf{\Omega} = k_1\mathbf{P}_r$, where k_1 is a positive proportional gain.

It is also verified that, being $\mathbf{t}_a$ a vector from point $\mathbf{p}_{a1}$ to point $\mathbf{p}_{a2}$ and $\mathbf{p}_{a1v}$, a vector from the frame origin to point $\mathbf{p}_{a1}$, the vector $\mathbf{P}_t = k_2\mathbf{t}_a + \mathbf{\Omega}\times\mathbf{p}_{a1v}$, with k_2 a positive proportional gain, is equal to the null vector when the handled cube assumes the reference pose. In this case the translation control velocity is given by $\mathbf{T}_r = \mathbf{P}_t$. By adequately adjusting k_1 and k_2 it is possible to improve the regulation velocity of position and orientation errors.

Figure 16 shows the final configuration of the vision-guided placing operation, a window shows the initial image as seen by the left camera. Another window shows the time evolution of the end-effector velocity components (Equation 20), in which case, due to the initial and desired cube pose, the angular components ω_x and ω_y are zero.

7. Conclusion

A software platform *RobSim* for analysis and design of robotic systems that includes image capturing devices was presented. It was developed within the Matlab environment to simulate kinematics of robotic structures and it allows implementing control strategies in order to follow trajectories, perform tasks, etc. Thus it is very suitable to implement robotic experiments before dealing with the real system. The platform is based on basic units called *primitives* that assembled together can simulate any robotic structure. Being modular it is expandable, another advantage is the inclusion of a video capturing device that allows implementing vision-guided robotic experiments. The platform was used here to model and simulate fixed and mobile robots. Image- and position-based servoing schemes were implemented for a robotic manipulator with a single and a two-camera arrangement and

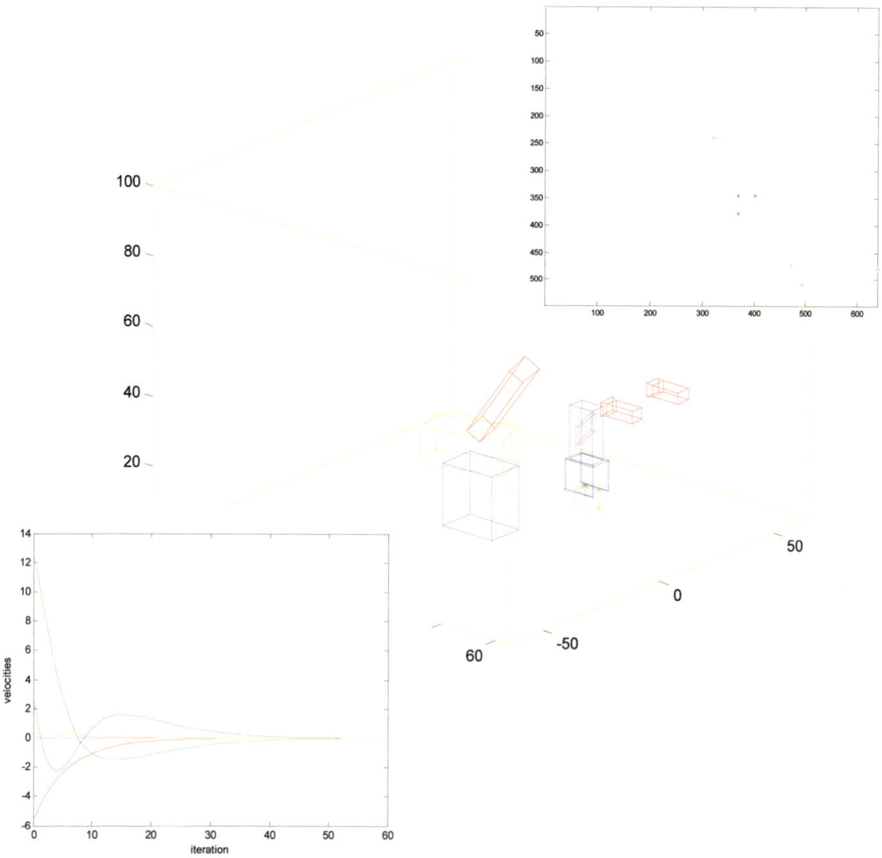

Fig. 16. Vision-guided placing operation - final configuration

simulations carried out within the *RobSim* platform. Further work is being addressed to introduce dynamical parameters into the primitives and simulation of more complex image features acquisition rather than image points.

8. References

Cervera, E. (2003) Visual Servoing Toolbox for Matlab/Simulink, http://vstoolbox.sourceforge.net/

Corke, P. I. (1996), A Robotics Toolbox for Matlab, *IEEE Robotics and Automation Magazine*, Vol. 3, No. 1, pp. 24-32.

Chaumette, F. and Hutchinson, S. (2006), Visual Servo Control - Part I: Basic Approach, *IEEE Robotics and Automation Magazine*, Vol. 13, No. 4, December 2006, pp. 82-90.

Chaumette, F. and Hutchinson, S. (2007), Visual Servo Control - Part II: Advanced Approaches, *IEEE Robotics and Automation Magazine*, Vol. 14, No. 1, March 2007, pp. 109-118.

Hutchinson, S.; Hager, D. & Corke, P. I. (1996), A Tutorial on Visual Servo Control, *IEEE Transactions on Robotics and Automation*, Vol. 12, No. 5, October 1996, pp. 651-670.

Legnani, G. (2005) Spacelib - Package for Matlab - User's Manual: http://www.bsing.ingunibst.it/glegnani.

Schilling, R. (1990), *Fundamentals of Robotics – Analysis and Control*, Prentice-Hall, ISBN-10: 0-1334-4433-9, NJ, USA.

Soares Jr., L. R. & Casanova Alcalde, V. H. (2006), Building Blocks for Simulation of Robotic Manipulators, *Proceedings of the 13th IASTED International Conference on Robotics and Applications*, pp. 408-413, ISBN 978-0-88986-685-0 , Würzburg, Germany, September, 2006.

Soares Jr., L. R. & Casanova Alcalde, V. H. (2006), An Educational Robotic Workstation based on the Rhino XR4 Robot, *Proceedings of the 36th ASEE/IEEE Frontiers in Education Conference*, pp. 7-12, ISBN 1-4244-0257-3 , San Diego, CA, USA, October 28-31, 2006.

Soares Jr., L. R. & Casanova Alcalde, V. H. (2008), *A Robotic Vehicle for Inspection and Maintenance Tasks over Transmission Lines*, University of Brasilia Technical Report.

2

Models and Control Strategies for Visual Servoing

Nils T Siebel, Dennis Peters and Gerald Sommer
Christian-Albrechts-University of Kiel
Germany

1. Introduction

Visual servoing is the process of steering a robot towards a goal using visual feedback in a closed control loop as shown in Figure 1. The *output* u_n of the controller is a *robot movement* which steers the robot towards the goal. The state x_n of the system cannot be directly observed. Instead a visual measurement process provides feedback data, the vector of *current image features* y_n. The *input* to the controller is usually the difference between desired ($y^\star$) and actual values of this vector—the *image error* vector Δy_n.

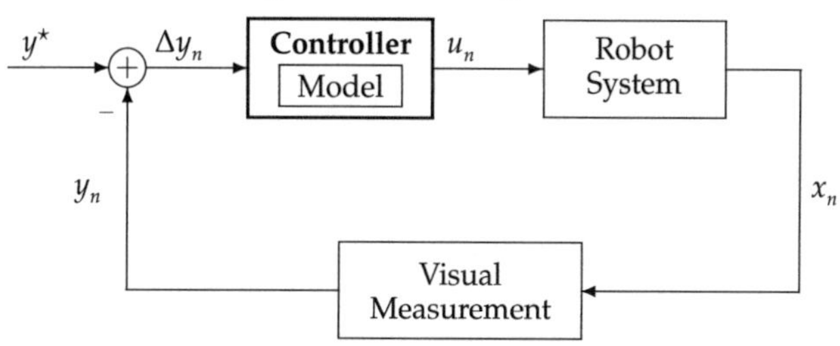

Fig. 1. Closed-loop image-based visual servoing control

In order for the controller to calculate the necessary robot movement it needs two main components:

1. a *model* of the environment—that is, a model of how the robot/scene will change after issuing a certain control commmand; and
2. a *control law* that governs how the next robot command is determined given current image measurements and model.

In this chapter we will look in detail on the effects different models and control laws have on the properties of a visual servoing controller. Theoretical considerations are combined with experiments to demonstrate the effects of popular models and control strategies on the behaviour of the controller, including convergence speed and robustness to measurement errors.

2. Building Models for Visual Servoing

2.1 Task Description

The aim of a visual servoing controller is to move the end-effector of one or more robot arms such that their configuration in relation to each other and/or to an object fulfils certain task-specific conditions. The feedback used in the controller stems from visual data, usually taken

Fig. 2. Robot Arm with Camera and Object

from one or more cameras mounted to the robot arm and/or placed in the environment. A typical configuration is shown in Figure 2. Here a camera is mounted to the robot's gripper ("eye-in-hand" setup), looking towards a glass jar. The controller's task in this case is to move the robot arm such that the jar can be picked up using the gripper. This is the case whenever the visual appearance of the object in the image has certain properties. In order to detect whether these properties are currently fulfilled a camera image can be taken and image processing techniques applied to extract the image positions of object markings. These image positions make up the *image feature vector.*

Since the control loop uses visual data the goal configuration can also be defined in the image. This can be achieved by moving the robot and/or the object in a suitable position and then acquiring a camera image. The image features measured in this image can act as *desired image features,* and a comparison of actual values at a later time to these desired values (*"image error"*) can be used to determine the degree of agreement with the desired configuration. This way of acquiring desired image features is sometimes called *"teaching by showing"*.

From a mathematical point of view, a successful visual servoing control process is equivalent to solving an optimisation problem. In this case a measure of the image error is minimised by moving the robot arm in the space of possible configurations. Visual servoing can also be regarded as practical feedback stabilisation of a dynamical system.

2.2 Modelling the Camera-Robot System

2.2.1 Preliminaries

The *pose* of an object is defined as its position and orientation. The *position* in 3D Euclidean space is given by the 3 Cartesian coordinates. The *orientation* is usually expressed by 3 angles, i.e. the rotation around the 3 coordinate axes. Figure 3 shows the notation used in this chapter, where *yaw, pitch* and *roll* angles are defined as the mathematically positive rotation around the x, y and z axis. In this chapter we will use the $\{\cdot\}$-notation for a *coordinate system,* for example $\{W\}$ will stand for the world coordinate system. A variable coordinate system—one which changes its pose to over time—will sometimes be indexed by the *time index* $n \in \mathbb{N} =$

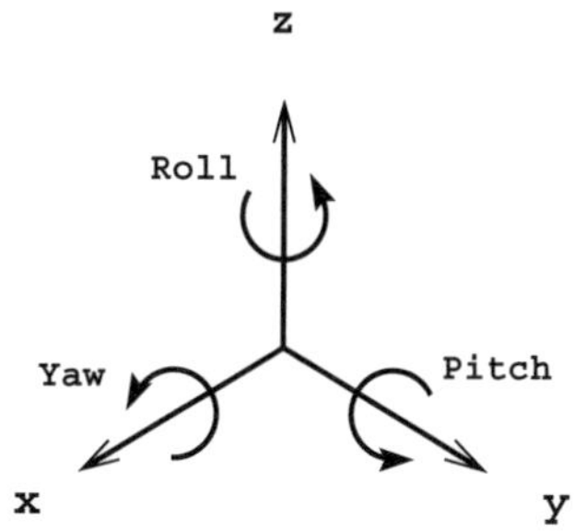

Fig. 3. Yaw, pitch and roll

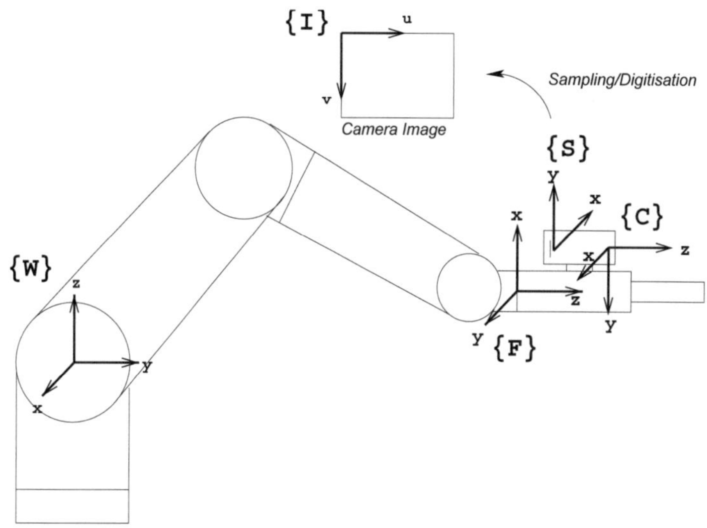

Fig. 4. World, Flange, Camera, Sensor and Image coordinate systems

$0, 1, 2, \ldots$. An example is the camera coordinate system $\{C_n\}$, which moves relative to $\{W\}$ as the robot moves since the camera is mounted to its hand.

Figure 4 lists the coordinate systems used for modelling the camera-robot system. The *world coordinate system* $\{W\}$ is fixed at the robot base, the *flange coordinate system* $\{F\}$ (sometimes called *"tool coordinate system"*, but this can be ambiguous) at the flange where the hand is mounted. The *camera coordinate system* $\{C\}$ (or $\{C_n\}$ at a specific time n) is located at the optical centre of the camera, the *sensor coordinate system* $\{S\}$ in the corner of its CCD/CMOS chip (sensor); their orientation and placement is shown in the figure. The *image coordinate system* which is used to describe positions in the digital image is called $\{I\}$. It is the only system to use pixel as its unit; all other systems use the same length unit, e.g. mm.

Variables that contain coordinates in a particular coordinate system will be marked by a superscript left of the variable, e.g. ${}^{A}x$ for a vector $x \in \mathbb{R}^n$ in $\{A\}$-coordinates. The coordinate transform which transforms a variable from a coordinate system $\{A\}$ to another one, $\{B\}$, will be written ${}^{B}_{A}\mathrm{T}$. If ${}^{A}x$ and ${}^{B}x$ express the pose of the same object then

$$ {}^{A}x = {}^{A}_{B}\mathrm{T}\;{}^{B}x, \qquad \text{and always} \qquad {}^{A}_{B}\mathrm{T} = \left({}^{B}_{A}\mathrm{T}\right)^{-1}. \tag{1} $$

The *robot's pose* is defined as the pose of $\{F\}$ in $\{W\}$.

2.2.2 Cylindrical Coordinates

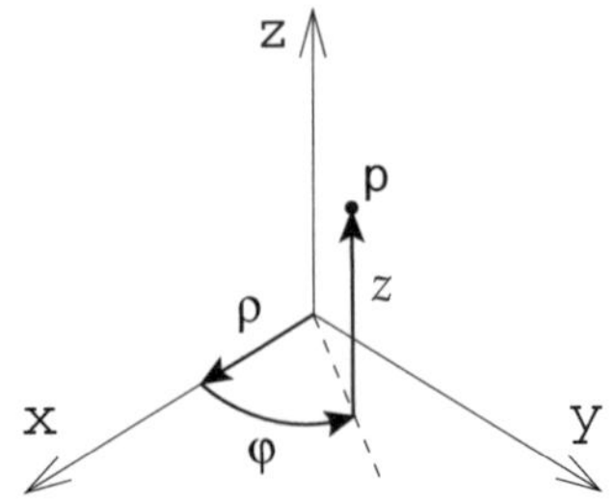

Fig. 5. A point $p = (\rho, \varphi, z)$ in cylindrical coordinates.

An alternative way to describe point positions is by using a cylindrical coordinate system as the one in Figure 5. Here the position of the point p is defined by the distance ρ from a fixed axis (here aligned with the Cartesian z axis), an angle φ around the axis (here $\varphi = 0$ is aligned with the Cartesian x axis) and a height z from a plane normal to the z axis (here the plane spanned by x and y). Using the commonly used alignment with the Cartesian axes as in Figure 5 converting to and from cylindrical coordinates is easy. Given a point $p = (x, y, z)$ in Cartesian coordinates, its cylindrical coordinates $p = (\rho, \varphi, z) \in \mathbb{R} \times]-\pi, \pi] \times \mathbb{R}$ are as follows:

$$\begin{aligned}
\rho &= \sqrt{x^2 + y^2} \\
\varphi &= \operatorname{atan2}(y, x) \\
&\stackrel{\star}{=} \begin{cases} 0 & \text{if } x = 0 \text{ and } y = 0 \\ \arcsin(\frac{y}{\rho}) & \text{if } x \geq 0 \\ \arcsin(\frac{y}{\rho}) + \pi & \text{if } x < 0 \end{cases} \\
z &= z,
\end{aligned} \tag{2}$$

($\star$ up to multiples of 2π), and, given a point $p = (\rho, \varphi, z)$ in cylindrical coordinates:

$$\begin{aligned}
x &= \rho \cos \varphi \\
y &= \rho \sin \varphi \\
z &= z.
\end{aligned} \tag{3}$$

2.2.3 Modelling the Camera

A simple and popular approximation to the way images are taken with a camera is the *pinhole camera model* (from the pinhole camera/camera obscura models by Ibn al-Haytham "Alhacen", 965–1039 and later by Gérard Desargues, 1591–1662), shown in Figure 6. A light ray from an object point passes an aperture plate through a very small hole ("pinhole") and arrives at the sensor plane, where the camera's CCD/CMOS chip (or a photo-sensitive film in the 17th century) is placed. In the digital camera case the sensor elements correspond to picture elements ("pixels"), and are mapped to the image plane. Since pixel positions are stored in the computer as unsigned integers the centre of the $\{I\}$ coordinate system in the image plane is shifted to the upper left corner (looking towards the object/monitor). Therefore the centre ${}^{I}c \neq (0, 0)^T$.

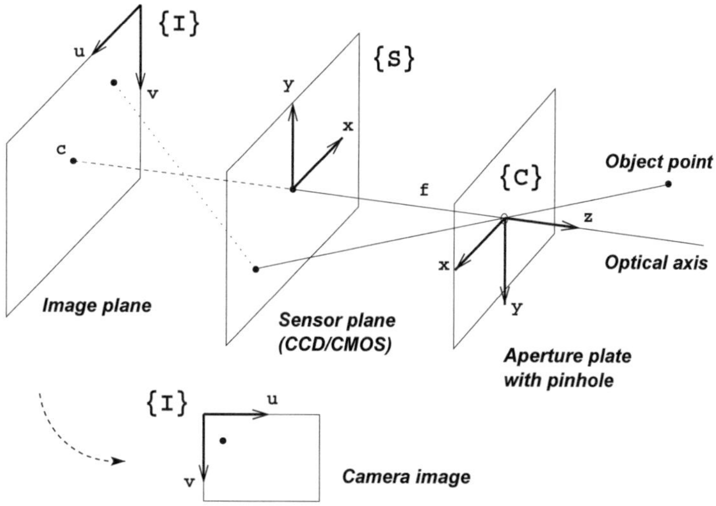

Fig. 6. Pinhole camera model

Sometimes the sensor plane is positioned in front of the aperture plate in the literature (e.g. in Hutchinson et al., 1996). This has the advantage that the x- and y-axis of $\{S\}$ can be (directionally) aligned with the ones in $\{C\}$ and $\{I\}$ while giving identical coordinates. However, since this alternative notation has also the disadvantage of being less intuitive, we use the one defined above.

Due to the simple model of the way the light travels through the camera the object point's position in $\{C\}$ and the coordinates of its projection in $\{S\}$ and $\{I\}$ are proportional, with a shift towards the new centre in $\{I\}$. In particular, the *sensor coordinates* ${}^{S}p = ({}^{S}x, {}^{S}y)^T$ of the image of an object point ${}^{C}p = ({}^{C}x, {}^{C}y, {}^{C}z)^T$ are given as

$$ {}^{S}x = \frac{{}^{C}x \cdot f}{{}^{C}z} \qquad \text{and} \qquad {}^{S}y = \frac{{}^{C}y \cdot f}{{}^{C}z}, \tag{4} $$

where f is the distance the aperture plate and the sensor plane, also called the *"focal length"* of the camera/lens.

The pinhole camera model's so-called *"perspective projection"* is not an exact model of the projection taking place in a modern camera. In particular, lens distortion and irregularities in the manufacturing (e.g. slightly tilted CCD chip or positioning of the lenses) introduce deviations. These modelling errors may need to be considered (or, corrected by a lens distortion model) by the visual servoing algorithm.

2.3 Defining the Camera-Robot System as a Dynamical System

As mentioned before, the camera-robot system can be regarded as a dynamical system. We define the *state* x_n of the robot system at a *time step* $n \in \mathbb{N}$ as the current robot pose, i.e. the pose of the flange coordinate system $\{F\}$ in world coordinates $\{W\}$. $x_n \in \mathbb{R}^6$ will contain the position and orientation in the x, y, z, yaw, pitch, roll notation defined above. The set of possible robot poses is $\mathcal{X} \subset \mathbb{R}^6$. The *output* of the system is the *image feature vector* y_n. It contains pairs of image coordinates of object markings viewed by the camera, i.e. $({}^{S}x_1, {}^{S}y_1, \ldots, {}^{S}x_M, {}^{S}y_M)^T$ for $M = \frac{m}{2}$ object markings (in our case $M = 4$, so $y_n \in \mathbb{R}^8$).

Let $\mathcal{Y} \subset \mathbb{R}^m$ be the set of possible output values. The *output (measurement) function* is $\eta : \mathcal{X} \to \mathcal{Y}, x_n \mapsto y_n$. It contains the whole measurement process, including projection onto the sensor, digitisation and image processing steps.
The *input (control) variable* $u_n \in \mathcal{U} \subset \mathbb{R}^6$ shall contain the desired pose change of the camera coordinate system. This robot movement can be easily transformed to a new robot pose $\tilde{u}_n$ in $\{W\}$, which is given to the robot in a move command. Using this definition of u_n an input of $(0,0,0,0,0,0)^T$ corresponds to no robot movement, which has advantages, as we shall see later. Let $\varphi : \mathcal{X} \times \mathcal{U} \to \mathcal{X}, (x_n, u_n) \mapsto x_{n+1}$ be the corresponding *state transition (next-state) function.*

With these definitions the camera-robot system can be defined as a time invariant, time discrete input-output system:

$$\begin{aligned} x_{n+1} &= \varphi\,(x_n, u_n) \\ y_n &= \eta\,(x_n). \end{aligned} \tag{5}$$

When making some mild assumptions, e.g. that the camera does not move relative to $\{F\}$ during the whole time, the state transition function φ can be calculated as follows:

$$\begin{aligned} \varphi(x_n, u_n) &= x_{n+1} = {}^{W}x_{n+1} = {}^{W}\tilde{u}_n \mathrel{\hat{=}} {}^{W}_{F_{n+1}}\mathrm{T} \\ &= \underbrace{{}^{W}_{F_n}\mathrm{T}}_{\hat{=}x_n} \circ \underbrace{{}^{F_n}_{C_n}\mathrm{T}}_{\star} \circ \underbrace{{}^{C_n}_{C_{n+1}}\mathrm{T}}_{\hat{=}u_n} \circ \underbrace{{}^{C_{n+1}}_{F_{n+1}}\mathrm{T}}_{\star}, \end{aligned} \tag{6}$$

where $\{F_n\}$ is the flange coordinate system at time step n, etc., and the $\hat{=}$ operator expresses the equivalence of a pose with its corresponding coordinate transform.
$\star =$ *external ("extrinsic") camera parameters;* ${}^{T_n}_{C_n}\mathrm{T} = {}^{T_{n+1}}_{C_{n+1}}\mathrm{T} = \left({}^{C_{n+1}}_{T_{n+1}}\mathrm{T}\right)^{-1} \quad \forall n \in \mathbb{N}.$

For $m = 2$ image features corresponding to coordinates $({}^{S}x, {}^{S}y)$ of a projected object point ${}^{W}p$ the equation for η follows analogously:

$$\begin{aligned} \eta(x) &= y = {}^{S}y = {}^{S}_{C}\mathrm{T}\ {}^{C}p \\ &= {}^{S}_{C}\mathrm{T} \circ {}^{C}_{T}\mathrm{T} \circ {}^{T}_{W}\mathrm{T}\ {}^{W}p, \end{aligned} \tag{7}$$

where ${}^{S}_{C}\mathrm{T}$ is the mapping of the object point ${}^{C}p$ depending on the focal length f according to the pinhole camera model / perspective projection defined in (4).

2.4 The Forward Model—Mapping Robot Movements to Image Changes

In order to calculate necessary movements for a given desired change in visual appearance the relation between a robot movement and the resulting change in the image needs to be modelled. In this section we will analytically derive a *forward model*, i.e. one that expresses image changes as a function of robot movements, for the eye-in-hand setup described above. This forward model can then be used to predict changes effected by controller outputs, or (as it is usually done) simplified and then inverted. An *inverse model* can be directly used to determine the controller output given actual image measurements.
Let $\Phi : \mathcal{X} \times \mathcal{U} \to \mathcal{Y}$ the function that expresses the system output y depending on the state x and the input u:

$$\Phi(x, u) := \eta \circ \varphi(x, u) = \eta(\varphi(x, u)). \tag{8}$$

For simplicity we also define the function which expresses the behaviour of $\Phi(x_n, \cdot)$ at a time index n, i.e. the dependence of image features on the camera movement u:

$$\Phi_n(u) := \Phi(x_n, u) = \eta(\varphi(x_n, u)). \tag{9}$$

This is the forward model we wish to derive.

Φ_n depends on the camera movement u and the current system state, the robot pose x_n. In particular it depends on the position of all object markings in the current camera coordinate system. In the following we need assume the knowledge of the camera's focal length f and the ${}^{C}z$ component of the positions of image markings in $\{C\}$, which cannot be derived from their image position $({}^{S}x, {}^{S}y)$. Then with the help of f and the image coordinates $({}^{S}x, {}^{S}y)$ the complete position of the object markings in $\{C\}$ can be derived with the pinhole camera model (4).

We will first construct the model Φ_n for the case of a single object marking, $M = \frac{m}{2} = 1$. According to equations (6) and (7) we have for an object point ${}^{W}p$:

$$\begin{aligned}\Phi_n(u) &= \eta \circ \varphi(x_n, u)\\ &= {}^{S}_{C_{n+1}}\mathrm{T} \circ {}^{C_{n+1}}_{C_n}\mathrm{T} \circ {}^{C_n}_{T}\mathrm{T} \circ {}^{T}_{W}\mathrm{T}\ {}^{W}p\\ &= {}^{S}_{C_{n+1}}\mathrm{T} \circ {}^{C_{n+1}}_{C_n}\mathrm{T}\ {}^{C_n}x,\end{aligned} \tag{10}$$

where ${}^{C_n}x$ are the coordinates of the object point in $\{C_n\}$.

In the system state x_n the position of an object point ${}^{C_n}x =: p = (p_1, p_2, p_3)^T$ can be derived with $({}^{S}x, {}^{S}y)^T$, assuming the knowledge of f and ${}^{C}z$, via (4). Then the camera changes its pose by ${}^{C}u =: u = (u_1, u_2, u_3, u_4, u_5, u_6)^T$; we wish to know the new coordinates $({}^{S}\tilde{x}, {}^{S}\tilde{y})^T$ of p in the image. The new position $\tilde{p}$ of the point in new camera coordinates is given by a translation by u_1 through u_3 and a rotation of the camera by u_4 through u_6. We have

$$\begin{aligned}\tilde{p} &= \mathrm{rot}_x(-u_4)\ \mathrm{rot}_y(-u_5)\ \mathrm{rot}_z(-u_6)\begin{pmatrix} p_1 - u_1 \\ p_2 - u_2 \\ p_3 - u_3 \end{pmatrix}\\ &= \begin{pmatrix} c_5 c_6 & c_5 s_6 & -s_5 \\ s_4 s_5 c_6 - c_4 s_6 & s_4 s_5 s_6 + c_4 c_6 & s_4 c_5 \\ c_4 s_5 c_6 + s_4 s_6 & c_4 s_5 s_6 - s_4 c_6 & c_4 c_5 \end{pmatrix}\begin{pmatrix} p_1 - u_1 \\ p_2 - u_2 \\ p_3 - u_3 \end{pmatrix}\end{aligned} \tag{11}$$

using the short notation

$$s_i := \sin u_i,\ c_i := \cos u_i \qquad \text{for } i = 4, 5, 6. \tag{12}$$

Again with the help of the pinhole camera model (4) we can calculate the $\{S\}$ coordinates of the projection of the new point, which finally yields the model Φ_n:

$$\begin{aligned}\begin{bmatrix} {}^{S}\tilde{x} \\ {}^{S}\tilde{y} \end{bmatrix} &= \Phi(x_n, u)\\ &= \Phi_n(u)\\ &= f \cdot \begin{bmatrix} \dfrac{c_5\, c_6\, (p_1 - u_1) + c_5\, s_6\, (p_2 - u_2) - s_5\, (p_3 - u_3)}{(c_4\, s_5\, c_6 + s_4\, s_6)\, (p_1 - u_1) + (c_4\, s_5\, s_6 - s_4\, c_6)\, (p_2 - u_2) + c_4\, c_5\, (p_3 - u_3)} \\[2ex] \dfrac{(s_4\, s_5\, c_6 - c_4\, s_6)\, (p_1 - u_1) + (s_4\, s_5\, s_6 + c_4\, c_6)\, (p_2 - u_2) + s_4\, c_5\, (p_3 - u_3)}{(c_4\, s_5\, c_6 + s_4\, s_6)\, (p_1 - u_1) + (c_4\, s_5\, s_6 - s_4\, c_6)\, (p_2 - u_2) + c_4\, c_5\, (p_3 - u_3)} \end{bmatrix}.\end{aligned} \tag{13}$$

2.5 Simplified and Inverse Models

As mentioned before, the controller needs to derive necessary movements from given desired image changes, for which an inverse model is beneficial. However, $\Phi_n(u)$ is too complicated to invert. Therefore in practice usually a linear approximation $\hat{\Phi}_n(u)$ of $\Phi_n(u)$ is calculated and then inverted. This can be done in a number of ways.

2.5.1 The Standard Image Jacobian

The simplest and most common linear model is the *Image Jacobian*. It is obtained by Taylor expansion of (13) around $u = 0$:

$$\begin{aligned} y_{n+1} &= \eta(\varphi(x_n, u)) \\ &= \Phi(x_n, u) \\ &= \Phi_n(u) \\ &= \Phi_n(0 + u) \\ &= \Phi_n(0) + J_{\Phi_n}(0)\, u + \mathcal{O}(\|u\|^2). \end{aligned} \tag{14}$$

With $\Phi_n(0) = y_n$ and the definition $J_n := J_{\Phi_n}(0)$ the image change can be approximated

$$y_{n+1} - y_n \approx J_n\, u \tag{15}$$

for sufficiently small $\|u\|_2$.

The Taylor expansion of the two components of (13) around $u = 0$ yields the Image Jacobian J_n for one object marking ($m = 2$):

$$J_n = \begin{pmatrix} -\dfrac{f}{{}^C z} & 0 & \dfrac{{}^S x}{{}^C z} & \dfrac{{}^S x\,{}^S y}{f} & -f - \dfrac{{}^S x^2}{f} & {}^S y \\ 0 & -\dfrac{f}{{}^C z} & \dfrac{{}^S y}{{}^C z} & f + \dfrac{{}^S y^2}{f} & -\dfrac{{}^S x\,{}^S y}{f} & -{}^S x \end{pmatrix} \tag{16}$$

where again image positions where converted back to sensor coordinates.

The Image Jacobian for M object markings, $M \in \mathbb{N}_{>1}$, can be derived analogously; the change of the $m = 2M$ image features can be approximated by

$$y_{n+1} - y_n \approx J_n\, u$$

$$= \begin{pmatrix} -\frac{f}{{}^{C}z_1} & 0 & \frac{{}^{S}x_1}{{}^{C}z_1} & \frac{{}^{S}x_1\,{}^{S}y_1}{f} & -f-\frac{{}^{S}x_1^2}{f} & {}^{S}y_1 \\ 0 & -\frac{f}{{}^{C}z_1} & \frac{{}^{S}y_1}{{}^{C}z_1} & f+\frac{{}^{S}y_1^2}{f} & -\frac{{}^{S}x_1\,{}^{S}y_1}{f} & -{}^{S}x_1 \\ \vdots & \vdots & \vdots & \vdots & \vdots & \vdots \\ -\frac{f}{{}^{C}z_M} & 0 & \frac{{}^{S}x_M}{{}^{C}z_M} & \frac{{}^{S}x_M\,{}^{S}y_M}{f} & -f-\frac{{}^{S}x_M^2}{f} & {}^{S}y_M \\ 0 & -\frac{f}{{}^{C}z_M} & \frac{{}^{S}y_M}{{}^{C}z_M} & f+\frac{{}^{S}y_M^2}{f} & -\frac{{}^{S}x_M\,{}^{S}y_M}{f} & -{}^{S}x_M \end{pmatrix} \begin{pmatrix} u_1 \\ \vdots \\ u_6 \end{pmatrix}, \tag{17}$$

for small $\|u\|_2$, where $({}^{S}x_i, {}^{S}y_i)$ are the sensor coordinates of the ith projected object marking and ${}^{C}z_i$ their distances from the camera, $i = 1, \ldots, M$.

2.5.2 A Linear Model in the Cylindrical Coordinate System

Iwatsuki and Okiyama (2005) suggest a formulation of the problem in cylindrical coordinates. This means that positions of markings on the sensor are given in polar coordinates, $(\rho, \varphi)^T$ where ρ and φ are defined as in Figure 5 ($z = 0$). The Image Jacobian J_n for one image point is given in this case by

$$J_n = \begin{pmatrix} -\frac{f c_\varphi}{{}^{C}z} & -\frac{f s_\varphi}{{}^{C}z} & \frac{{}^{C}y s_\varphi + {}^{C}x c_\varphi}{{}^{C}z} & \left(f + \frac{{}^{C}y^2}{f}\right) s_\varphi + \frac{{}^{C}x\,{}^{C}y c_\varphi}{f} & \left(-f - \frac{{}^{C}x^2}{f}\right) c_\varphi - \frac{{}^{C}x\,{}^{C}y s_\varphi}{f} & {}^{C}y c_\varphi - {}^{C}x s_\varphi \\ \frac{f s_\varphi}{{}^{C}z} & -\frac{f c_\varphi}{{}^{C}z} & \frac{{}^{C}y c_\varphi + {}^{C}x s_\varphi}{{}^{C}z} & \left(f + \frac{{}^{C}y^2}{f}\right) c_\varphi - \frac{{}^{C}x\,{}^{C}y s_\varphi}{f} & \left(f + \frac{{}^{C}x^2}{f}\right) s_\varphi - \frac{{}^{C}x\,{}^{C}y c_\varphi}{f} & -{}^{C}y s_\varphi - {}^{C}x c_\varphi \end{pmatrix} \tag{18}$$

with the short notation

$$s_\varphi := \sin\varphi \quad \text{and} \quad c_\varphi := \cos\varphi. \tag{19}$$

and analogously for $M > 1$ object markings.

2.5.3 Quadratic Models

A quadratic model, e.g. a quadratic approximation of the system model (13), can be obtained by a Taylor expansion; a resulting approximation for $M = 1$ marking is

$$y_{n+1} = \begin{bmatrix} {}^{S}\tilde{x} \\ {}^{S}\tilde{y} \end{bmatrix} = \Phi_n(0) + J_{\Phi_n}(0)\, u + \frac{1}{2} \begin{bmatrix} u^T H_{S_x} u \\ u^T H_{S_y} u \end{bmatrix} + \mathcal{O}(\|u\|^3). \tag{20}$$

where again $\Phi_n(0) = y_n$ and $J_{\Phi_n}(0) = J_n$ from (16), and the Hessian matrices are

$$H_{S_x} = \begin{pmatrix} 0 & 0 & \frac{-f}{{}^{C}z^2} & \frac{-{}^{S}y}{{}^{C}z} & \frac{2{}^{S}x}{{}^{C}z} & 0 \\ 0 & 0 & 0 & \frac{-{}^{S}x}{{}^{C}z} & 0 & \frac{-f}{{}^{C}z} \\ \frac{-f}{{}^{C}z^2} & 0 & \frac{2{}^{S}x}{{}^{C}z^2} & \frac{2{}^{S}x{}^{S}y}{f{}^{C}z} & \frac{-2{}^{S}x^2}{f{}^{C}z} & \frac{{}^{S}y}{{}^{C}z} \\ \frac{-{}^{S}y}{{}^{C}z} & \frac{-{}^{S}x}{{}^{C}z} & \frac{2{}^{S}x{}^{S}y}{f{}^{C}z} & {}^{S}x\left(1+2\left(\frac{{}^{S}y}{f}\right)^2\right) & -{}^{S}y\left(1+2\left(\frac{{}^{S}x}{f}\right)^2\right) & \frac{{}^{S}y^2-{}^{S}x^2}{f} \\ \frac{2{}^{S}x}{{}^{C}z} & 0 & \frac{-2{}^{S}x^2}{f{}^{C}z} & -{}^{S}y\left(1+2\left(\frac{{}^{S}x}{f}\right)^2\right) & 2{}^{S}x\left(1+\frac{{}^{S}x^2}{f}\right)^2 & -2{}^{S}x\frac{{}^{S}y}{f} \\ 0 & \frac{-f}{{}^{C}z} & \frac{{}^{S}y}{{}^{C}z} & \frac{{}^{S}y^2-{}^{S}x^2}{f} & \frac{-2{}^{S}x{}^{S}y}{f} & -{}^{S}x \end{pmatrix} \tag{21}$$

as well as

$$H_{S_y} = \begin{pmatrix} 0 & 0 & 0 & 0 & \frac{{}^{S}y}{{}^{C}z} & \frac{f}{{}^{C}z} \\ 0 & 0 & \frac{-f}{{}^{C}z^2} & \frac{-2{}^{S}y}{{}^{C}z} & \frac{{}^{S}x}{{}^{C}z} & 0 \\ 0 & \frac{-f}{{}^{C}z^2} & \frac{2{}^{S}y}{{}^{C}z^2} & \frac{2{}^{S}y^2}{f{}^{C}z} & \frac{-2{}^{S}x{}^{S}y}{f{}^{C}z} & \frac{-{}^{S}x}{{}^{C}z} \\ 0 & \frac{-2{}^{S}y}{{}^{C}z} & \frac{2{}^{S}y^2}{f{}^{C}z} & 2{}^{S}y\left(1+\frac{{}^{S}y^2}{f}\right)^2 & \left(\frac{{}^{S}y}{f}\right)\left(-2{}^{S}x\frac{{}^{S}y}{f}\right) & \frac{-2{}^{S}x{}^{S}y}{f} \\ \frac{{}^{S}y}{{}^{C}z} & \frac{{}^{S}x}{{}^{C}z} & \frac{-2{}^{S}x{}^{S}y}{f{}^{C}z} & \left(\frac{{}^{S}y}{f}\right)\left(-2{}^{S}x\frac{{}^{S}y}{f}\right) & {}^{S}y\left(1+2\left(\frac{{}^{S}x}{f}\right)^2\right) & \frac{{}^{S}x^2-{}^{S}y^2}{f} \\ \frac{f}{{}^{C}z} & 0 & \frac{-{}^{S}x}{{}^{C}z} & \frac{-2{}^{S}x{}^{S}y}{f} & \frac{{}^{S}x^2-{}^{S}y^2}{f} & -{}^{S}y \end{pmatrix}. \tag{22}$$

2.5.4 A Mixed Model

Malis (2004) proposes a way of constructing a mixed model which consists of different linear approximations of the target function Φ. Let x_n again be the current robot pose and $x^\star$ the teach pose. For a given robot command u we set again $\Phi_n(u) := \Phi(x_n, u)$ and now also $\Phi^\star(u) := \Phi(x^\star, u)$ such that $\Phi_n(0) = y_n$ und $\Phi^\star(0) = y^\star$. Then Taylor expansions of Φ_n and $\Phi^\star$ at $u = 0$ yield

$$y_{n+1} = y_n + J_{\Phi_n}(0)u + \mathcal{O}(\|u\|^2) \tag{23}$$

and

$$y_{n+1} = y_n + J_{\Phi^\star}(0)u + \mathcal{O}(\|u\|^2). \tag{24}$$

In other words, both Image Jacobians, $J_n := J_{\Phi_n}(0)$ and $J^\star := J_{\Phi^\star}(0)$ can be used as linear approximations of the behaviour of the robot system. One of these models has its best validity at the current pose, the other at the teach pose. Since we are moving the robot from one towards the other it may be useful to consider both models. Malis proposes to use a mixture of these two models, i.e.

$$y_{n+1} - y_n \approx \frac{1}{2}(J_n + J^\star)\, u. \tag{25}$$

In his control law (see Section 3 below) he calculates the pseudoinverse of the Jacobians, and therefore calls this approach "Pseudo-inverse of the Mean of the Jacobians", or short "PMJ". In a variation of this approach the computation of mean and pseudo-inverse is exchanged, which results in the "MPJ" method. See Section 3 for details.

2.5.5 Estimating Models

Considering the fact that models can only ever approximate the real system behaviour it may be beneficial to use measurements obtained during the visual servoing process to update the model "online". While even the standard models proposed above use current measurements to estimate the distance ${}^{C}z$ from the object to use this estimate in the Image Jacobian, there are also approaches that estimate more variables, or construct a complete model from scratch. This is most useful when no certain data about the system state or setup are available. The following aspects need to be considered when estimating the Image Jacobian—or other models:

- How precise are the measurements used for model estimation, and how large is the sensitivity of the model to measurement errors?
- How many measurements are needed to construct the model? For example, some methods use 6 robot movements to measure the 6-dimensional data within the Image Jacobian. In a static look-and-move visual servoing setup which may reach its goal in 10-20 movements with a given Jacobian the resulting increase in necessary movements, as well as possible mis-directed movements until the estimation process converges, need to be weighed against the flexibility achieved by the automatic model tuning.

The most prominent approach to estimation methods of the whole Jacobian is the *Broyden approach* which has been used by Jägersand (1996). The Jacobian estimation uses the following update formula for the current estimate $\hat{J}_n$:

$$\hat{J}_n := {}^{C_n}_{C_{n-1}}\mathrm{T} \left(\hat{J}_{n-1} + \frac{(y_n - y_{n-1} - \hat{J}_{n-1}\, u_n)\, u_n^T}{u_n^T u_n} \right), \tag{26}$$

with an additional weighting of the correction term

$$J_n := \gamma\, \hat{J}_{n-1} + (1-\gamma)\, \hat{J}_n, \quad 0 \le \gamma < 1 \tag{27}$$

to reduce the sensitivity of the estimate to measurement noise.
In the case of Jägersand's system using an estimation like this makes sense since he worked with a dynamic visual servoing setup where many more measurements are made over time compared to our setup ("static look-and-move", see below).
In combination with a model-based measurement a non-linear model could also make sense. A number of methods for the estimation of quadratic models are available in the optimisation literature. More on this subject can be found e.g. in Fletcher (1987, chapter 3) and Sage and White (1977, chapter 9).

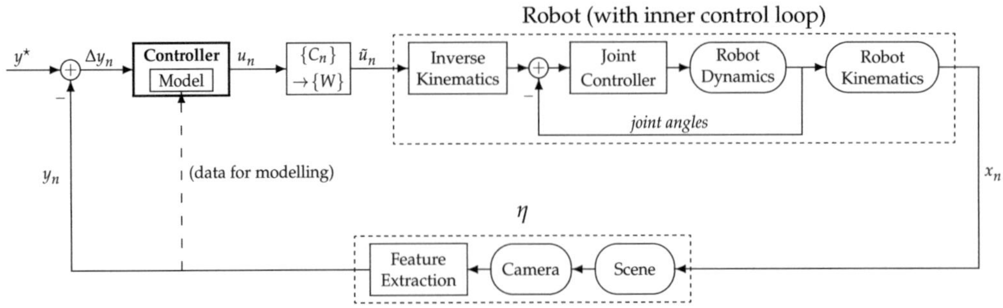

Fig. 7. Typical closed-loop image-based visual servoing controller

3. Designing a Visual Servoing Controller

Using one of the models defined above we wish to design a controller which steers the robot arm towards an object of unknown pose. This is to be realised in the visual feedback loop depicted in Figure 7. Using the terminology defined by Weiss et al. (1987) the visual servoing controller is of the type "*Static Image-based Look-and-Move*". "Image-based" means that goal and error are defined in image coordinates instead of using positions in normal space (that would be "position-based"). "Static Look-and-Move" means that the controller is a sampled data feedback controller and the robot does not move while a measurement is taken. This traditionally implies that the robot is controlled by giving world coordinates to the controller instead of directly manipulating robot joint angles (Chaumette and Hutchinson, 2008; Hutchinson et al., 1996).

The object has 4 circular, identifiable markings. Its appearance in the image is described by the *image feature vector* $y_n \in \mathbb{R}^8$ that contains the 4 pairs of image coordinates of these markings in a fixed order. The desired pose relative to the object is defined by the object's appearance in that pose by measuring the corresponding *desired image features* $y^\star \in \mathbb{R}^8$ ("*teaching by showing*"). Object and robot are then moved so that no Euclidean position of the object or robot is known to the controller. The *input* to the controller is the *image error* $\Delta y_n := y^\star - y_n$. The current image measurements y_n are also given to the controller for adapting its internal model to the current situation. The *output* of the controller is a relative movement of the robot in the camera coordinate system, a 6-dimensional vector $(x, y, z, yaw, pitch, roll)$ for a 6 DOF movement.

Controllers can be classified into approaches where the control law (or its parameters) are adapted over time, and approaches where they are fixed. Since these types of controllers can exhibit very different controlling behaviour we will split our considerations of controllers into these two parts, after some general considerations.

3.1 General Approach

Generally, in order to calculate the necessary camera movement u_n for a given desired image change $\Delta\tilde{y}_n := \tilde{y}_{n+1} - y_n$ we again use an approximation $\hat{\Phi}_n$ of Φ_n, for example the image Jacobian J_n. Then we select

$$u_n \in \underset{u \in \mathcal{U}(x_n)}{\operatorname{argmin}} \left\| \Delta\tilde{y}_n - \hat{\Phi}_n(u) \right\|_2^2. \tag{28}$$

where a given algorithm may or may not enforce a restriction $u \in \mathcal{U}(x_n)$ on the admissible movements when determining u. If this restriction is inactive and we are using a Jacobian, $\hat{\Phi}_n = J_n$, then the solution to (28) with minimum norm $\|u_n\|_2$ is given by

$$u_n = J_n^+ \, \Delta\tilde{y}_n \tag{29}$$

where J_n^+ is the *pseudo-inverse* of J_n.
With 4 coplanar object markings $m = 8$ and thereby $J_n \in \mathbb{R}^{8\times 6}$. One can show that J_n has maximum rank[1], so $\operatorname{rk} J_n = 6$. Then the pseudo-inverse $J_n^+ \in \mathbb{R}^{6\times 8}$ of J_n is given by:

$$J_n^+ = (J_n^T J_n)^{-1} J_n^T \tag{30}$$

(see e.g. Deuflhard and Hohmann, 2003, chapter 3).

When realising a control loop given such a controller one usually sets a fixed error threshold $\varepsilon > 0$ and repeats the steps

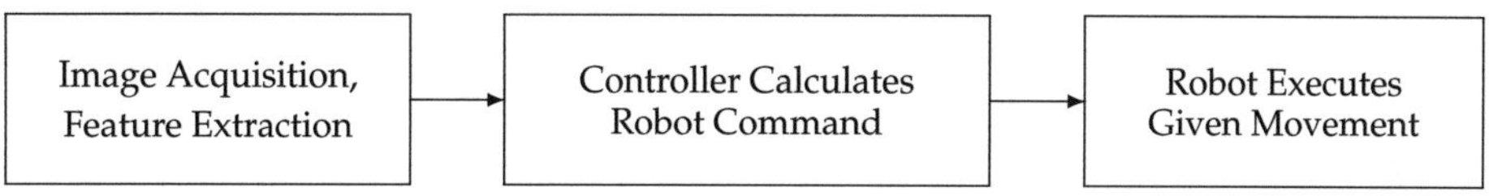

until

$$\|\Delta y_n\|_2 = \|y^\star - y_n\|_2 < \varepsilon, \tag{31}$$

or until

$$\|\Delta y_n\|_\infty = \|y^\star - y_n\|_\infty < \varepsilon \tag{32}$$

if one wants to stop only when the maximum deviation in *any* component of the image feature vector is below ε. Setting $\varepsilon := 0$ is not useful in practice since measurements even in the same pose tend to vary a little due to small movements of the robot arm or object as well as measurement errors and fluctuations.

3.2 Non-Adaptive Controllers

3.2.1 The Traditional Controller

The most simple controller, which we will call the *"Traditional Controller"* due to its heritage, is a straightforward proportional controller as known in engineering, or a dampened Gauss-Newton algorithm as it is known in mathematics.
Given an Image Jacobian J_n we first calculates the full Gauss-Newton step Δu_n for a complete movement to the goal in one step (desired image change $\Delta\tilde{y}_n := \Delta y_n$):

$$\Delta u_n := J_n^+ \, \Delta y_n \tag{33}$$

without enforcing a restriction $u \in \mathcal{U}(x_n)$ for the admissibility of a control command.
In order to ensure a convergence of the controller the resulting vector is then scaled with a dampening factor $0 < \lambda_n \leq 1$ to get the controller output u_n. In the traditional controller the factor λ_n is constant over time and the most important parameter of this algorithm. A typical value is $\lambda_n = \lambda = 0.1$; higher values may hinder convergence, while lower values also significantly slow down convergence. The resulting controller output u_n is given by

[1] One uses the fact that no 3 object markings are on a straight line, ${}^{C}z_i > 0$ for $i = 1, \ldots, 4$ and all markings are visible (in particular, neither all four ${}^{C}x_i$ nor all four ${}^{C}y_i$ are 0).

$$u_n := \lambda \cdot J_n^+ \, \Delta y_n. \tag{34}$$

3.2.2 Dynamical and Constant Image Jacobians

As mentioned in the previous section there are different ways of defining the Image Jacobian. It can be defined in the current pose, and is then calculated using the current distances to the object, ${}^{C}z_i$ for marking i, and the current image features. This is the *Dynamical Image Jacobian* J_n. An alternative is to define the Jacobian in the teach (goal) pose $x^\star$, with the image data $y^\star$ and distances at that pose. We call this the *Constant Image Jacobian* $J^\star$. Unlike J_n, $J^\star$ is constant over time and does not require image measurements for its adaptation to the current pose.

From a mathematical point of view the model J_n has a better validity in the current system state and should therefore yield better results. We shall later see whether this is the case in practice.

3.2.3 The Retreat-Advance Problem

Fig. 8. Camera view in the start pose with a pure rotation around the ${}^{C}z$ axis

When the robot's necessary movement to the goal pose is a pure rotation around the optical axis (${}^{C}z$, approach direction) there can be difficulties when using the standard Image Jacobian approach (Chaumette, 1998). The reason is that the linear approximation J_n models the relevant properties of Φ_n badly in these cases. This is also the case with $J^\star$ if this Jacobian is used. The former will cause an unnecessary movement away from the object, the latter a movement towards the goal. The larger the roll angle, the more pronounced is this phenomenon, an extreme case being a roll error of $\pm\pi$ (all other pose elements already equal to the teach pose) where the Jacobians suggest a pure movement along the ${}^{C}z$ axis. Corke and Hutchinson (2001) call this the *"Retreat-Advance Problem"* or the *"Chaumette Conundrum"*.

3.2.4 Controllers using the PMJ and MPJ Models

In order to overcome the Retreat-Advance Problem the so-called *"PMJ Controller"* (Malis, 2004) uses the pseudo-inverse of the mean of the two Jacobians J_n and $J^\star$. Using again a dampening factor $0 < \lambda \leq 1$ the controller output is given by

$$u_n = \lambda \cdot \left(\frac{1}{2} \left(J_n + J^\star \right) \right)^+ \Delta y_n. \tag{35}$$

Analogously, the *"MPJ Controller"* works with the mean of the pseudo-inverse of the Jacobians:

$$u_n = \lambda \cdot \left(\frac{1}{2} \left(J_n^+ + J^{\star +} \right) \right) \Delta y_n. \tag{36}$$

Otherwise, these controllers work like the traditional approach, with a constant dampening λ.

3.2.5 Defining the Controller in the Cylindrical Coordinate System

Using the linear model by Iwatsuki and Okiyama (2005) in the cylindrical coordinate system as discussed in Section 2.5.2 a special controller can also be defined. The authors define the image error for the ith object marking as follows:

$$e_i := \begin{pmatrix} \rho^\star - \rho \\ \rho(\varphi^\star - \varphi) \end{pmatrix} \tag{37}$$

where $(\rho, \varphi)^T$ is the current position and $(\rho^\star, \varphi^\star)$ the teach position. The control command u is then given by

$$u = \lambda \tilde{J}^+ e, \tag{38}$$

$\tilde{J}^+$ being the pseudo-inverse of the Image Jacobian in cylindrical coordinates from equation (18). e is the vector of pairs of image errors in the markings, i.e. a concatenation of the e_i vectors.

It should be noted that even if e is given in cylindrical coordinates, the output u of the controller is in Cartesian coordinates.

Due to the special properties of cylindrical coordinates, the calculation of the error and control command is very much dependent on the definition of the origin of the coordinate system. Iwatsuki and Okiyama (2005) therefore present a way to shift the origin of the coordinate system such that numerical difficulties are avoided.

One approach to select the origin of the cylindrical coordinate system is such that the current pose can be transformed to the desired (teach) pose with a pure rotation around the axis normal to the sensor plane, through the origin. For example, the general method given by Kanatani (1996) can be applied to this problem.

Let $l = (l_x, l_y, l_z)^T$ be the unit vector which defines this rotation axis, and $o = (o_x, o_y)^T$ the new origin, obtained by shifting the original origin $(0,0)^T$ in $\{S\}$ by $(\eta, \xi)^T$.

If $|l_z|$ is very small then the rotation axis l is almost parallel to the sensor. Then η and ξ are very large, which can create numerical difficulties. Since the resulting cylindrical coordinate system approximates a Cartesian coordinate system as $\eta, \xi \to \infty$, the standard Cartesian Image Jacobian J_n from (17) can therefore used if $|l_z| < \delta$ for a given lower limit δ.

3.3 Adaptive Controllers

Using adaptive controllers is a way to deal with errors in the model, or with problems resulting from the simplification of the model (e.g. linearisation, or the assumption that the camera works like a pinhole camera). The goal is to ensure a fast convergence of the controller in spite of these errors.

3.3.1 Trust Region-based Controllers

Trust Region methods are known from mathematics as globally convergent optimisation methods (Fletcher, 1987). In order to optimise "difficult" functions one uses a model of its properties, like we do here with the Image Jacobian. This model is adapted to the current state/position in the solution space, and therefore only valid within some region around the current state. The main idea in trust region methods is to keep track of the validity of the current system model, and adapt a so-called "*Trust Region*", or "*Model Trust Region*" around the current state within which the model does not exhibit more than a certain pre-defined "acceptable error".

To our knowledge the first person to use trust region methods for a visual servoing controller was Jägersand (1996). Since the method was adapted to a particular setup and cannot be used here we have developed a different trust region-based controller for our visual servoing scenario (Siebel et al., 1999). The main idea is to replace the constant dampening λ for Δu_n with a variable dampening λ_n:

$$u_n := \lambda_n \cdot \Delta u_n = \lambda_n \cdot J_n^+ \, \Delta y_n. \tag{39}$$

The goal is to adapt λ_n before each step to balance the avoidance of model errors (by making small steps) and the fast movement to the goal (by making large steps).

In order to achieve this balance we define an *actual model error* e_n which is set in relation to a *desired (maximum) model error* e_{des}[2] to adapt a bound α_n for the movement of projected object points on the sensor. Using this purely image-based formulation has advantages, e.g. having a measure to avoid movements that lead to losing object markings from the camera's field of view.

Our algorithm is explained in Figure 9 for one object marking. We wish to calculate a robot command to move such that the current point position on the sensor moves to its desired position. In step ①, we calculate an undampened robot movement Δu_n to move as close to this goal as possible ($\Delta \tilde{y}_n := \Delta y_n$) according to an Image Jacobian J_n:

$$\Delta u_n := J_n^+ \, \Delta y_n. \tag{40}$$

This gives us a predicted movement ℓ_n on the sensor, which we define as the maximum movement on the sensor for all M markings:

$$\ell_n := \max_{i=1,\dots,M} \left\| \begin{bmatrix} (J_n \, \Delta u_n)_{2i-1} \\ (J_n \, \Delta u_n)_{2i} \end{bmatrix} \right\|_2, \tag{41}$$

where the subscripts to the vector $J_n \, \Delta u_n$ signify a selection of its components.

Before executing the movement we restrict it in step ② such that the distance on the sensor is less or equal to a current limit α_n:

$$\begin{aligned} u_n &:= \lambda_n \cdot \Delta u_n \\ &= \min \left\{ 1, \frac{\alpha_n}{\ell_n} \right\} \cdot J_n^+ \, \Delta y_n. \end{aligned} \tag{42}$$

[2] While the name "desired error" may seem unintuitive the name is chosen intentionally since the α adaptation process (see below) can be regarded as a control process to have the robot system reach exactly this amount of error, by controlling the value of α_n.

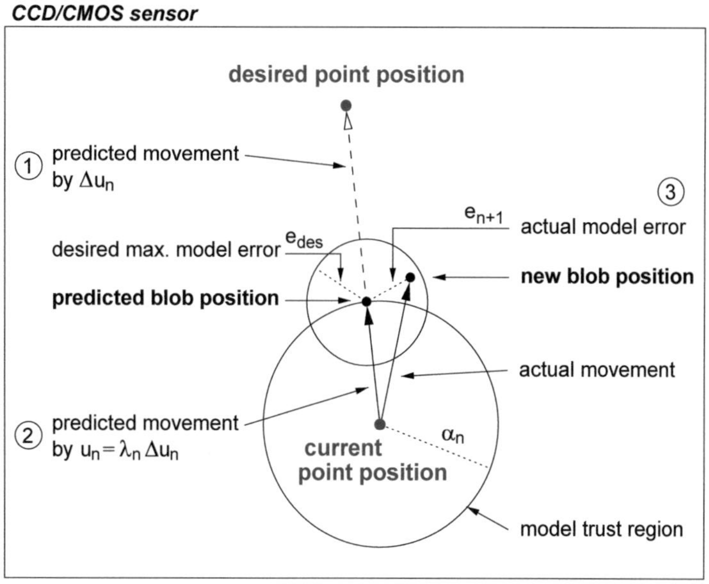

Fig. 9. Generation of a robot command by the trust region controller: view of the image sensor with a projected object marking

After this restricted movement is executed by the robot we obtain new measurements y_{n+1} and thereby the actual movement and model (prediction) error e_{n+1} ③, which we again define as the maximum deviation on the sensor for $M > 1$ markings:

$$e_{n+1} := \max_{i=1,\ldots,M} \left\| \begin{bmatrix} (\hat{y}_{n+1})_{2i-1} \\ (\hat{y}_{n+1})_{2i} \end{bmatrix} - \begin{bmatrix} (y_{n+1})_{2i-1} \\ (y_{n+1})_{2i} \end{bmatrix} \right\|_2 . \tag{43}$$

where $\hat{y}_{n+1}$ is the vector of predicted positions on the sensor,

$$\hat{y}_{n+1} := y_n + J_n\, u_n. \tag{44}$$

The next step is the adaptation of our restriction parameter α_n. This is done by comparing the model error e_{n+1} with a given desired (maximum admissible) error e_{des}:

$$r_{n+1} := \frac{e_{n+1}}{e_{\text{des}}} \tag{45}$$

where r_n is called the *relative model error*. A small value signifies a good agreement of model and reality. In order to balance model agreement and a speedy control we adjust α_n so as to achieve $r_n = 1$. Since we have a linear system model we can set

$$\alpha_{n+1} := \alpha_n \cdot \frac{e_{\text{des}}}{e_{n+1}} = \frac{\alpha_n}{r_{n+1}} \tag{46}$$

with an additional restriction on the change rate, $\frac{\alpha_{n+1}}{\alpha_n} \leq 2$. In practice, it may make sense to define minimum and maximum values $\alpha_{\min}$ and $\alpha_{\max}$ and set $\alpha_0 := \alpha_{\min}$.
In the example shown in Figure 9 the actual model error is smaller than e_{des}, so α_{n+1} can be larger than α_n.

```
Let n := 0; α_0 := α_start; y* given
Measure current image features y_n and calculate Δy_n := y* − y_n
WHILE ||Δy_n||_∞ ≥ ε
  Calculate J_n
  IF n > 0
    Calculate relative model error r_n via (43)
    Adapt α_n by (46)
  END IF
  Calculate u_sd_n := J_n^T Δy_n, λ_n := ||u_sd_n|| / ℓ_sd_n and u_gn_n := J_n^+ Δy_n
  Calculate u_dl_n via (52)
  Send control command u_dl_n to the robot
  Measure y_{n+1} and calculate Δy_{n+1}; let n := n + 1
END WHILE
```

Fig. 10. Algorithm: Image-based Visual Servoing with the Dogleg Algorithm

3.3.1.1 Remark:

By restricting the movement on the sensor we have implicitly defined the set $\mathcal{U}(x_n)$ of admissible control commands in the state x_n as in equation (33). This $\mathcal{U}(x_n)$ is the trust region of the model J_n.

3.3.2 A Dogleg Trust Region Controller

Powell (1970) describes the so-called *Dogleg Method* (a term known from golf) which can be regarded as a variant of the standard trust region method (Fletcher, 1987; Madsen et al., 1999). Just like in the trust region method above, a current model error is defined and used to adapt a trust region. Depending on the model error, the controller varies between a Gauss-Newton and a gradient (steepest descent) type controller.

The undampened Gauss-Newton step u_{gn_n} is calculated as before:

$$u_{\text{gn}_n} = J_n^{+} \Delta y_n, \tag{47}$$

and the steepest descent step u_{sd_n} is given by

$$u_{\text{sd}_n} = J_n^{T} \Delta y_n. \tag{48}$$

The dampening factor λ_n is set to

$$\lambda_n := \frac{\|u_{\text{sd}_n}\|_2^2}{\ell_{\text{sd}_n}} \tag{49}$$

where again

$$\ell_{\text{sd}_n} := \max_{i=0,\dots,M} \left\| \begin{pmatrix} (\Delta\hat{y}_{\text{sd}_n})_{2i-1} \\ (\Delta\hat{y}_{\text{sd}_n})_{2i} \end{pmatrix} \right\|_2^2 \tag{50}$$

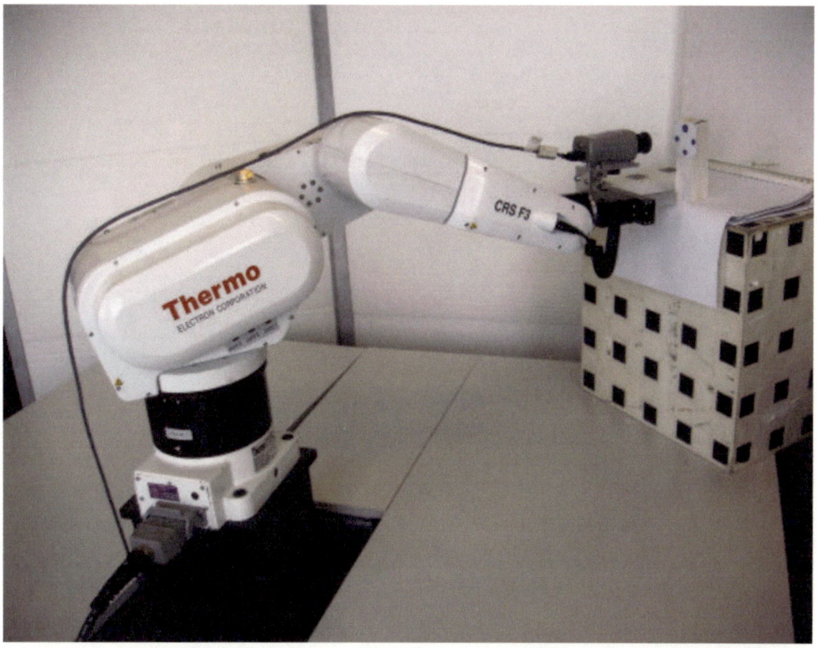

Fig. 11. Experimental setup with Thermo CRS F3 robot, camera and marked object

is the maximum predicted movement on the sensor, here the one caused by the steepest descent step u_{sd_n}. Analogously, let

$$\ell_{\mathrm{gn}_n} := \max_{i=0,\ldots,M} \left\| \begin{pmatrix} (\Delta\hat{y}_{\mathrm{gn}_n})_{2i-1} \\ (\Delta\hat{y}_{\mathrm{gn}_n})_{2i} \end{pmatrix} \right\|_2^2 \tag{51}$$

be the maximum predicted movement by the Gauss Newton step. With these variables the dog leg step $u_n = u_{\mathrm{dl}_n}$ is calculated as follows:

$$u_{\mathrm{dl}_n} := \begin{cases} u_{\mathrm{gn}_n} & \text{if } \ell_{\mathrm{gn}_n} \leq \alpha_n \\ \alpha_n \frac{u_{\mathrm{sd}n}}{\|u_{\mathrm{sd}n}\|_2} & \text{if } \ell_{\mathrm{gn}_n} > \alpha_n \text{ and } \ell_{\mathrm{sd}_n} \geq \alpha_n \\ \lambda_n u_{\mathrm{sd}_n} + \beta_n (u_{\mathrm{gn}_n} - \lambda_n u_{\mathrm{sd}_n}) & \text{else} \end{cases} \tag{52}$$

where in the third case β_n is chosen such that the maximum movement on the sensor has length α_n.

The complete dogleg algorithm for visual servoing is shown in Figure 10.

4. Experimental Evaluation

4.1 Experimental Setup and Test Methods

The robot setup used in the experimental validation of the presented controllers is shown in Figure 11. Again a eye-in-hand configuration and an object with 4 identifiable markings are used. Experiments were carried out both on a Thermo CRS F3 (pictured here) and on a Unimation Stäubli RX-90 (Figure 2 at the beginning of the chapter). In the following only

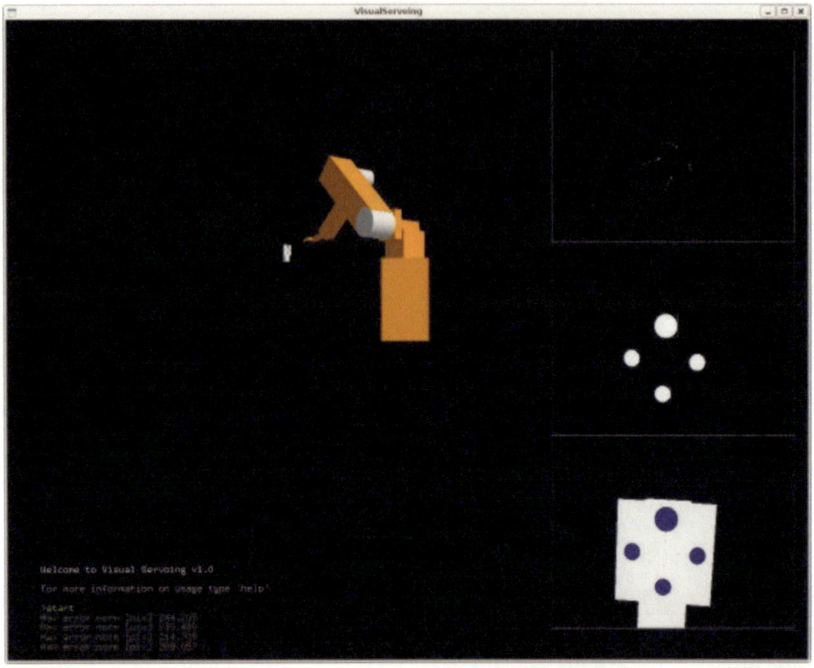

Fig. 12. OpenGL Simulation of camera-robot system with simulated camera image (bottom right), extracted features (centre right) and trace of objects markings on the sensor (top right)

the CRS F3 experiments are considered; the results with the Stäubli RX-90 were found to be equivalent. The camera was a Sony DFW-X710 with IEEE1394 interface, 1024×768 pixel resolution and an $f = 6.5\,\text{mm}$ lens.

In addition to the experiments with a real robot two types of simulations were used to study the behaviour of controllers and models in detail. In our *OpenGL Simulation*[3], see Figure 12, the complete camera-robot system is modelled. This includes the complete robot arm with inverse kinematics, rendering of the camera image in a realistic resolution and application of the same image processing algorithms as in the real experiments to obtain the image features. Arbitrary robots can be defined by their Denavit-Hartenberg parameters (cf. Spong et al., 2005) and geometry in an XML file. The screenshot above shows an approximation of the Stäubli RX-90.

The second simulation we use is the *Multi-Pose Test*. It is a system that uses the exact model as derived in Section 2.2, without the image generation and digitisation steps as in the OpenGL Simulation. Instead, image coordinates of objects points as seen by the camera are calculated directly with the pinhole camera model. Noise can be added to these measurements in order to examine how methods react to these errors. Due to the small computational complexity of the Multi-Pose Test it can be, and has been applied to many start and teach pose combinations (in our experiments, 69,463 start poses and 29 teach poses). For a given algorithm and parameter set the convergence behaviour (success rate and speed) can thus be studied on a statistically relevant amount of data.

[3] The main parts of simulator were developed by Andreas Jordt and Falko Kellner when they were students in the Cognitive Systems Group.

4.2 List of Models and Controllers Tested

In order to test the advantages and disadvantages of the models and controllers presented above we combine them in the following way:

Short Name	Controller	Model	Parameters
Trad const	Traditional	$\Delta y_n \approx J^\star u$	$\lambda = 0.2$
Trad dyn	Traditional	$\Delta y_n \approx J_n u$	$\lambda = 0.1$, sometimes $\lambda = 0.07$
Trad PMJ	Traditional	$\Delta y_n \approx \frac{1}{2}(J_n + J^\star)\, u$	$\lambda = 0.25$
Trad MPJ	Traditional	$u \approx \frac{1}{2}\,(J_n^+ + J^{\star+})\,\Delta y_n$	$\lambda = 0.15$
Trad cyl	Traditional	$\Delta y_n \approx \tilde{J}_n u$ (cylindrical)	$\lambda = 0.1$
TR const	Trust-Region	$\Delta y_n \approx J^\star u$	$\alpha_0 = 0.09, e_{\text{des}} = 0.18$
TR dyn	Trust-Region	$\Delta y_n \approx J_n u$	$\alpha_0 = 0.07, e_{\text{des}} = 0.04$
TR PMJ	Trust-Region	$\Delta y_n \approx \frac{1}{2}(J_n + J^\star)\, u$	$\alpha_0 = 0.07, e_{\text{des}} = 0.09$
TR MPJ	Trust-Region	$u \approx \frac{1}{2}\,(J_n^+ + J^{\star+})\,\Delta y_n$	$\alpha_0 = 0.05, e_{\text{des}} = 0.1$
TR cyl	Trust-Region	$\Delta y_n \approx \tilde{J}_n u$ (cylindrical)	$\alpha_0 = 0.04, e_{\text{des}} = 0.1$
Dogleg const	Dogleg	$u \approx J^{\star+}\Delta y_n$ and $u \approx J_n^T \Delta y_n$	$\alpha_0 = 0.22, e_{\text{des}} = 0.16, \lambda = 0.5$
Dogleg dyn	Dogleg	$u \approx J_n^+ \Delta y_n$ and $u \approx J_n^T \Delta y_n$	$\alpha_0 = 0.11, e_{\text{des}} = 0.28, \lambda = 0.5$
Dogleg PMJ	Dogleg	$\Delta y_n \approx \frac{1}{2}(J_n + J^\star)\, u$ and $u \approx J_n^T \Delta y_n$	$\alpha_0 = 0.29, e_{\text{des}} = 0.03, \lambda = 0.5$
Dogleg MPJ	Dogleg	$u \approx \frac{1}{2}\,(J_n^+ + J^{\star+})\,\Delta y_n$ and $u \approx J_n^T \Delta y_n$	$\alpha_0 = 0.3, e_{\text{des}} = 0.02, \lambda = 0.5$

Here we use the definitions as before. In particular, J_n is the dynamical Image Jacobian as defined in the current pose, calculated using the current distances to the object, ${}^{C}z_i$ for marking i, and the current image features in its entries. The distance to the object is estimated in the real experiments using the known relative distances of the object markings, which yields a fairly precise estimate in practice. $J^\star$ is the constant Image Jacobian, defined in the teach (goal) pose $x^\star$, with the image data $y^\star$ and distances at that pose. $\Delta y_n = y_{n+1} - y_n$ is the change in the image predicted by the model with the robot command u.

The values of the parameters detailed above were found to be useful parameters in the Multi-Pose Test. They were therefore used in the experiments with the real robot and the OpenGL Simulator. See below for details on how these values were obtained.

λ is the constant dampening factor applied as the last step of the controller output calculation. The Dogleg controller did not converge in our experiments without such an additional dampening which we set to 0.5. The Trust-Region controller works without additional dampening. α_0 is the start and minimum value of α_n. These, as well as the desired model error e_{des} are given in mm on the sensor. The sensor measures 4.8×3.6 mm which means that at its 1024×768 pixel resolution 0.1 mm $\approx$ 22 pixels after digitisation.

4.3 Experiments and Results

The Multi-Pose Test was run first in order to find out which values of parameters are useful for which controller/model combination. 69,463 start poses and 29 teach poses were combined randomly into 69,463 fixed pairs of tasks that make up the training data. We studied the following two properties and their dependence on the algorithm parameters:

1. *Speed:* The number of iterations (steps/robot movements) needed for the algorithm to reach its goal. The mean number of iterations over all successful trials is measured.
2. *Success rate:* The percentage of experiments that reached the goal. Those runs where an object marking was lost from the camera view by a movement that was too large and/or mis-directed were considered not successful, as were those that did not reach the goal within 100 iterations.

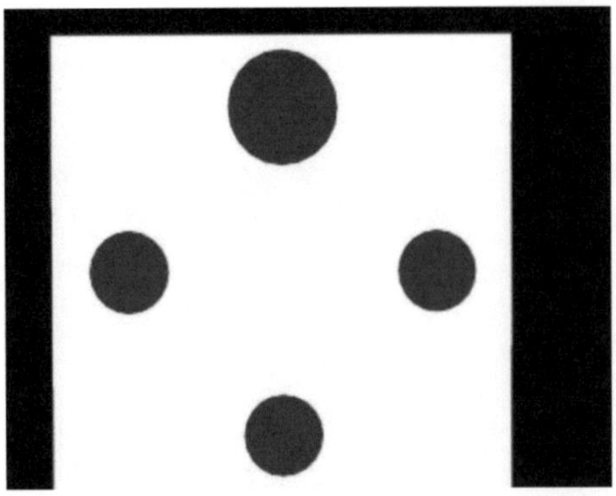

(a) Teach pose

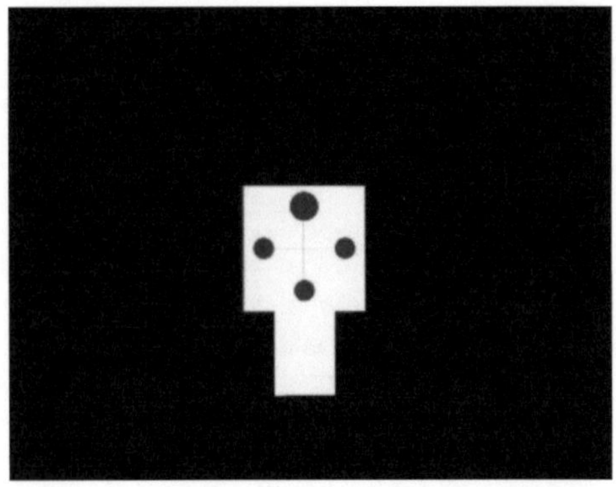

(b) Pose 1 (0,0,-300,0°,0°,0°)

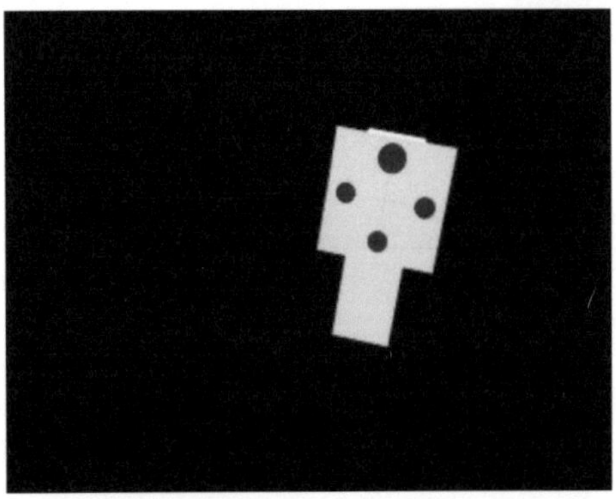

(c) Pose 2 (20,-50,-300,-10°,-10°,-10°)

(d) Pose 3 (0,0,0,-5°,-3°,23°)

(e) Pose 4 (150,90,-200,10°,-15°,30°)

(f) Pose 5 (0,0,0,0°,0°,45°)

Fig. 13. Teach and start poses used in the experiments; shown here are simulated camera images in the OpenGL Simulator. Given for each pose is the relative movement in {C} from the teach pose to the start pose. Start pose 4 is particularly difficult since it requires both a far reach and a significant rotation by the robot. Effects of the linearisation of the model or errors in its parameters are likely to cause a movement after which an object has been lost from the camera's field of view. Pose 5 is a pure rotation, chosen to test for the retreat-advance problem.

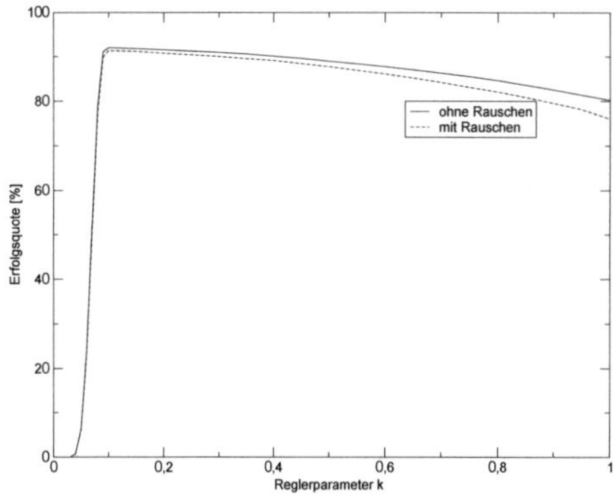

(a) Trad const, success rate

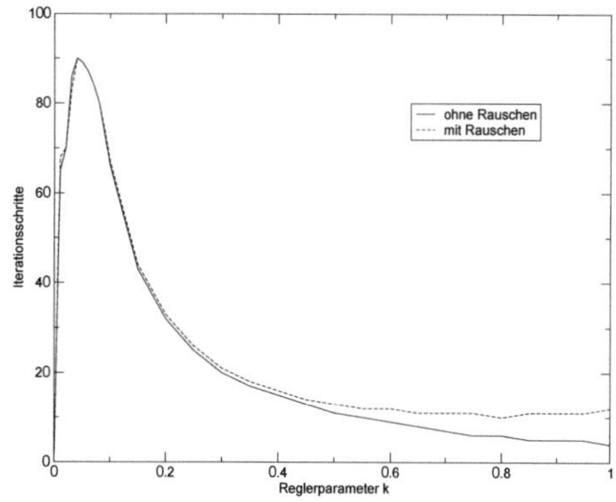

(b) Trad const, speed

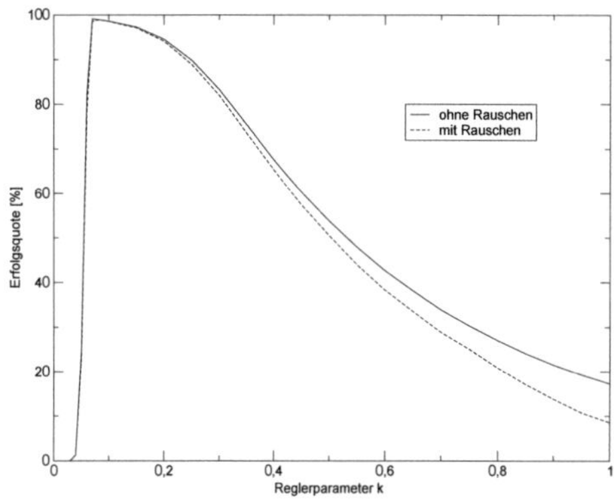

(c) Trad dyn, success rate

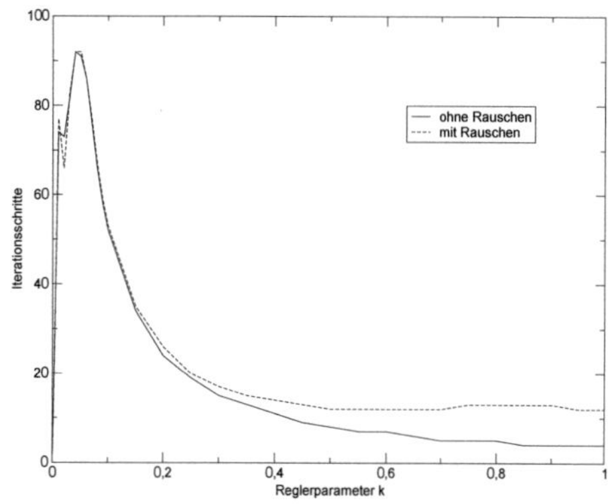

(d) Trad dyn, speed

Fig. 14. Multi-Pose Test: Traditional Controller with const. and dyn. Jacobian. Success rate and average speed (number of iterations) are plotted as a function of the dampening parameter λ.

Using the optimal parameters found by the Multi-Pose Test we ran experiments on the real robot. Figure 13 shows the camera images (from the OpenGL simulation) in the teach pose and five start poses chosen such that they cover the most important problems in visual servoing. The OpenGL simulator served as an additional useful tool to analyse why some controllers with some parameters would not perform well in a few cases.

4.4 Results with Non-Adaptive Controllers

Figures 14 and 15 show the results of the Multi-Pose Test with the Traditional Controller using different models. For the *success rates* it can be seen that with λ-values below a certain value $\approx$ 0.06–0.07 the percentages are very low. On the other hand, raising λ above $\approx$ 0.08–0.1 also significantly decreases success rates. The reason is the proportionality of image error and (length of the) robot movement inherent in the control law with its constant factor λ. During the course of the servoing process the norm of the image error may vary by as much as a factor of 400. The controller output varies proportionally. This means that at the beginning of the control process very large movements are carried out, and very small movements at the end.

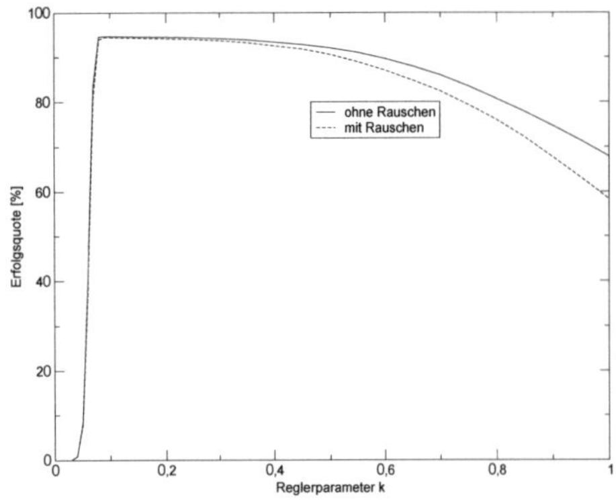

(a) Trad PMJ, success rate

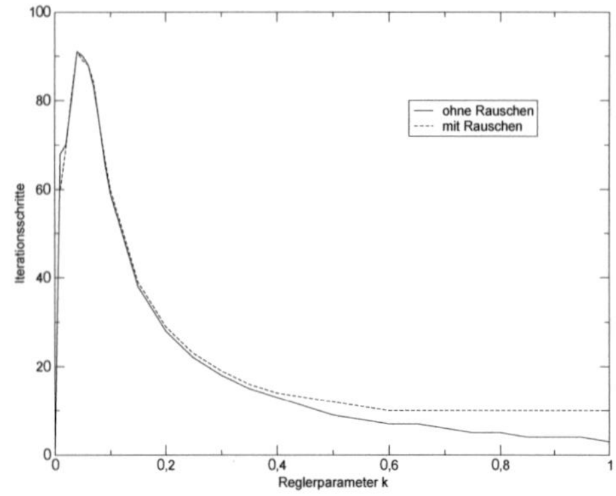

(b) Trad PMJ, speed

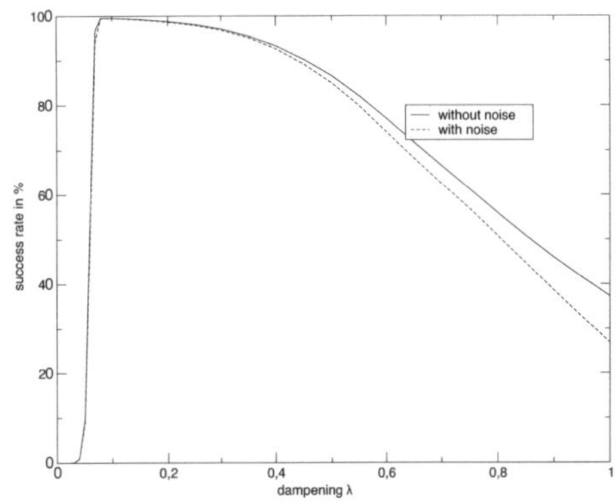

(c) Trad MJP, success rate

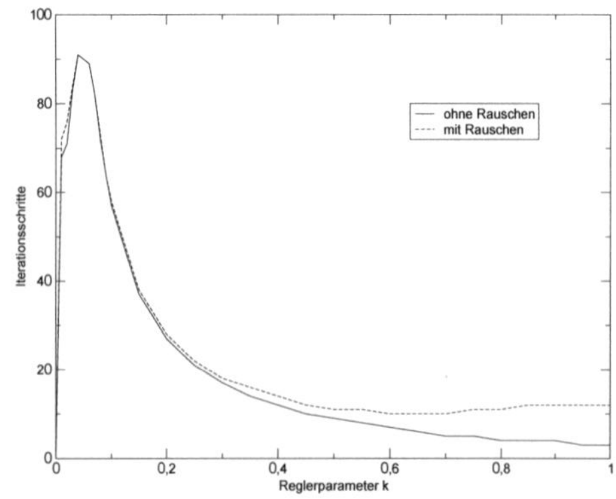

(d) Trad MJP, speed

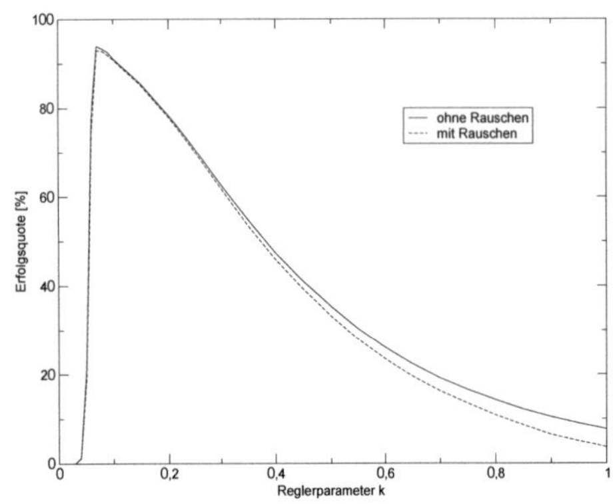

(e) Trad cyl, success rate

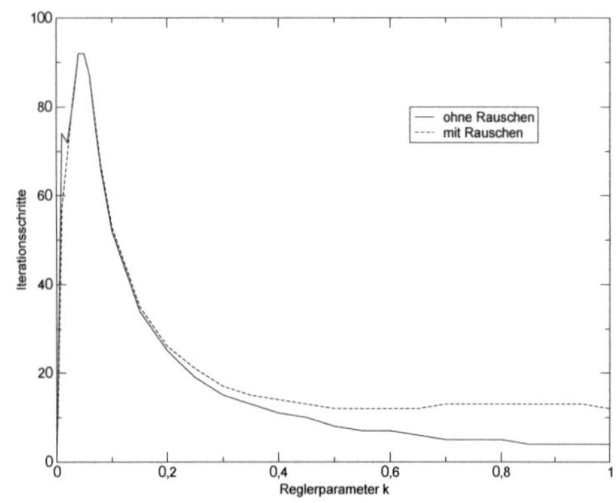

(f) Trad cyl, speed

Fig. 15. Multi-Pose Test: Traditional Controller with PMJ, MPJ and cylindrical models. Shown here are again the success rate and speed (average number of iterations of successful runs) depending on the constant dampening factor λ. As before, runs that did not converge in the first 100 steps were considered unsuccessful.

Controller	param.	Real Robot start pose					OpenGL Sim. start pose					Multi-Pose	
	λ	1	2	3	4	5	1	2	3	4	5	speed (iter.)	success (%)
Trad const	0.2	49	55	21	46	31	44	44	23	44	23	32	91.53
Trad dyn	0.1	63	70	48	∞	58	46	52	45	∞	47	52	98.59
	0.07				121					81		76	99.11
Trad MJP	0.15	41	51	33	46	37	35	39	31	41	32	37	99.27
Trad PMJ	0.25	29	29	17	∞	35	26	26	18	∞	32	38	94.52
Trad cyl	0.1	59	∞	50	70	38	46	49	49	58	49	52	91.18

Table 1. All results, Traditional Controller, optimal value of λ. "∞" means no convergence

The movements at the beginning need strong dampening (small λ) in order to avoid large misdirected movements (Jacobians usually do not have enough validity for 400 mm movements), those at the end need little or no dampening (λ near 1) when only a few mm are left to move. The version with the constant image Jacobian has a better behaviour for larger (≥ 0.3) values of λ, although even the optimum value of $\lambda = 0.1$ only gives a success rate of 91.99 %. The behaviour for large λ can be explained by $J^{\star}$'s smaller validity away from the teach pose; when the robot is far away it suggests smaller movements than J_n would. In practise this acts like an additional dampening factor that is stronger further away from the object.
The adaptive Jacobian gives the controller a significant advantage if λ is set well. For $\lambda = 0.07$ the success rate is 99.11 %, albeit with a speed penalty, at as many as 76 iterations. With $\lambda = 0.1$ this decreases to 52 at 98.59 % success rate.
The use of the PMJ and MJP models show again a more graceful degradation of performance with increasing λ than J_n. The behaviour with PMJ is comparable to that with $J^{\star}$, with a maximum of 94.65 % success at $\lambda = 0.1$; here the speed is 59 iterations. Faster larger λ, e.g. 0.15 which gives 38 iterations, the success rate is still at 94.52 %. With MJP a success rate of 99.53 % can be achieved at $\lambda = 0.08$, however, the speed is slow at 72 iterations. At $\lambda = 0.15$ the controller still holds up well with 99.27 % success and significantly less iterations: on average 37.
Using the cylindrical model the traditional controller's success is very much dependant on λ. The success rate peaks at $\lambda = 0.07$ with 93.94 % success and 76 iterations; a speed 52 can be achieved at $\lambda = 0.1$ with 91.18 % success. Overall the cylindrical model does not show an overall advantage in this test.
Table 1 shows all results for the traditional controller, including real robot and OpenGL results. It can be seen that even the most simple pose takes at least 29 steps to solve. The Trad MJP method is the clearly the winner in this comparison, with a 99.27 % success rate and on average 37 iterations. Pose 4 holds the most difficulties, both in the real world and in the OpenGL simulation. In the first few steps a movement is calculated that makes the robot lose the object from the camera's field of view. The Traditional Controller with the dynamical Jacobian achieves convergence only when λ is reduced from 0.1 to 0.07. Even then the object marking comes close to the image border during the movement. This can be seen in Figure 16 where the trace of the centre of the object markings on the sensor is plotted. With the cylindrical model the controller moves the robot in a way which avoids this problem. Figure 16(b) shows that there is no movement towards the edge of the image whatsoever.

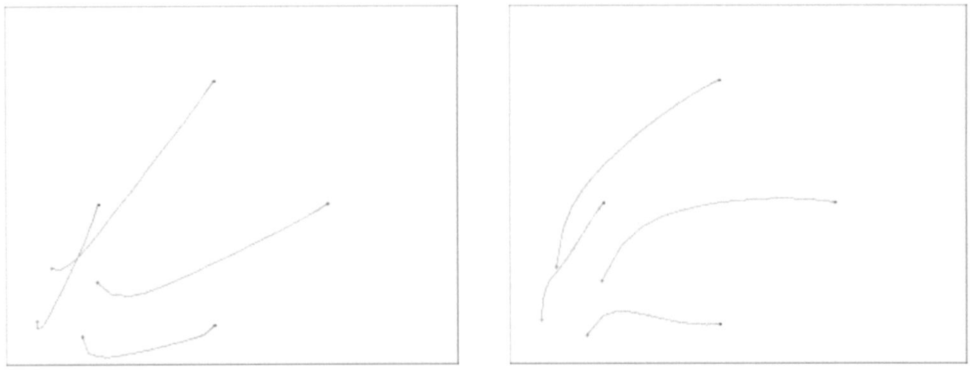

(a) Trad dyn, $\lambda = 0.07, 81$ steps (b) Trad cyl, $\lambda = 0.1, 58$ steps

Fig. 16. Trad. Controller, dyn. and cyl. model, trace of markings on sensor, pose 4 (OpenGL).

4.5 Results with Adaptive Controllers

In this section we wish to find out whether the use of dynamical dampening by a limitation of the movement on the sensor (image-based trust region methods) can speed up the slow convergence of the traditional controller. We will examine the Trust-Region controller first, then the Dogleg controller.

Figure 17 shows the behaviour for the constant and dynamical Jacobians as a function of the main parameter, the desired maximum model error e_{des}. The success rate for both variants is only slightly dependent on e_{des}, with rates over 91 % (Trust const) and 99 % (Trust dyn) for the whole range of values from 0.01 to 0.13 mm when run without noise. The speed is significantly faster than with the Traditional Controller at 13 iterations ($e_{\text{des}} = 0.18, 91.46$ % success) and 8 iterations ($e_{\text{des}} = 0.04, 99.37$ % success), respectively. By limiting the step size dynamically the Trust Region methods calculate smaller movements than the Traditional Controller at the beginning of the experiment but significantly larger movements near the end. This explains the success rate (no problems at beginning) and speed advantage (no active dampening towards the end). The use of the mathematically more meaningful dynamical model J_n helps here since the Trust Region method avoids the large mis-directed movements far away from the target without the need of the artificial dampening through $J^{\star}$. The Trust/dyn. combination shows a strong sensitivity to noise; this is mainly due to the amplitude of the noise (standard deviation 1 pixel) which exceeds the measurement errors in practice when the camera is close to the object. This results in convergence problems and problems detecting convergence when the robot is very close to its goal pose. In practise (see e.g. Table 2 below) the controller tends to have fewer problems. In all five test poses, even the difficult pose 4 the controller converges with both models without special adjustment (real world and OpenGL), with a significant speed advantage of the dynamical model. In pose 5 both are delayed by the retreat-advance problem but manage to reach the goal successfully.

The use of the MJP model helps the Trust-Region Controller to further improve its results. Success rates (see Figure 18) are as high as 99.68 % at $e_{\text{des}} = 0.01$ (on average 16 iterations), with a slightly decreasing value when e_{des} is increased: still 99.58 % at $e_{\text{des}} = 0.1$ (7 iterations, which makes it the fastest controller/model combination in our tests).

As with the Traditional Controller the use of the PMJ and cylindrical model do not show overall improvements for visual servoing over the dynamical method. The results, are also

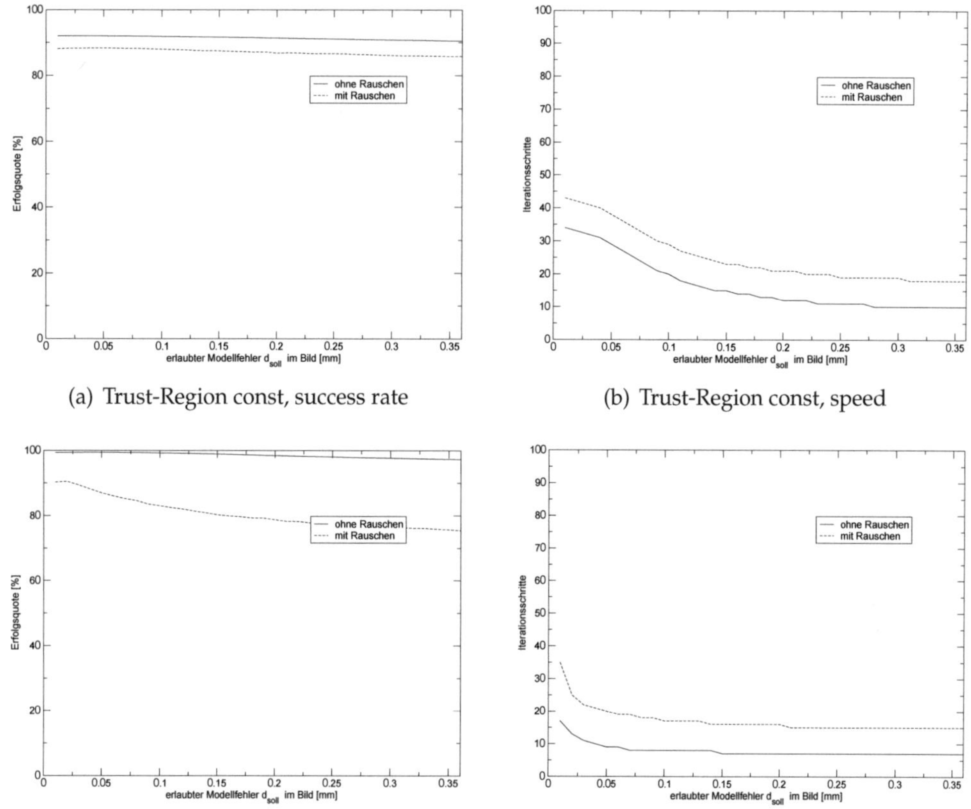

(a) Trust-Region const, success rate

(b) Trust-Region const, speed

(c) Trust-Region dyn, success rate

(d) Trust-Region dyn, speed

Fig. 17. Multi-Pose Test: Trust-Region Controller with const. and dyn. Jacobian

shown also in Figure 18. Table 2 details the results for all three types of tests. It can be seen that while both models have on average better results than with the constant Jacobian they do have convergence problems that show in the real world. In pose 2 (real robot) the cylindrical model causes the controller to calculate an unreachable pose for the robot at the beginning, which is why the experiment was terminated and counted as unsuccessful.

The Dogleg Controller shows difficulties irrespective of the model used. Without an additional dampening with a constant $\lambda = 0.5$ no good convergence could be achieved. Even with dampening its maximum success rate is only 85 %, with $J^\star$ (at an average of 10 iterations). Details for this combination are shown in Figure 19 where we see that the results cannot be improved by adjusting the parameter e_{des}. With other models only less than one in three poses can be solved, see results in Table 2.

A thorough analysis showed that the switching between gradient descent and Gauss-Newton steps causes the problems for the Dogleg controller. This change in strategy can be seen in Figure 20 where again the trace of projected object markings on the sensor is shown (from the real robot system). The controller first tries to move the object markings towards the centre of the image, by applying gradient descent steps. This is achieved by changing yaw and pitch angles only. Then the Dogleg step, i.e. a combination of gradient descent and Gauss-Newton

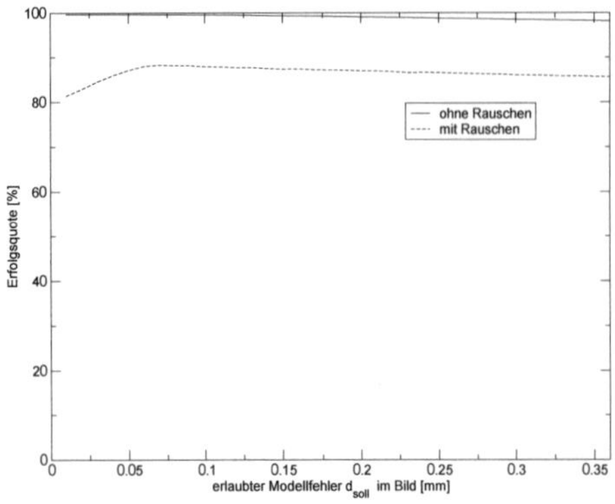

(a) Trust-Region MJP, success rate

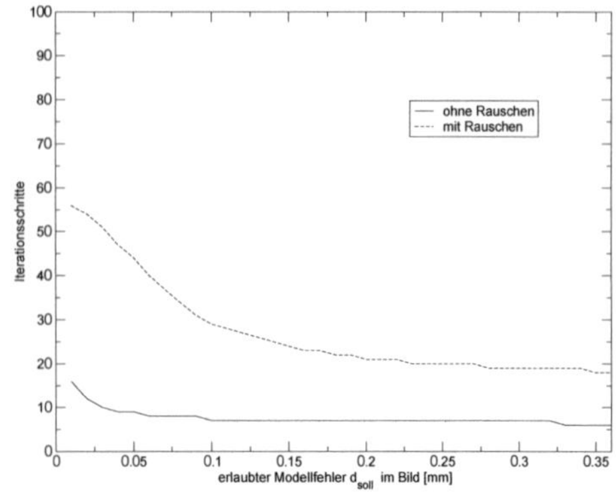

(b) Trust-Region MJP, speed

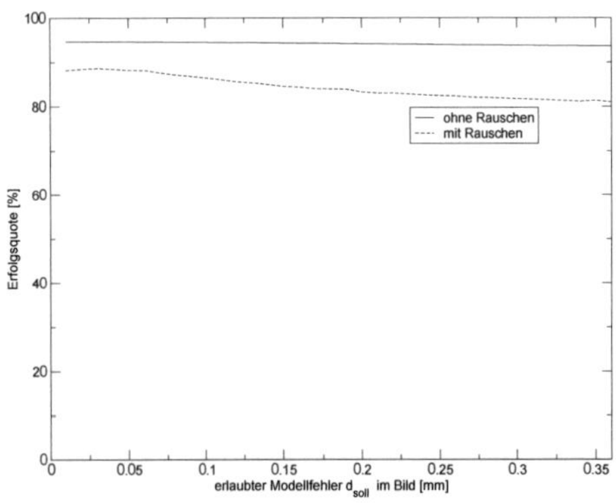

(c) Trust-Region PMJ, success rate

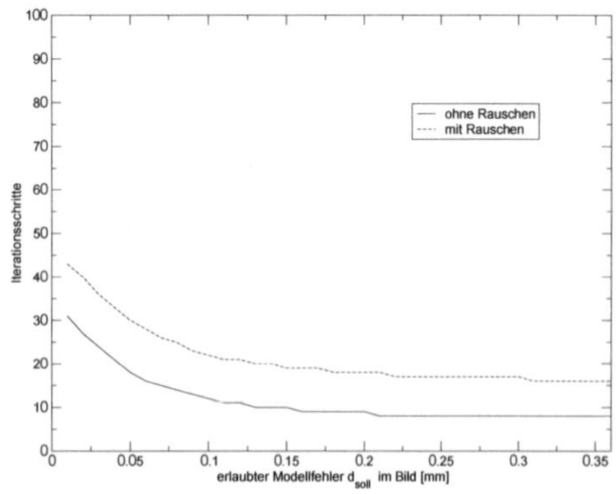

(d) Trust-Region PMJ, speed

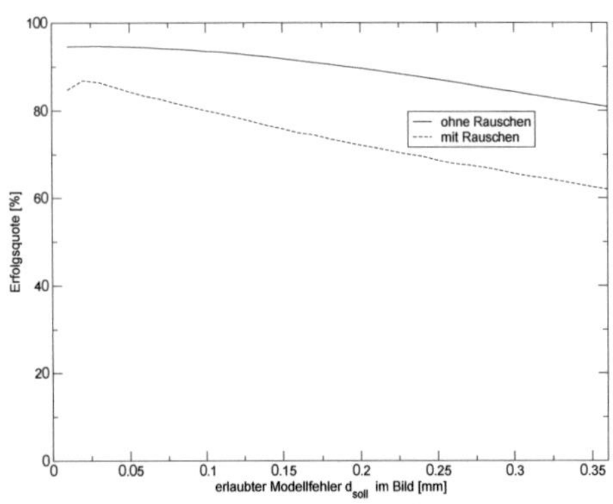

(e) Trust-Region cyl, success rate

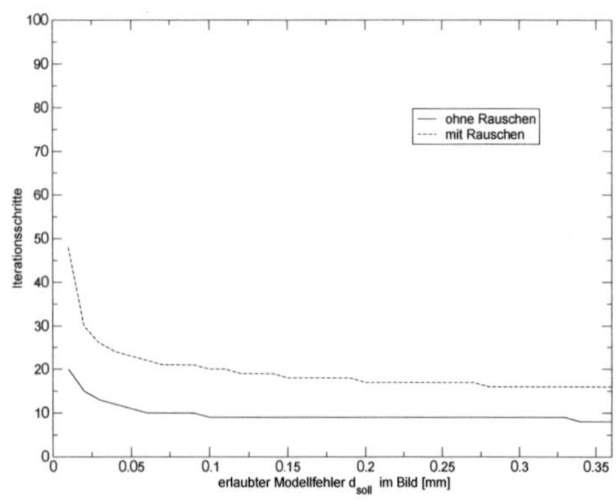

(f) Trust-Region cyl, speed

Fig. 18. Multi-Pose Test: Trust-Region Controller with PMJ, MPJ and cylindrical model. Plotted are the success rate and the speed (average number of iterations of successful runs) depending on the desired (maximum admissible) error, e_{des}.

Controller	param.		Real Robot start pose					OpenGL Sim. start pose					Multi-Pose speed	Multi-Pose success
	α_{start}	e_{des}	1	2	3	4	5	1	2	3	4	5	(iter.)	(%)
Trust const	0.09	0.18	22	29	11	39	7	20	26	6	31	7	13	91.46
Trust dyn	0.07	0.04	10	15	9	17	17	9	12	7	14	6	8	99.37
Trust MJP	0.05	0.1	8	9	11	13	7	7	9	6	11	5	7	99.58
Trust PMJ	0.07	0.09	21	28	7	∞	13	20	25	6	∞	5	13	94.57
Trust cyl	0.04	0.1	10	∞	7	11	15	8	18	6	11	6	9	93.5
Dogleg const	0.22	0.16	19	24	8	∞	12	17	25	4	21	9	10	85.05
Dogleg dyn	0.11	0.28	13	∞	∞	∞	13	8	∞	6	∞	16	9	8.4
Dogleg MJP	0.3	0.02	∞	∞	10	∞	13	∞	∞	5	∞	7	8	26.65
Dogleg PMJ	0.29	0.03	14	13	5	∞	12	9	13	5	14	7	8	31.47

Table 2. All results, Trust-Region and Dogleg Controllers. "∞" means no success.

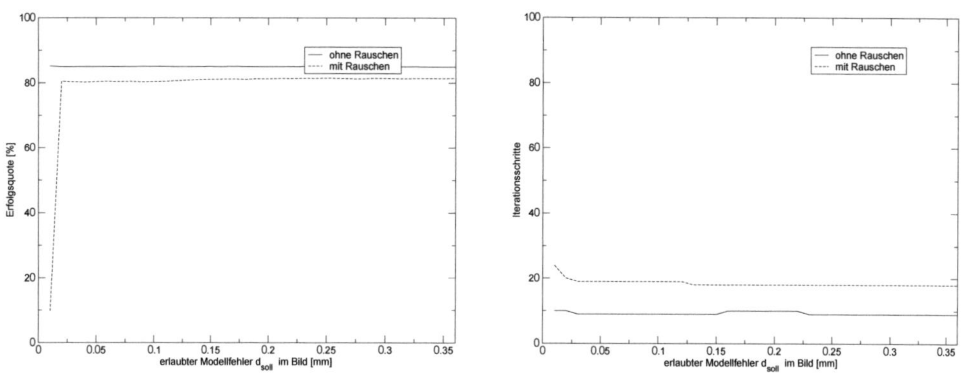

(a) Dogleg const, success rate (b) Dogleg const, speed

Fig. 19. Multi-Pose Test: Dogleg Controller with constant Image Jacobian

step (with the respective Jacobian), is applied. This causes zigzag movements on the sensor. These are stronger when the controller switches back and forth between the two approaches, which is the case whenever the predicted and actual movements differ by a large amount.

5. Analysis and Conclusion

In this chapter we have described and analysed a number of visual servoing controllers and models of the camera-robot system used by these controllers. The inherent problem of the traditional types of controllers is the fact that these controllers do not adapt their controller output to the current state in which the robot is: far away from the object, close to the object, strongly rotated, weakly rotated etc. They also cannot adapt to the strengths and deficiencies of the model, which may also vary with the current system state. In order to guarantee successful robot movements towards the object these controllers need to restrict the steps the robot takes, and they do so by using a constant scale factor ("dampening"). The constancy of this scale factor is a problem when the robot is close to the object as it slows down the movements too much.

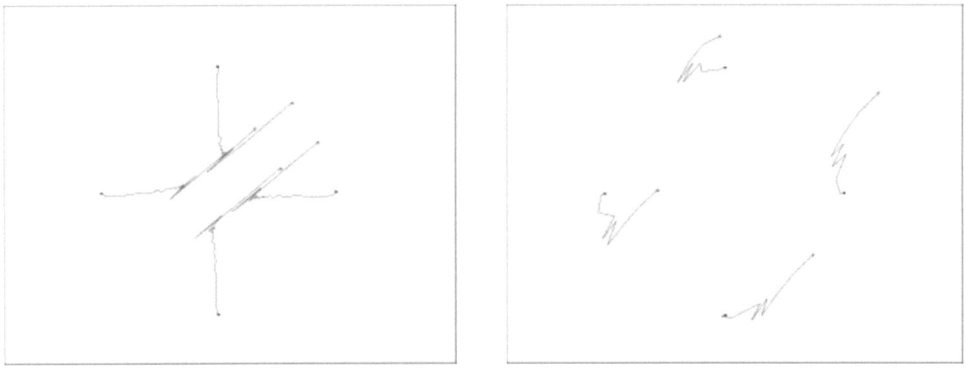

(a) Dogleg const, pose 2, 24 steps (b) Dogleg MJP, pose 3, 10 steps

Fig. 20. Dogleg, const and MJP model, trace of markings on sensor, poses 2 and 3 (real robot).

Trust-region based controllers successfully overcome this limitation by adapting the dampening factor in situations where this is necessary, but only in those cases. Therefore they achieve both a better success rate and a significantly higher speed than traditional controllers.
The Dogleg controller which was also tested does work well with some poses, but on average has much more convergence problems than the other two types of controllers.
Overall the Trust-Region controller has shown the best results in our tests, especially when combined with the MJP model, and almost identical results when the dynamical image Jacobian model is used. These models are more powerful than the constant image Jacobian which almost always performs worse.
The use of the cylindrical and PMJ models did not prove to be helpful in most cases, and in those few cases where they have improved the results (usually pure rotations, which is unlikely in most applications) the dynamical and MJP models also achieved good results.
The results found in experiments with a real robot and those carried out in two types of simulation agree on these outcomes.

Acknowledgements

Part of the visual servoing algorithm using a trust region method presented in this chapter was conceived in 1998–1999 while the first author was at the University of Bremen. The advice of Oliver Lang and Fabian Wirth at that time is gratefully acknowledged.

6. References

François Chaumette. Potential problems of stability and convergence in image-based and position-based visual servoing. In David J Kriegmann, Gregory D Hager, and Stephen Morse, editors, *The Confluence of Vision and Control*, pages 66–78. Springer Verlag, New York, USA, 1998.

François Chaumette and Seth Hutchinson. Visual servoing and visual tracking. In Bruno Siciliano and Oussama Khatib, editors, *Springer Handbook of Robotics*, pages 563–583. Springer Verlag, Berlin, Germany, 2008.

Peter I. Corke and Seth A. Hutchinson. A new partioned approach to image-based visual servo control. *IEEE Transactions on Robotics and Automation*, 237(4):507–515, August 2001.

Peter Deuflhard and Andreas Hohmann. *Numerical Analysis in Modern Scientific Computing: An Introduction*. Springer Verlag, New York, USA, 2nd edition, 2003.

Roger Fletcher. *Practical Methods of Optimization*. John Wiley & Sons, New York, Chichester, 2nd edition, 1987.

Seth Hutchinson, Gregory D Hager, and Peter Corke. A tutorial on visual servo control. Tutorial notes, Yale University, New Haven, USA, May 1996.

Masami Iwatsuki and Norimitsu Okiyama. A new formulation of visual servoing based on cylindrical coordinate system. *IEEE Transactions on Robotics*, 21(2):266–273, April 2005.

Martin Jägersand. Visual servoing using trust region methods and estimation of the full coupled visual-motor Jacobian. In *Proceedings of the IASTED Applications of Control and Robotics, Orlando, USA*, pages 105–108, January 1996.

Kenichi Kanatani. *Statistical Optimization for Geometric Computation: Theory and Practice*. Elsevier Science, Amsterdam, The Netherlands, 1996.

Kaj Madsen, Hans Bruun Nielsen, and Ole Tingleff. Methods for non-linear least squares problems. Lecture notes, Department of Informatics and Mathematical Modelling, Technical University of Denmark, Lyngby, Denmark, 1999.

Ezio Malis. Improving vision-based control using efficient second-order minimization techniques. In *Proceedings of 2004 International Conference on Robotics and Automation (ICRA 2004), New Orleans, USA*, pages 1843–1848, April 2004.

Michael J D Powell. A hybrid method for non-linear equations. In Philip Rabinowitz, editor, *Numerical Methods for Non-linear Algebraic Equations*, pages 87–114. Gordon and Breach, London, 1970.

Andrew P Sage and Chelsea C White. *Optimum Systems Control*. Prentice-Hall, Englewood Cliffs, USA, 2nd edition, 1977.

Nils T Siebel, Oliver Lang, Fabian Wirth, and Axel Gräser. Robuste Positionierung eines Roboters mittels Visual Servoing unter Verwendung einer Trust-Region-Methode. In *Forschungsbericht Nr. 99-1 der Deutschen Forschungsvereinigung für Meß-, Regelungs- und Systemtechnik (DFMRS) e.V.*, pages 23–39, Bremen, Germany, November 1999.

Mark W Spong, Seth Hutchinson, and Mathukumalli Vidyasagar. *Robot Modeling and Control*. John Wiley & Sons, New York, Chichester, 2005.

Lee E Weiss, Arthur C Sanderson, and Charles P Neuman. Dynamic sensor-based control of robots with visual feedback. *IEEE Journal of Robotics and Automation*, 3(5):404–417, October 1987.

3

The Uncalibrated Microscope Visual Servoing for Micromanipulation Robotic System

Xinhan Huang, Xiangjin Zeng and Min Wang
Dep. Of Control Science & Engineering, Huazhong University of Science & Technology
P. R. China

1. Introduction

MEMS technology exploits the existing microelectronics infrastructure to create complex machines with micron feature sizes. These machines can perform complex functions including communication, actuation and sensing. However, micron sized devices with incompatible processes, different materials, or complex geometries, have to be 'assembled'. Manual assembly tasks need highly skilled operator to pick and place micro-parts manually by means of microscopes and micro-tweezers. This is a difficult, tedious and time consuming work. Visual feedback is an important approach to improve the control performance of micro manipulators since it mimics the human sense of vision and allows for operating on the noncontact measurement environment.

The image jacobian matrix model has been proved to be an effective tool to approach the robotic visual servoing problem theoretically and practically. It directly bridges the visual sensing and the robot motion with linear relations, without knowing the calibration model of the visual sensor such as cameras. However, image jacobian matrix is a dynamic time-varying matrix, which cannot be calibrated by fix robotic or CCD camera parameters, especially for micro-manipulation based on micro vision. So, it is an exigent request for us to estimate parameters of image jacobian matrix on-line.

Many papers about image jacobian matrix online estimation have been reported. Clearly, Performance of the online estimation of the image jacobian matrix is the key issue for the quality of the uncalibrated micro-vision manipulation robotic. Unfortunately, the current estimation methods have problems such as estimation-lag, singularity, convergence and its speed. Especially in dynamic circumstances, these problems become more serious. There are other efforts to deal with the online estimation of the image jacobian matrix and the uncalibrated coordination control. Piepmeier et al. present a moving target tracking task based on the quasi-Newton optimization method. In order to compute the control signal, the jacobian of the objective function is estimated on-line with a broyden's update formula (equivalent to a RLS algorithm). This approach is adaptive, but cannot guarantee the stability of the visual servoing. Furthermore, the cost function using RLS is restricted by prior knowledge for obtaining some performance.

To deal with those problems discussed above, we apply an improved broyden's method to estimate the image jacobian matrix. Without prior knowledge, the method employs chebyshev polynomial as a cost function to approximate the best value. Our results show that, when calibration information is unavailable or highly uncertain, chebyshev polynomial algorithm can achieve a satisfactory result, which can bring additional performance and flexibility for the control of complex robotic systems. To verify the effectiveness of the method, both the simulations and experiments are carried out, and its jacobian estimation results show that our proposed method can attain a good performance.

2. Overview of micromanipulation robotic system

IRIS have developed the autonomous embryo pronuclei DNA injection system, visual servoing and precision motion control are combined in a hybrid control scheme. Experimental results demonstrate that the success rate of automatic injection is 100%. The time required of performing the injections is comparable with manual operation by a proficient technician. Nagoya University Fukuda Professor's research team have developed nano manipulation hybrid system based on the scanning electron microscope (FE-SEM) and the emission electron microscopy, which has been used for operating the single-cell and the individual biological cells. Columbia University Dr. Tie Hu and Dr. Allen have developed the medical micro-endoscopic imaging system. Georgiev has employed the micro-robotic system to manipulate the protein crystal and his team have established the planting robot - automatic stripe planting robotics. Ferreira has presented a self-assembly method based on the microscopy visual servoing and virtual reality technology. Kemal has proposed a visual feedback method based on the closed-loop control.

The complete micromanipulation system in our lab consists of micromanipulation stage, microscopes vision and micro-gripper. The system construction is showed in Fig.1.

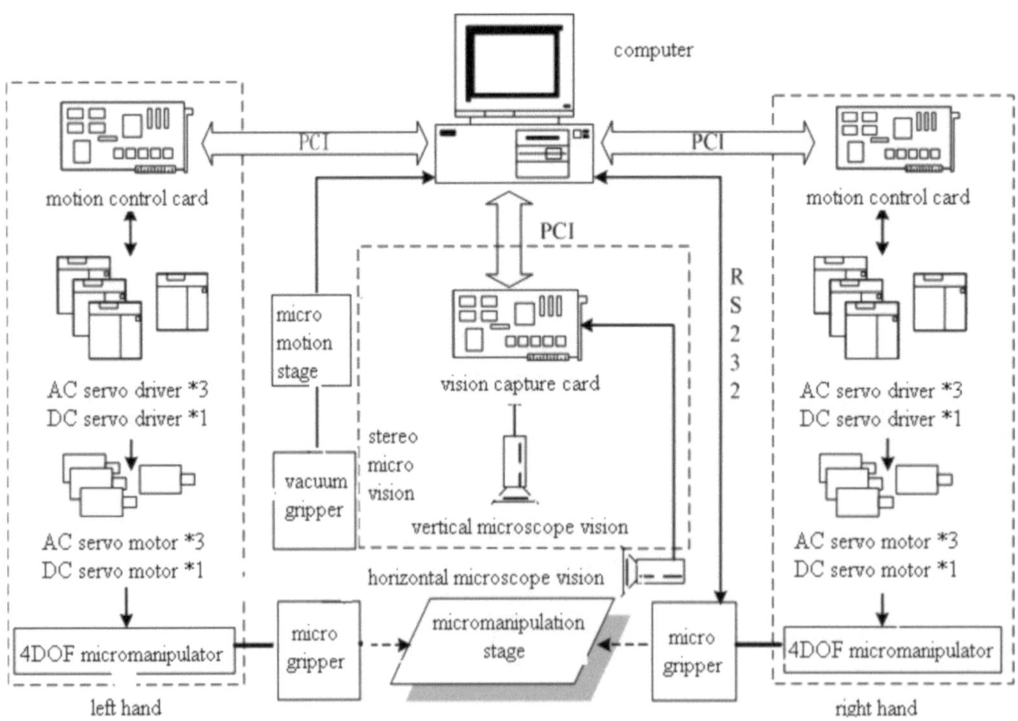

Fig. 1. The system construction of three hands cooperation micromanipulation stage

Micromanipulation hands

The left and right hands consist of 3D high precise micro-motion stage driven by the AC servoing motor and one DOF pose adjust joint driven by the DC servoing motor. The motion range of the 3D micro-motion stage is 50 x 50 x 50mm and the position of precise is 2.5μm. The rotary range of the pose adjust joint motor is ±180° and the resolution is 0.01. The third hand consists of three DOF motors driven by the DC servoing motor and the operation range is 20 x 20 x 20mm.

Microscope vision

Microscope vision is a main method that the micromanipulation robotic obtains environment information. A microscopic vision unit with two perpendicular views was developed to reduce the structural complexity of mechanism. The vision system consists of vertical crossed two rays, which can monitor micro-assembly space by stereo method and obtain the space and pose information of objects and end-effector, providing control and decision-making information for robotic.

End-effector

There are two driven types micro-gripper developed by us. One is driven by vacuum and the other one is driven by piezoelectricity ceramic, which can operate the micro parts with different size, shape and material.

3. Problems statement

Obtaining accurate camera parameters is crucial to the performance of an image-based visual controller because the image Jacobian or interaction matrix is widely used to map the image errors into the joint space of the manipulator. It is well known that camera calibration is tedious and time consuming. To avoid this, tremendous efforts have been made for on-line estimation of the image Jacobian.

The micro-assembly technology based on microscope visual servoing can be used to obtain good performance in micro size parts assembly. To obtain this level of performance and precision, one need is to identify and position the multi microsize objects. Therefore, we must consider the effective of the identifying algorithm and the position algorithm.

For an autonomous micro-assembly under microscope, it is difficult to maintain the identifying precisely since there are no reliable microsize objects sources for the reason of poor shine. So, the feature attributes reduce can become necessary for enhancing the identifying preciseness. After identifying the multi microsize parts, it is a very important issue we faced that converts the image space coordinate into robotic space coordinate, namely, how we compute the image jacobian matrix. Focusing on the microscope environment, calibrating the parameters may be not meet the requirement. So, using the uncalibrated microscope visual servoing method becomes the best path for us. Then, the uncalibrated microscope visual servoing systems can be built without considering the real-time performance and the stability of system. It means that we employ the time-consuming for exchanging the good performance. Nevertheless, the real-time performance and the stability of system are also important for micromanipulation. So the new algorithms have to be developed to cope with this information. The visual control law is essential to successfully produce high-resolution micro-assembly tasks. Its role is to improve control system performance. Are the classical control laws such as PID and intelligence control law all adapted for the micromanipulation system? Therefore, we must consider this status. Now, we will discuss all the above problems.

4. Multi-objects identifying and recognizing

In order to assemble the multi micro objects under microscope, it is necessary that identifies firstly these objects. In pattern recognition field, the moment feature is one of the shape feature that be used in extensive application. Invariant moments are the statistical properties of the image, meeting that the translation, reduction and rotation are invariance. Hu (Hu, 1962)has presented firstly invariant moments to be used for regional shape recognition. For closed structure and not closed structure, because the moment feature can not be calculated directly, it needs to construct firstly regional structure. Besides, because the moment involves in the calculation of all the pixels of intra-regional and border, it means that it can be more time-consuming. Therefore, we apply the edge extraction algorithm to process image firstly, and then calculate the edge image's invariant moments to obtain the feature attribute, which solves the problem discussed above.

After feature attribute extraction, the classification algorithm should be provided during the final target identification. The main classifier used at present can be divided into three categories: one is the statistics-based method and its representatives are such as the bayes method, KNN method like centre vector and SVM (Emanuela B et al., 2003), (Jose L R et al., 2004), (Yi X C & James Z W, 2003), (Jing P et al., 2003), (Andrew H S & Srinivas M, 2003), (Kaibo D et al., 2002) ; One is the rule-based method and its representatives are decision tree and rough sets; the last one is the ANN-based method. Being SVM algorithm is a convex optimization problem, its local optimal solution must be global optimal solution, which is better than the other learning algorithms. Therefore, we employ SVM classification algorithm to classify the targets. However, the classic SVM algorithm is established on the basis of the quadratic planning. That is, it can not distinguish the attribute's importance from training sample set. In addition, it is high time to solve the large volume data classification and time series prediction, which must improve its real-time data processing and shorten the training time and reduce the occupied space of the training sample set.

For the problem discussed above, we present an improved support vector machine classification, which applies edge extraction's invariant moments to obtain object's feature attribute. In order to enhance operation effectiveness and improve classification performance, a feature attribute reduction algorithm based on rough set (Richard Jensen & Qiang Shen, 2007), (Yu chang rui et al., 2006) has been developed, with the good result to distinguish training data set's importance.

Invariant moments theory

Image $(p+q)$ order moments: we presume that $f(i,j)$ represents the two-dimensional continuous function. Then, it's $(p+q)$ order moments can be written as (1).

$$M_{pq} = \iint i^p j^q f(i,j)\,di\,dj \qquad (p,q = 1,2,\ldots) \tag{1}$$

In terms of image computation, we use generally the sum formula of $(p+q)$ order moments shown as (2).

$$M_{pq} = \sum_{i=1}^{M}\sum_{j=1}^{N} f(i,j) i^p j^q \quad (p,q = 1,2,...) \tag{2}$$

Where p and q can choose all of the non-negative integer value, they create infinite sets of the moment. According to Papulisi's theorem, the infinite sets can determine completely a two-dimensional image $f(i,j)$.

In order to ensure location invariance of the shape feature, we must compute the image $(p+q)$ order center moment. That is, calculates the invariant moments using the center of object as the origin of the image. The center of object $(i^{'}, j^{'})$ can be obtained from zero-order moment and first-order moment. The centre-moment formula can be shown as (3).

$$M_{pq} = \sum_{i=1}^{M}\sum_{j=1}^{N} f(i,j)(i-i^{'})^p (j-j^{'})^q \tag{3}$$

At present, most studies about the two-dimensional invariant moments focus on extracting the moment from the full image. This should increase the computation amount and can impact on the real-time of system. Therefore, we propose the invariant moments method based on edge extraction, which gets firstly the edge image and then achieves the invariant moments feature attribute. Obviously, it keeps the region feature of moment using the proposed method. In addition, being the role of edge detection, the data that participate calculation have made a sharp decline, reducing greatly the computation amount.The invariant moments are the functions of the seven moments, meeting the invariance of the translation, rotation and scale.

Improved support vector machine and target identify

1) Support Vector Machine: The basic idea of SVM is that applies a nonlinear mapping Φ to map the data of input space into a higher dimensional feature space, and then does the linear classification in this high-dimensional space.

Presumes that the sample set (x_i, y_i), $(i = 1,...,n)$, $x \in R_d$ can be separated linearly, where x is d dimensional feature vector and $y \in \{-1,1\}$ is the class label. The general form of judgement function in its linear space is $f(x) = w \bullet x + b$, Then, the classification hyperplane equation can be shown as (4).

$$w \bullet x + b = 0 \tag{4}$$

If class m and n can be separated linearly in the set, there exists (w, b) to meet formula as (5).

$$\begin{aligned} w \bullet x_i + b > 0, (x_i \in m) \\ w \bullet x_i + b < 0, (x_i \in n) \end{aligned} \tag{5}$$

Where w is weight vector and b is the classification threshold. According to (4), if w and b are zoomed in or out at the same time, the classification hyperplane in (4) will keep invariant . We presume that the all sample data meet $|f(x) \geq 1|$, and the samples that is closest classification hyperplane meet $|f(x) = 1|$, then, this classification gap is equivalent to $2/\|w\|$. So the classification gap is biggest when $\|w\|$ is minimum.

Although the support vector machine with a better classification performance, but it can only classify two types of samples, and the practical applications often require multiple categories of classification. As a result, SVM need to be extended to more categories of classification issues, For the identification of a number of small parts in micromanipulation, we applied the Taiwan scholar Liu presented method based on the "one-to-many" method of fuzzy support vector machine for multi-target classification.

2) Improved Support Vector Machine: For the completion of the sample training, it is a usual method that all the feature attribute values after normalization have been used for modeling, which will increase inevitably the computation amount and may lead to misjudge the classification system being some unnecessary feature attributes. Therefore, bringing a judgement method to distinguish the attribute importance may be necessary for us. So we employ rough set theory to complete the judgement for samples attribute's importance. Then, we carry out SVM forecast classification based on the reduction attributes.

Now, we introduce rough set theory. The decision-making system is $S=(U,A,V,f)$, where U is the domain with a non-null limited set and $A=C\bigcup D$. C, D represents conditions and decision-making attributes set respectively. V is the range set of attributes ($V=\bigcup_{a\in A} V_a$), V_a is the range of attribute a. f is information function ($f:UXA\rightarrow V$). If exists $f(x,a)\in V_a$ under $\forall x\in U \ a\in A$ and $\forall B\subseteq A$ is a subset of the conditions attributes set, we call that $Ind(B)$ is S's un-distinguish relationship. Formula $Ind(B)=\{(x,y)\in UXU \mid \forall a\in B, f(x,a)=f(y,a)\}$ represents that x and y is indivisible under subset B. Given $X\subseteq U$, $B(x_i)$ is the equivalent category including x_i in term of the equivalent relationship $Ind(B)$. We can define the next approximate set $\underline{B(X)}$ and the last approximate set $\overline{B(X)}$ of subset X as follows:

$$\underline{B(X)}=\{x_i\in U \mid B(x_i)\subseteq X\}$$

$$\overline{B(X)}=\{x_i\in U \mid B(x_i)\bigcap X\neq\phi\}$$

If there is $\overline{B(X)}-\underline{B(X)}=\phi$, the set X is able to define set based on B. Otherwise, call X is the rough set based on B. The positive domain of X based on B are the objects set that can be determined to belong to X based knowledge B. Namely, $POS_B(X)=\underline{B(X)}$. The dependence of decision-making attributes D and conditions attributes C can be defined as follows.

$$\gamma(C,D)=card(POS_C(D))/card(U)$$

Where $card(X)$ is the base number of the set X.

The attributes reduction of rough set is that the redundant attributes have been deleted but there is not loss information. The formula $R=\{R \quad R\subseteq C, \ \gamma(R,D)=\gamma(C,D)\}$ is the reduction attributes set. Therefore, we can use equation attributes dependence as conditions for terminating iterative computing.

In order to complete the attribute reduction, we present a heuristic attribute reduction algorithm based-on rough set's discernibility matrix, which applies the frequency that attributes occurs in matrix as the heuristic rules and then obtains the minimum attributes's relative reduction.

The discernibility matrix was introduced by Skowron and has been defined as (6):

$$(c_{ij}) = \begin{cases} a \in A : r(x_i) \neq r(x_j) & D(x_i) \neq D(x_j) \\ \varnothing & D(x_i) = D(x_j) \\ -1 & \forall r, \exists r(x_i) = r(x_j) \quad D(x_i) \neq D(x_j) \end{cases} \tag{6}$$

According to formula (6), The value of elements is the different attributes combination when the attributes for the decision-making are different and the attributes for the conditions are different. The values of elements are null when the attributes for the decision-making are the same. The values of elements are -1 when the attributes for the decision-making are the same and the attributes for the conditions are different.

If p (a) is the attribute importance formula of attribute a, we can propose the formula as (7) according to the frequency that attribute occurs:

$$p(a) = \gamma \frac{1}{|U^2|} \sum_{a \in c_{ij}} \frac{1}{|c_{ij}|} \tag{7}$$

Where γ is the general parameter and c_{ij} are the elements of the discernibility matrix. Obviously, the greater the frequency that attribute occurs, the greater its importance is. Therefore, we can compute the importance of attributes and eliminate the attributes that its importance is the smallest using the heuristic rules in formula (7). And then, we can obtain the relative reduction attributes.

Now, we give the heuristics attribute reduction algorithm based-on rough set's discernibility matrix.

Input: the decision-making table $(U, A \cup D, V, f)$

Output: the relative attribute reduction

Algorithm steps:

Step I computes the identification discernibility matrix M.

Step II determines the core attributes and find the attributes combination that the core attributes is not included.

Step III obtains conjunctive normal form $P=\wedge(\vee\ c_{ij}:(i = 1, 2, 3 \ldots s;\ j = 1, 2, 3 \ldots m))$ of the attributes combination by step II, where c_{ij} are elements of each attribute combination. And then converts the conjunctive normal form to disjunctive normal form.

Step IV determines the importance of attribute according to formula (7).

Step V computes the smallest importance of attributes by steps IV and then eliminate the less importance attribute to obtain the attributes reduction.

After reducing the attribute, the samples feature attributes will be sent to SVM for establishing model. Support vector machines uses Gaussian kernel function, and Gaussian kernel function shows a good performance in practical applications of learning. Finally, we can finish the classification of the final prediction data.

Feature extraction and data pretreat

The main task of classification is to identify and classify the manipulator (microgripper, vacuum suction) and operation targets (cylindrical metal part, glass ball), which can provide convenience for follow-up visual servo task. Fig.2 shows the original image of operation targets and manipulator in microscopic environment.Fig.3 is the image after edge extraction of operation targets and manipulator in microscopic environment.

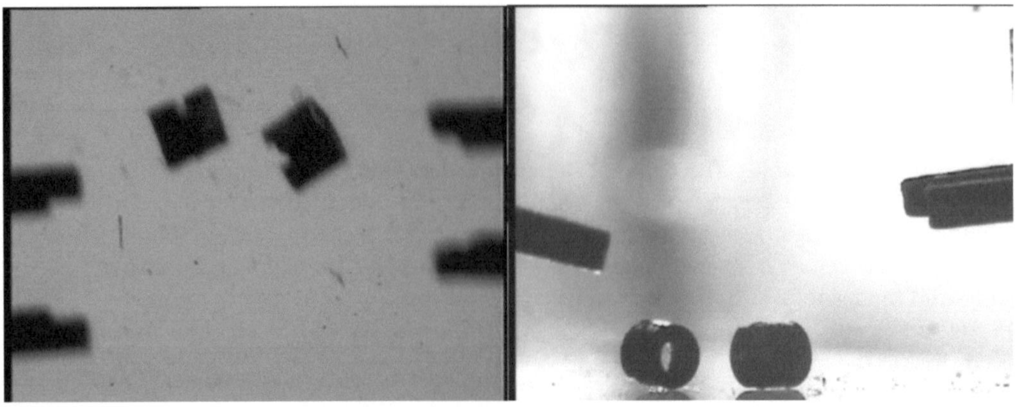

(a) Microscopic images in vertical view field (b) Microscopic images in horizontal view field

Fig. 2. The original microscopic image of object and the endeffector in vertical (a) and horizontal (b) view fields

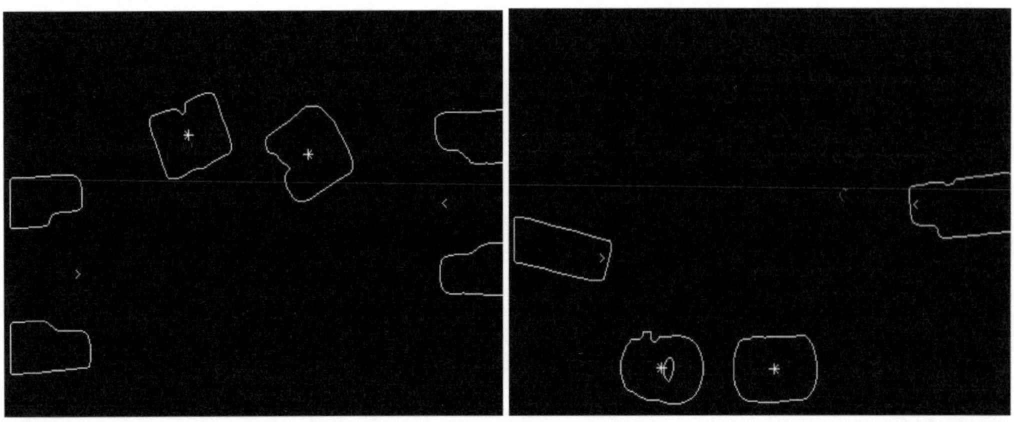

(a) Microscopic images in vertical view field (b) Microscopic images in horizontal view field

Fig. 3. The object centre image and the end centre of the endeffector after processing in vertical (a) and horizontal (b) view fields

Table 1 gives the feature attribute's normalization value of four different objectives using invariant moments algorithm. We compute the feature attribute of objects in all directions and only list one of the feature attribute.

Category	F1	F 2	F 3	F4	F5	F6	F7
Cyl. metal part	1.0000	-0.9910	0.9935	-0.1600	0.1076	1.0000	-0.5762
Glass Small Ball	1.0000	0.9900	-0.9946	0.1822	0.1178	0.9952	-0.5486
Micro Gripper	-0.9897	-0.7610	-1.0000	-1.0000	-0.9999	0.9554	-1.0000
Vacuum Suction	0.1673	0.9993	0.3131	0.9915	0.9857	-0.9577	0.9861

Table 1. The feature attribute's normalization value of different objects using invariant moments algorithm

Result of identification and analysis

We compare firstly the data classification effectiveness on a number of micro objects using the traditional support vector machines algorithm and rough set + SVM, and the results are shown in table 2.

	SVM		SVM +Rough set	
	Correction rate	Classification time (*ms*)	Correction rate	Classification time (*ms*)
Micro Object	93.45%	2108.24	95.89%	357.65

Table 2. The comparison results of using two classification methods

According to table 2, the correction rate of classification based on the proposed SVM classification algorithm has been over 95 pre cent, being higher than the single SVM algorithm's correction rate. So, we can draw the conclusion that the attribute reduction improves the classification ability. Besides, compared with the single SVM algorithm's calculation time,it can be seen clearly from Table 2 that the calculation time of the proposed algorithm is less than about five times, meaning that the system becomes more effective.
Then, Table 3 provides the comparison results of classification accuracy using SVM classification and SVM+rough set classification with joining the other 25 feature attributes (gray, area, perimeter, texture, etc.). In table 3, The first column is the times of data sets; second column is the number of conditions attributes after attribute reduction; third column is the classification accuracy using the SVM; fourth column is the classification accuracy using SVM and rough set algorithm. The number of conditions attributes of the final classification for entering to SVM is 14.25, less than 25 features attribute. Thus it simplifies the follow-up SVM forecast classification process.

Times	Property	classification accurateness	
		SVM	SVM + rough set
1	10	90.00 %	95.10 %
2	15	90.25 %	96.00 %
3	9	89.00 %	92.87 %
4	21	92.15 %	97.08 %
5	15	90.80 %	92.33 %
6	12	90.00 %	93.50 %
7	12	94.00 %	95.22 %
8	20	92.16 %	97.40 %

Table 3. The comparison results of classification accurateness using SVM and SVM + Rough set classification

5. The uncalibrated microscope visual servoing

As a result of the particularity of micro-manipulation and micro-assembly environment, we can not calibrate the parameter of micromanipulation robotic as the industrial robots calibration. So, we employ the uncalibrated visual servoing method. The uncalibrated visual servoing is a hot issue in the field of robot vision research over the past decade, which estimates the image jacobian matrix elements on-line, increasing the system's adaptability for environmental change.

Many scholars in this area have done a lot of researches. Piepmeier developed a dynamic quasi-Newton method. Using the least square method, Lu developed an algorithm for on-line calculating the exterior orientation. Chen proposed a homography based adaptive tracking controller by estimating the unknown depth and object parameters. Yoshimi and Allen proposed an estimator of the image Jacobian for a peg-in-hole alignment task. Hosoda and Asada employed the Broyden updating formula to estimate the image Jacobian. Ruf presented an on-line calibration algorithm for position-based visual servoing. Papanikolopoulos developed an algorithm for on-line estimating the relative distance of the target with respect to the camera.

Visual-servo architecture of the micro manipulator

The dynamic image-based look-and-move system is the most suitable visual servoing architecture for the micromanipulation operation, and some commercial software is available. In the micro-vision system based optic-microscope, a camera can only be mounted on the microscope. This control system has both the end-effector feedback and its joint level feedback. A classical proportional control scheme is given by:

$$V = -\lambda \hat{L}^{+} e$$

Where L_e is defined by

$$\dot{e} = L_e V$$

In order to finish three-dimensional small object positioning task, in the actual operation, micro-manipulation tasks will be divided into horizontal direction (XY plane) movement and the vertical direction (YZ plane) movement. The manipulator in the XY plane moves first, positioning small parts in the above, then does so in the YZ plane movement, positioning small parts at the centre. Therefore, we apply two image jacobian matrixs, including horizontial view field of image jacobian matrix and vertical view field of image jacobian matrix, which can complete the positioning and tracking three-dimensional objects. The change of robot movement $[dx, dy]^T$ and the change of image characteristics $[du, dv]^T$ can be wirte as (8):

$$\begin{bmatrix} dx \\ dy \end{bmatrix} = \dot{J} \begin{bmatrix} du \\ dv \end{bmatrix} \tag{8}$$

According to the online estimtion image Jacobian matrix J, set the position of the error $e = f^d - f^c$, which f^d is the expectations of position of objects (small cylindrical parts, 600 um diameter) and f^c is the centre of endeffector. Then, the control law of PD controller u (k) is:

$$u(k) = K_p(\dot{J}^T\dot{J})^{-1}\dot{J}^T e(k) + K_d(\dot{J}^T\dot{J})^{-1}\dot{J}^T \frac{\Delta e(k)}{T_S} \tag{9}$$

Where T_s is the time interval, and K_p is proportional gain and K_d is differential gain. Its control structure is shown in Fig.4.

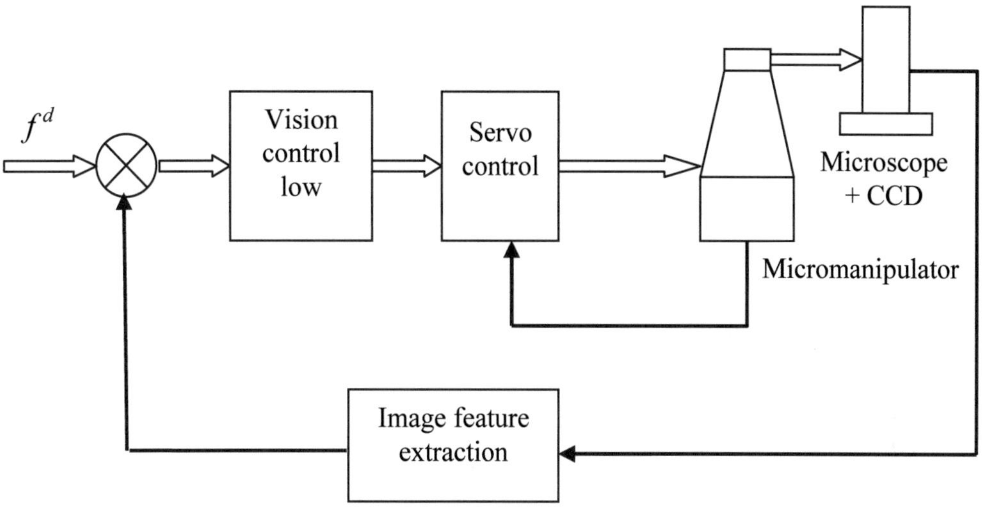

Fig. 4. Micromanipulator servo control structure

In the next section, the pseudo-inverse of image Jacobian will be addressed. In order to meet the request of the high precise micro-manipulation task, robotic must employ the visual servoing method. The methods of visual servoing need calibrate precisely intrinsic parameter of camera. However, the system calibration is the complicated and difficult problem, especially for micro-manipulation based on microscope vision. So, we present the uncalibrated method to estimate image jacobian matrix online.

Image jacobian

The image jacobian defines the relationship between the velocity of a robot end-effector and the change of an image feature. Considering q = [q1, q2…qm]^R represents the coordinates of robot end-effector in the task space. An n-dimensional vector: f = [f1, f2…fn]^T is corresponding position in image feature. Then, the image jacobian matrix J_q is defined as

$$\dot{f} = J_q(q)\dot{q} \tag{10}$$

where

$$J_q(q) = \left[\frac{\partial f}{\partial q}\right] = \begin{bmatrix} \frac{\partial f_1(q)}{\partial q_1} & \cdots & \frac{\partial f_1(q)}{\partial q_m} \\ \cdots & & \cdots \\ \frac{\partial f_n(q)}{\partial q_1} & \cdots & \frac{\partial f_n(q)}{\partial q_m} \end{bmatrix} \tag{11}$$

Broyden's method for image jacobian matrix estimation

The image jacobian matrix can be calculated by calibrating the inner and outer parameter of robotic system & sensor system. However, it is impossible to obtain precise system parameter under a dynamic or uncertainty environment. Considering those, we employ broyden's method to estimate the image jacobian matrix.

According to equation (10), Provided that two image feature error function $e_f(q) = f - f^*$, the Taylor series expansion of e_f is shown as

$$e_f(q) = e_f(q_m) + \frac{\partial e(q_m)}{\partial q}(q - q_m) + \ldots + R_n(x) \tag{12}$$

Where $R_n(x)$ is Lagrange remaining. We define $\overset{*}{J}_q(q_n)$ as the Nth image jacobian to be estimated, then

$$\overset{*}{J}_q(q) = \frac{\partial e(q_n)}{\partial q} \tag{13}$$

Ignoring the high order term and Lagrange remaining $R_n(x)$, Equation (14) can be obtained from (12) and (13), which is shown as

$$e_f(q) = e_f(q_m) + \overset{*}{J}_q(q_n)(q - q_m) \tag{14}$$

The broyden algorithm is described as

$$A_{k+1} = A_k + \frac{(y^{(k)} - A_k s^{(k)})s^{(k)^T}}{\left\|s^{(k)}\right\|_2^2} \qquad (k = 0,1,2,\ldots) \tag{15}$$

Therefore, we can obtain image jacobian estimation $\overset{*}{J}_q(q_{k+1})$ as shown in (16)

$$\overset{*}{J}_q(q_{k+1}) = \overset{*}{J}_q(q_k) + \frac{(\Delta e - \overset{*}{J}_q(q_k)\Delta q)\Delta q^T}{\Delta q^T \Delta q} \tag{16}$$

In (16), We will apply the cost function to minimize $\overset{*}{J}_q(q_{k+1}) - \overset{*}{J}_q(q_k)$.

Chebyshev polynomial approximation algorithm

Provided that

$$N_K(q) = e_f(q_k) + \overset{*}{J}_q(q)(q - q_k) \tag{17}$$

If $N_k(q) \in c[-1,1]$, for Chebyshev polynomial serial $\{T_n, n = 0,1,\ldots\}$ with weight $\rho(x) = (1 - x^2)^{-\frac{1}{2}}$, it's optimization square approximation polynomial can be shown as

$$s^*_n(x) = \frac{a_0}{2} + \sum_{i=1}^{n} a_i T_i(x) \tag{18}$$

where

$$a_i = \frac{2}{\pi}\int_{-1}^{1} \frac{N_k(x)T_i(x)}{\sqrt{1-x^2}}dx \quad (k = 0,1,2...n) \tag{19}$$

then

$$N(q) = \lim_{n\to\infty}(\frac{a_0}{2} + \sum_{i=1}^{n} a_i T_i(q)) \tag{20}$$

if we use part sum s^*_n as $N(q)$'s approximation, under some conditions, there is a fast speed for $a_n \to 0$.

Theoretically, Compared with RLS algorithm, Chebyshev polynomial approximation algorithm is independent of the prior knowledge of system, and it has fast approximate speed than that of other methods. Experiments will prove its correctness. Surely, The unsatisfied thing of chebyshev polynomial approximation algorithm, we encountered, lies in that it require N(q)'s good smoothness . It is a difficulty for us to meet this need for most conditions.

Chebyshev polynomial approximation algorithm implementation

Let's consider firstly the chebyshev polynomial approximation algorithm implementation. Usually, $N_K(q) = e_f(q_k) + \overset{*}{J}_q(q)(q - q_k)$ is a function whose variable interval lies in [a, b], it means that we need to convert variable interval of [a, b] into [-1, 1]. Thus, as shown in equation (21), it can finish this conversion

$$t = \frac{b-a}{2}x + \frac{b+a}{2} \tag{21}$$

Following task is that how to obtain parameter a_i (i = 0, 1, 2...) from formula (11). It presumes that we apply the zero point of $T_{n+1}(x)$ as discrete point set, namely, $x_i = \cos\frac{2i-1}{2(n+1)}\pi$ (i=1, 2...n+1), so a_i can be calculated as follows

$$a_i = \frac{2}{n+1}\sum_{i=1}^{n+1} N(x_i)T(x_i) \quad (i = 0,1,2...) \tag{22}$$

Comparison chebyshev polynomial approximation with RLS

Some papers [4][5] provide RLS algorithm to approximate best value for minimum cost function. The cost function using RLS is shown as equation (23).

$$Min(k) = \sum_{i=1}^{n} \lambda^{k-i} \|N_k(q_{i-1}) - N_{i-1}(q_{i-1})\|^2 \tag{23}$$

Where λ is a rate of dependency for prior data. As shown in equation (23), In order to obtain some performance, the cost function using RLS algorithm depends on the data of the several past steps, it mean that the prior knowledge must be obtained for finishing the task. Similarly, the cost function using chebyshev polynomial is shown as equation (24).

$$M(k)=\sum_{i=1}^{n}\|N_k(q_{i-1})-N_{i-1}(q_{i-1})\|^2 \tag{24}$$

Clearly, the cost function using chebyshev polynomial is independent of the prior data.

Jacobian estimator with improved broyden's method

As discussed in the above two sections, an improved broyden with chebyshev polynomial approximate algorithm estimator of image jacobian is developed. A graphical representation of the estimate process is shown in Fig.5. Firstly, the broyden estimator starts with initial endeffector position q^0 and precision ε. Then, Camera captures an image of endeffector for extracting corresponding image coordinate feature f^k, Which provides the possibility for calculating $\overset{*}{J}(q^k)$ by formula $\overset{*}{J}(q^k)=[f'(q^k)]^{-1}$. Secondly, Camera captures an image of target to obtain expectative image coordinate feature f^{k+1}. With the obtained $\overset{*}{J}(q^k)$, the servoing control law can be deduced in equation (25). Finally, Program judges whether precision ε satisfies system requirement or not. If precision ε arrives the requirement, system will be ended, otherwise system will be executed repeat processing.

$$u(k)=K\Delta q=K\overset{*+}{J}(k)(f^{k+1}-f^k) \tag{25}$$

Where K is proportion gain.

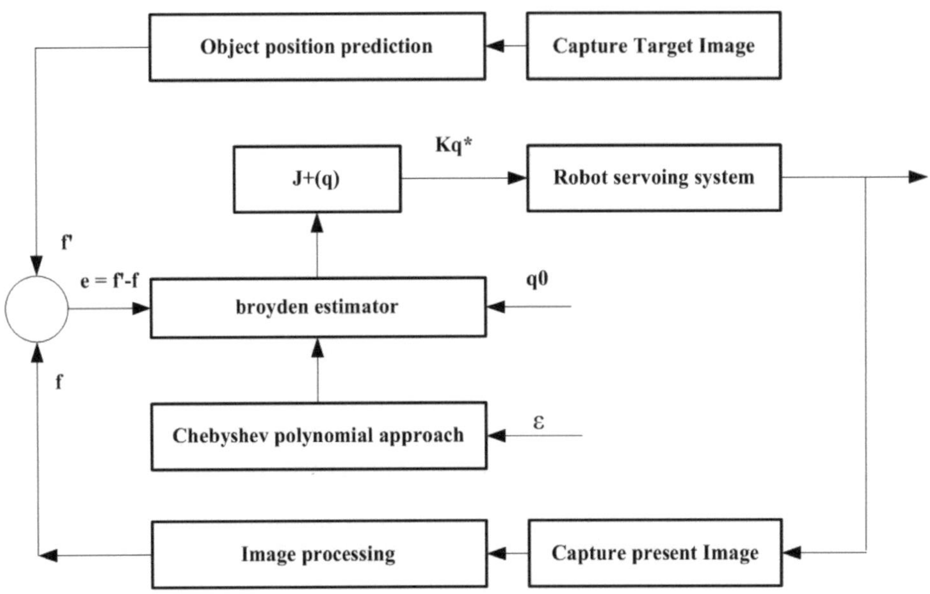

Fig. 5. A broyden with chebyshev polynomial approximation estimator of image jacobian

6. Experiments and simulations

Micro manipulation system

Microscopic visual servoing is the sensor-based control strategy in microassembly. The microscopic vision feedback has been identified as one of the more promising approaches to improve the precision and efficiency of micromanipulation tasks. A robotic microassembly system has been developed in our lab. Fig.6 is three hands coordination micro-manipulation system.

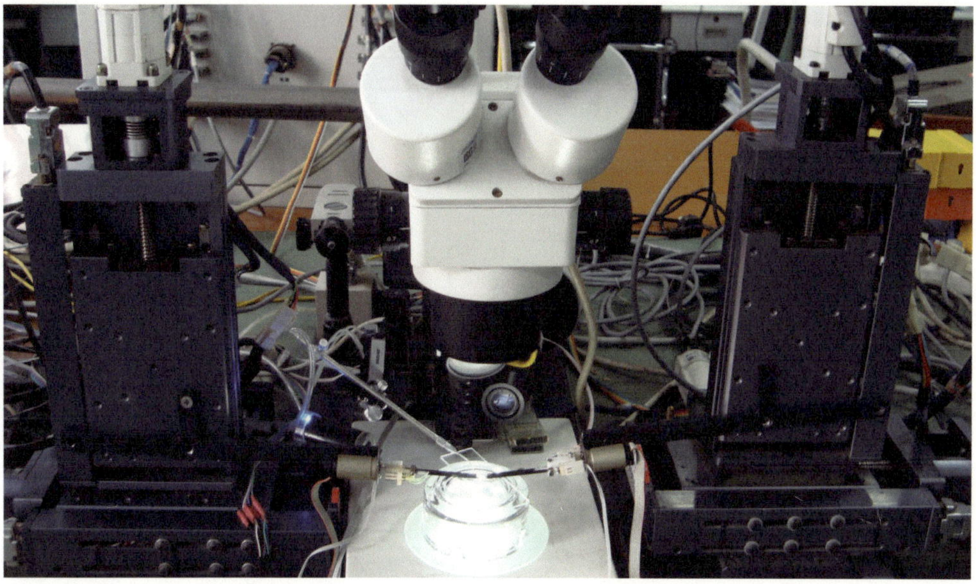

Fig. 6. Three hands coordination micro-manipulation system

Image jacobian estimation results

As the micromanipulator performs a continuous 4D movement with translation step of 10um and rotation step of 0.20, the broyden's method with chebyshev polynomial approximation algorithm executes an online estimation of the jacobian matrix elements. The manipulator kinematic parameters and microscopic vision parameters are not known in the estimation. The image size adopted in image processing is 400 X 300 pixels.
We test firstly the endeffector moving trajectory according to the online estimation method of the jacobian matrix. Fig.7 shows the endeffector moving trajectory in the vertical direction camera (left) and in horizontal direction camera (right).

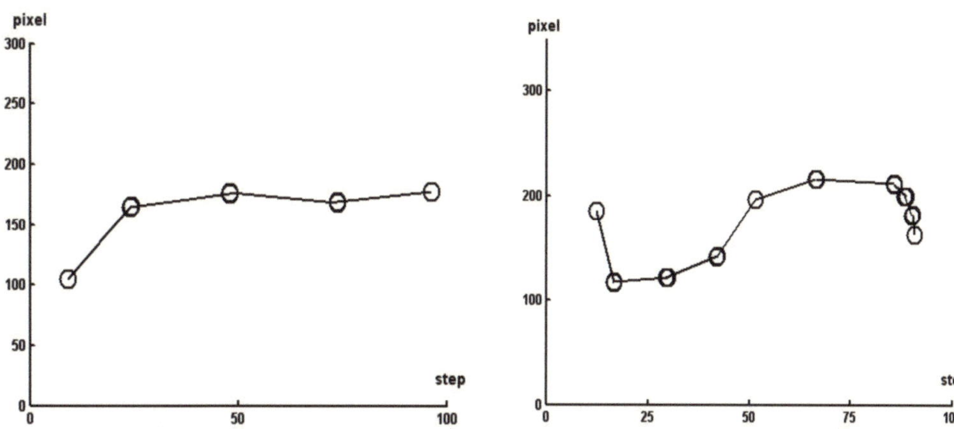

Fig. 7. The endeffector moving trajectory in vertical and horizontal direction camera

Next, we demonstrate the microscopic visual servoing experiment based on the improved broyden's method of image jacobian for a moving target. The initial position of micro gripper is (0.0, 0.0) and the moving target initial position is (x, y) = (0.8, -0.3) with the velocity of about 4mm/s. The task is done at the time of 10s with the tracking error between the target and the micro-gripper about 25 pixels. Fig.8 gives the trajectory of target and gripper of in vertical direction camera (left) and in horizontal direction camera (right).

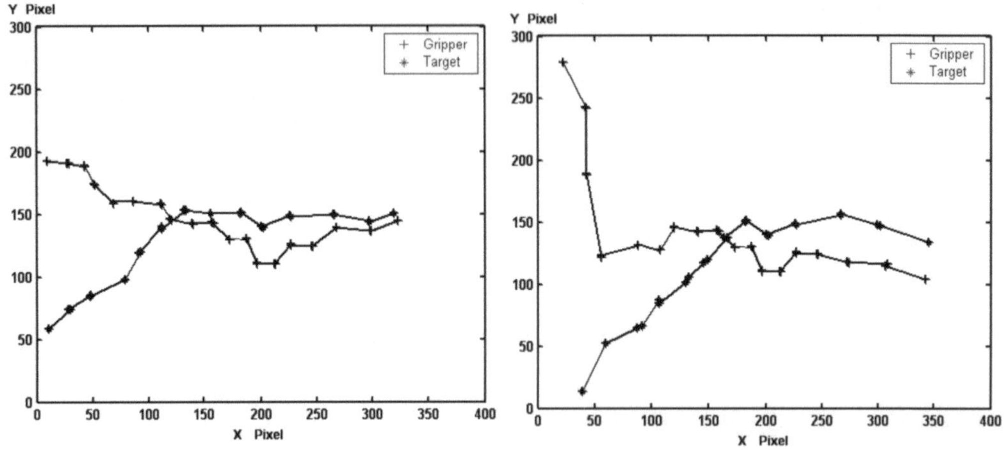

Fig. 8. Trajectory of target and gripper in vertical direction camera in horizontal direction camera

As shown in Fig.8, we can find that micro-gripper and the target have a large tracking error at initial stages. The reason for the large error is that there are a lot of noises and a small control output to step motor. With the progression of time, the error decreases to 25 pixels, it satisfies the tracking task requirement.

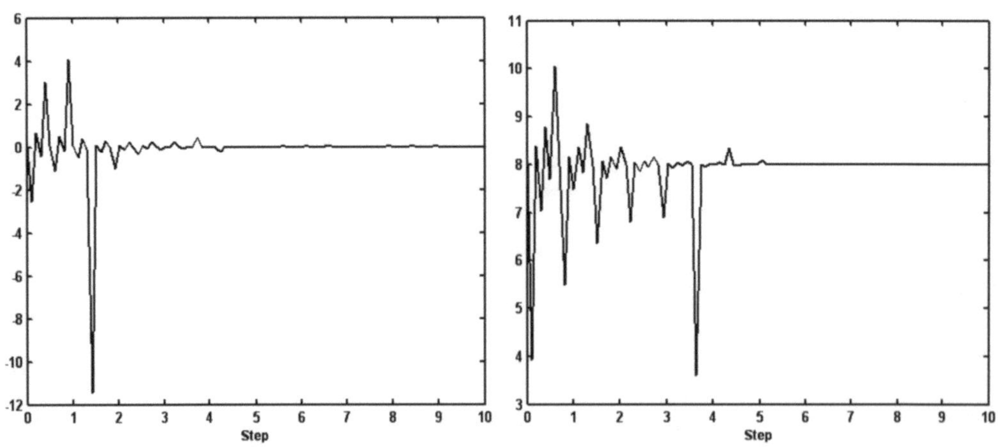

Fig. 9. Convergence speed of chebyshev algorithm (left) and Convergence speed of RLS (right)

Then, we have finished the experiment using chebyshev polynomial and RLS as cost function to estimate image jacobian matrix. The comparisons of convergence speed of two cost functions are shown in Fig.9. Clearly, compared with the RLS algorithm, it achieves a good performance in speed and stability when we apply chebyshev polynomial as a cost function.

Automatic position test

To accomplish micromanipulator positioning and girpping small parts, we must firstly obtain the centre of object and the centre of the end of endeffector. The centre of object and the end of endeffector can be accessed by a series of image processing (gray, de-noising, filter, canny operator, edge extraction, fuzzy c-means clustering). Fig.10 shows the original microscopic image of object and the endeffector in vertical and horizontal view fields. Fig.11 shows the object centre image and the end centre of the endeffector after processing in vertical and horizontal view fields. In Fig.11, the XY image plane coordinates of the center of the object is (147,99) and the centre of the end of the endeffector is (343,77).

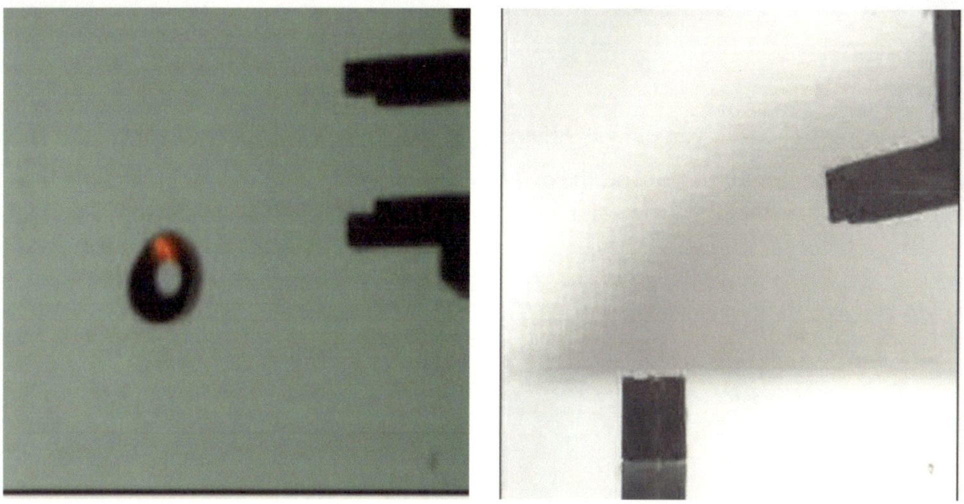

(a) (b)

(a) Microscopic images in vertical view field (b) Microscopic images in horizontal view field

Fig. 10. The original microscopic image of object and the endeffector in vertical (a) and horizontal (b) view fields

Assuming that the initial parameters of PD controller Kp is 10 and Kd is 0, that is, only joined proportional control, control effect is shown in Fig.12. we can see the implementation of automatic positioning objects to the target center, a greater oscillation and overshoot. When Kp is 10 and Kd is 1.5, which incorporates proportional and differential control, control result is shown in Fig.13. Differential joined inhibits apparently the system overshoot, and the system meets the rapid and smooth. Finally, the implementation of micro-manipulator positioning and automatic gripping operations is given, it can be obtained the satisfied implementation with the results to the system application requirements.

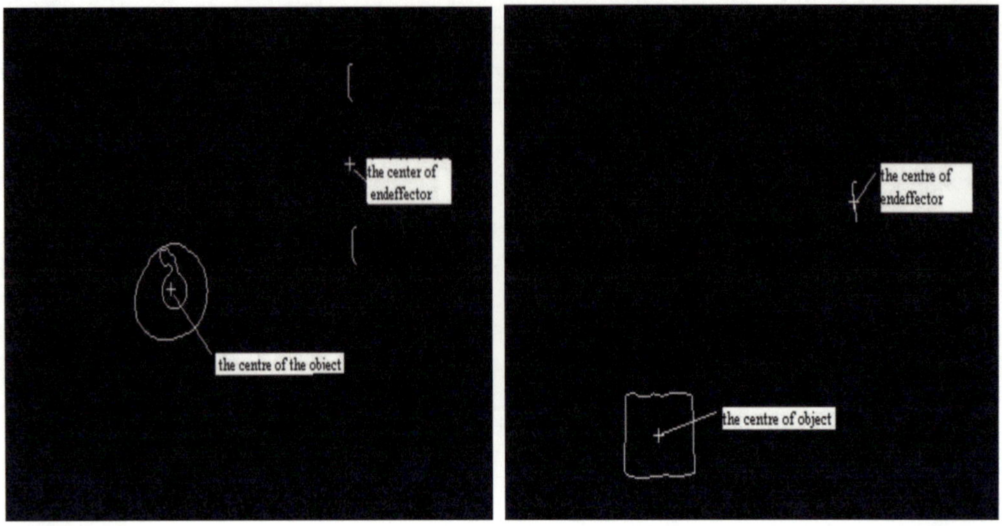

(a) (b)

(a) Microscopic images in vertical view field (b) Microscopic images in horizontal view field

Fig. 11. The object centre image and the end centre of the endeffector after processing in vertical (a) and horizontal (b) view fields

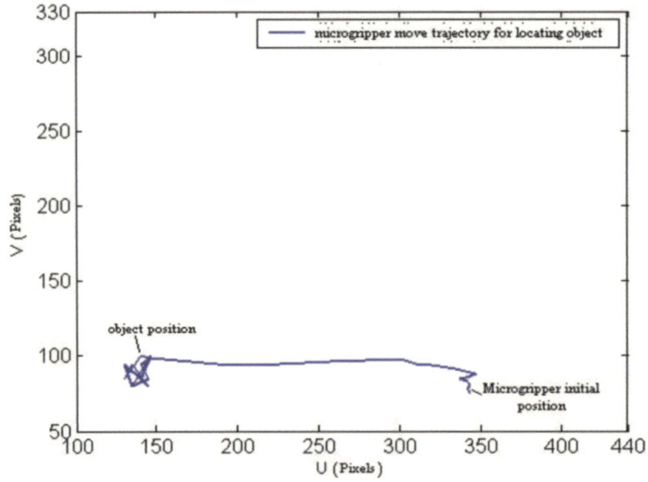

Fig. 12. The trajectories of micromanipulator approaching goal objects with only proportional control (XY plane)

Finally, In order to verify the effective of uncalibrated visual servoing method, we test the experiments of single microgripper hand to position automatic and grip micro objects. The flow chart of single microgripper hand to position automatic and grip micro objects is shown in Fig.14.

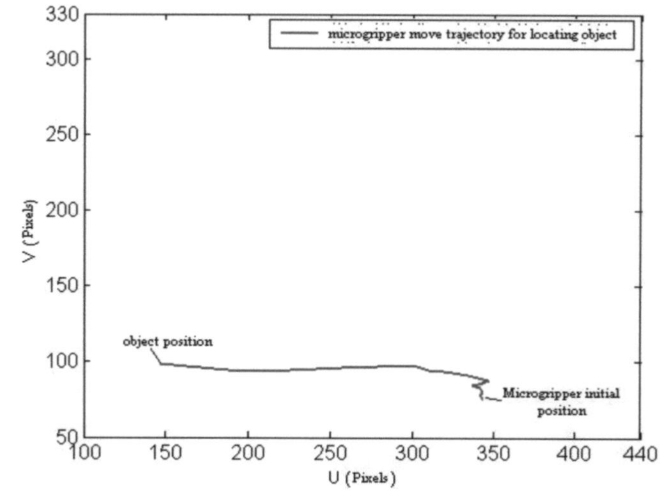

Fig. 13. The trajectories of micromanipulator approaching goal objects with proportional and differential control (XY plane)

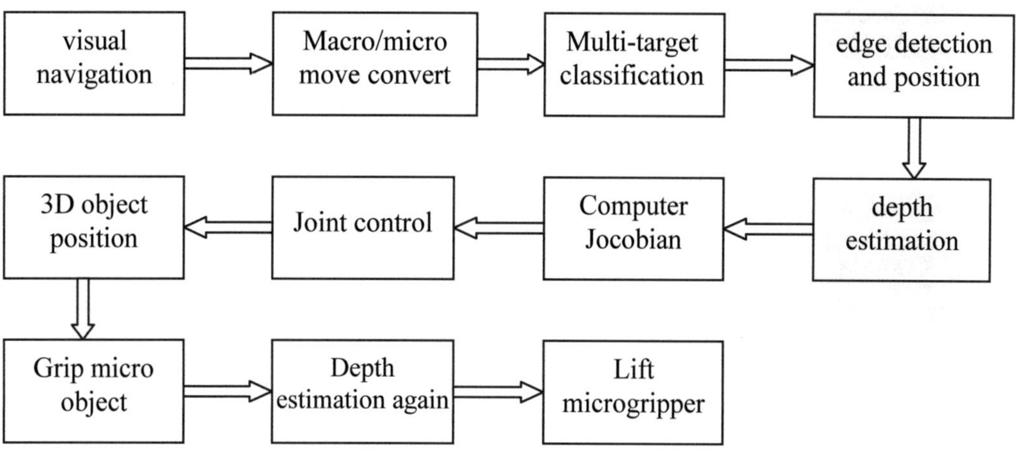

Fig. 14. The flow chart of single microgripper hand to position automatic and grip micro objects under microscope visual information

Fig.15 shows the process of the piezoelectric microgripper automatically locating and gripping the micro-target in the vertical view field. The time-consuming of process is about one minute:

(a) to (c) is the process of piezoelectric microgripper close to the target micro-target;

(d) is the process of the end of piezoelectric microgripper positioning the center of the micro target;

(e) is the process of the piezoelectric microgripper gripping the micro target;

(f) is the process of the piezoelectric microgripper lifting the designated height for follow-up of the micro-target assembly.

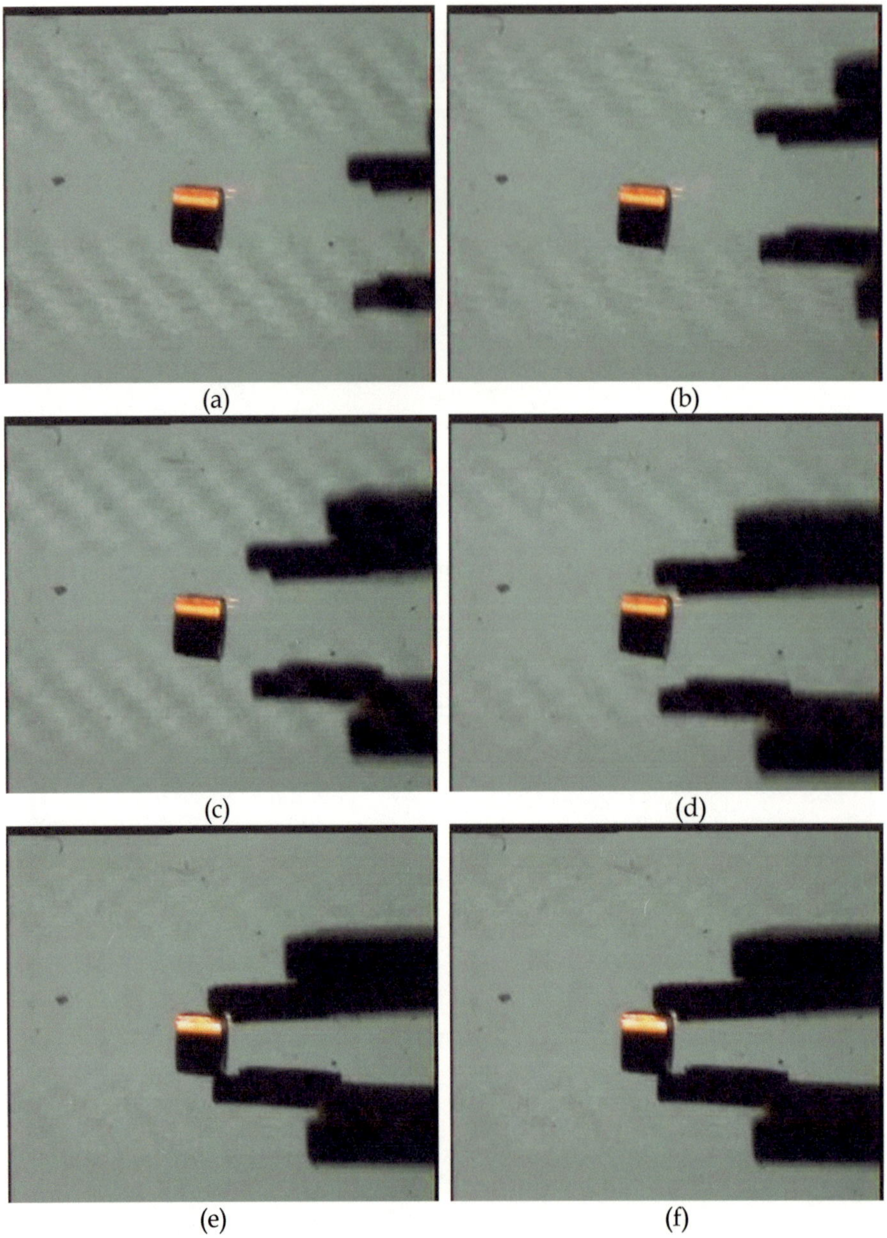

Fig. 15. The process of the piezoelectric microgripper automatically locateing and gripping the micro-target in the vertical view field

Fig.16 shows the process of the piezoelectric microgripper automatically locating and gripping the micro-target in the horizontal view field. The time-consuming of process is about one minute:

(a) to (c) is the process of piezoelectric microgripper close to the target micro-target;

(d) is the process of the end of piezoelectric microgripper positioning the center of the micro target;

(e) is the process of the piezoelectric microgripper gripping the micro target;
(f) is the process of the piezoelectric microgripper lifting the designated height for follow-up of the micro-target assembly.

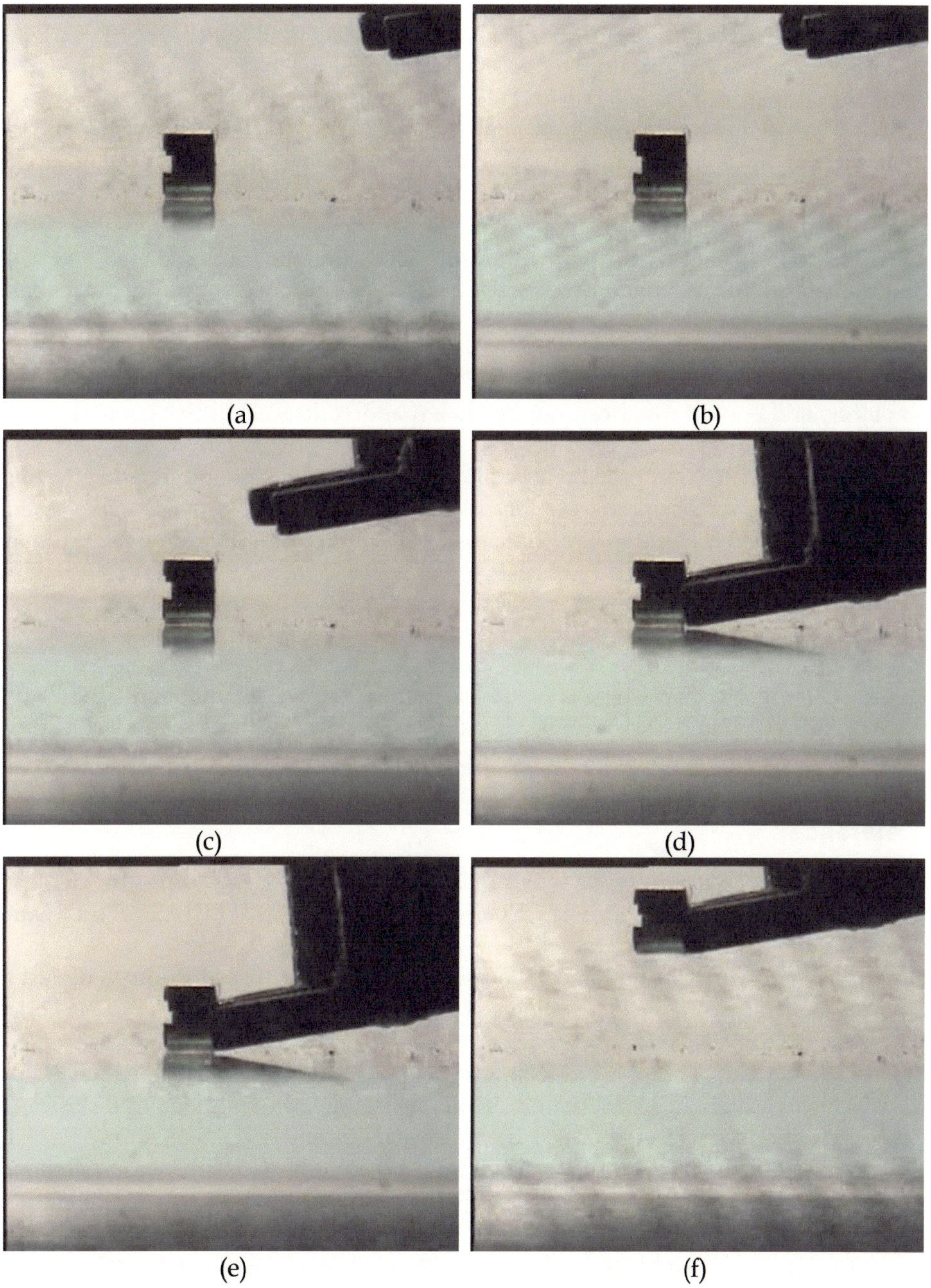

Fig. 16. The process of the piezoelectric microgripper automatically locateing and gripping the micro-target in the horizontal view field

7. Conclusion

For the completion of three-dimensional micro-sized components assembly, an improved support vector machine algorithm is presented, which is employed to identify multi micro objects. Then apply an improved broyden's method to estimate the image jacobian matrix on line. Classical RLS algorithm can provide an optimal estimate to a well-prior knowleage in image jacobian model for uncalibrated visual servoing. However the method has a strict requirement on the prior knowledge and shows a poor adaptability on convergence speed and stability to unknown dynamic applications. A novel improved broyden's method using chebyshev polynomial approximation algorithm for jacobian matrix estimation has been presented. Finally, design a PD controller to control micro-robot. In the microscopic visual environment, the visual servo task of micromanipulator positioning and automatic gripping micro-parts are completed. The experiment results show that the proposed method can meet the requirements of micro-assembly tasks.

8. Future work

Micro-objects and end-effectors can not be shown and controlled at the same time with a single zoom threshold or focus ratio because of the non-uniform light intensity. Therefore, the study of multi-scale's multi-objects classification algorithm is important and effective for improving the accuracy of micro-assembly tasks. Besides, Research on the micro-assembly control strategy based on multi-sensor data fusion is an important technique to improve the micro-robot system performance.

9. Acknowledgments

This work is supported by Chinese National Natural Science Foundation (CNSF) under Grant No. 60275013 and No. 60873032, as well as by National High-tech Research Developing Plan of China under Grant No. 2008AA8041302

10. Reference

J.A. Piepmeier, GV. McMurray,H. Lipkin, "A Dynamic Jacobian Estimation Method for Uncalibrated Visual Servoing," Proceedings on the 1999 IEEE/ASME International Conference on Advanced Intelligent Mechatronics. 944-949,1999.

Jianbo Su, Yugeng Xi, "Uncalibrated Hand Eye Coordination based on auto disturbance rejection controller," Proceedings of the 41st IEEE conference on Decision and control, 923-924, 2002.

Kang Qing sheng, Hao Ting, MENG Zhang_da et al, "Pseudo inverse Estimation of Image jacobian Matrix in Uncalibration Visual Servoing, " Proceedings of the 2006 IEEE International conference on Mechatronics and Automation, 25-28, 2006.

J. A. Piepmeier,G. V. MacMurray, H. Lipkin, "A Dynamic Quasi-Newton Method for Uncalibrated Visual Servoing," IEEE International Conference on Robotics and Automation, 1595-1600, 1999.

J. A. Piepmeier, G. V. MacMurray, H. Lipkin, "Uncalibrated Dynamic Visual Servoing, " IEEE Transactions on Robotics and Automation, l.20(1), 143-147, 2004.

J. Su, H. Ma, W. Qiu et al, " Task-independent robotic uncalibrated hand-eye coordination based on the extended state observer, " IEEE Trans. on Systems, Man, and Cybernetics, 34(4), 1917-1922, 2004.

S. Hutchinson, G.D. Hager, P.I. Corke, "A Tutorial on Visual Servo Control, " IEEE Transactions on Robotics and Automation, 12(5),651-670,1996.

H.Sutanto, R.Sharma,V Varma, "Image based Auto docking without Calibration, "Proceedings of the 1997 IEEE Internolionol Conference on Robotics and Automation, 974-979, 1997.

Y. Shen, D. Song, Y. H. Liu et al, "Asymptotic trajectory tracking of manipulators using uncalibrated visual feedback, "IEEE/ASME Trans. on Mechatronics, 8(1), 87-98, 2003.

C.-P. Lu, E. Mjolsness,G. D. Hager, "Online computation of exterior orientation with application to hand-eye calibration," Mathematical and Computer Modeling, 24(5), 121-143, 1996.

A. Astolfi, L. Hsu, M. Netto, et al, "Two solutions to the adaptive visual servoing problem, "IEEE Trans. on Robotics and Automation, 18(3), 387-392, 2002.

H. Wang, Y. H. Liu, "Adaptive image-based trajectory tracking of robots," Proc. Of IEEE Int. Conf.on Robotics and Automation, 564-569, 2005.

B. H. Yoshimi, P. K. Allen," Alignment using an uncalibrated camera system, "IEEE. Trans. on Robotics and Automation,11(4), 516-521, 1995.

Hu M K.. (1962). Visual attribute recognition by moment invariants. IEEE Trans on Information Theory, vol. 8, pp.179-187.

Emanuela B, Andrea B and Salvatore C.(2003). An innovative real time technique for buried object detection. IEEE Trans on Geoscience and Remote Sensing, vol.40, no.4, pp. 927-931.

Jose L R, Manel M and Mario D P.(2004). Support vector method for robot ARMA system identification. IEEE trans on signal processing, vol.52, no.1, pp. 155-164.

Yi X C & James Z W.(2003). Support vector learning for fuzzy rule based classification systems. IEEE trans on fuzzy system, vol.11, no.1, pp. 716-727.

Jing P, Douglas R H and Dai H K.(2003). LDA/SVM driven nearest neighbor classification. IEEE trans on neural networks, vol.14, no.4, pp. 940-942.

Andrew H S & Srinivas M. (2003). Identifying important features for intrusion detection using support vector machines and neural networks. Proceedings of the 2003 symposium on applications and internet.

Kaibo D, Keerthi S S and Aun N P.(2002). Evaluation of simple performance measures for tuning SVM hyperparameters. Neurocomputing, pp.1-19.

Richard Jensen & Qiang Shen.(2007). Fuzzy rough sets assisted attribute selection. IEEE transactions on fuzzy systems, vol.15, no.1, pp.73-89.

Yu chang rui, Wang hong wei and Luo yan.(2006). A heuristic algorithm for attribute reduction of decision making problem based on rough set. Proceedings of the sixth international conference on intelligent systems design and applications.

Kang Q. S., Hao T., Meng Z.D and Dai,X,Z. (2006). Pseudo inverse Estimation of Image jacobian Matrix in Uncalibration Visual Servoing. Proceedings of the 2006 IEEE International conference on Mechatronics and Automation ,pp.25-28.

Malik A. S. & Choi T. S.(2007). Consideration of illumination effects and optimization of window size for accurate calculation of depth map for 3D shape recovery. Pattern Recognition, vol.40, no.1,pp.154-170.

Shen, Song D., Liu Y. H and K Li. (2003). Asymptotic trajectory tracking of manipulators using uncalibrated visual feedback. IEEE/ASME Trans. on Mechatronics, vol.8, no.1,pp.87-98.

Piepmeier J. A., MacMurray G. V and Lipkin H. (1999). A Dynamic Quasi-Newton Method for Uncalibrated Visual Servoing. IEEE International Conference on Robotics and Automation , pp.1595-1600.

Piepmeier J. A. & MacMurray G. V.(2004). Lipkin H., Uncalibrated Dynamic Visual Servoing. IEEE Transactions on Robotics and Automation, vol.20, no.1, pp. 143-147.

Malis E. (2004). Visual servoing invariant to changes in camera-intrisic parameters. IEEE Trans. Robot. Autom, vol.20,no.1, pp.72–81.

Su J., Ma H., Qiu W and Xi Y.(2004). Task-independent robotic uncalibrated hand-eye coordination based on the extended state observer. IEEE Trans. on Systems, Man, and Cybernetics, vol.34, no.4, pp. 1917-1922.

4

Human-in-the-Loop Control for a Broadcast Camera System

Rares Stanciu and Paul Oh
Drexel University
USA

1. Introduction

There are many tools that carry cameras. Their working domain is usually surveillance, surface inspection, and broadcasting. Devices like rovers, gantries, and aircrafts often possess video cameras. The task is usually to maneuver the vehicle and position the camera to obtain the desired fields-of-view. A platform widely used in the broadcasting industry can be seen in Figure 1. The specific parts are usually the tripod, the boom, and the motorized pan-tilt unit (PTU).

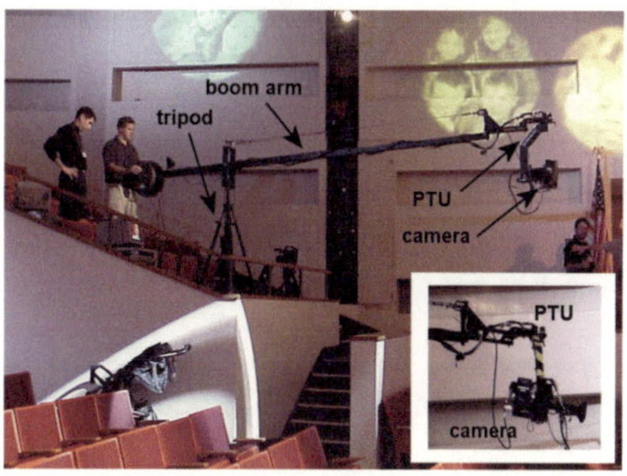

Fig. 1. The operator can move the boom horizontally and vertically to position the camera. The pan-tilt (*lower right inset*) head provides additional DOFs.

Manual operation of such a tool requires two skilled operators. Typically, one person will handle the boom while the second operator will coordinate the PTU camera to track the subjects using two joysticks. Tracking the moving objects is difficult because there are many degrees-of-freedom (DOFs) to be coordinated simultaneously. Increasing the target's speed increases the tracking difficulty. Using computer vision and control techniques ensures the automatic camera tracking and reduces the number of DOFs the operator has to coordinate. This way the platform can be operated by one person concentrating only on the booming. The use of such techniques enables the tracking of faster moving objects.
Searching through the literature on this subject reveals that there is a wealth of existing research in the visual servoing domain. An excellent starting point in the literature search is (Hutchinson et all., 1996). Extensive research is described in (Corke & Good, 1996); (Hill &

Park, 1973); (Oh & Allen, 2001); (Oh, 2002); (Sanderson & Weiss, 1980); (Stanciu & Oh, 2002); (Stanciu & Oh, 2003); (Papanikolopoulos et all., 1993). It is to be noted that in these publications, researchers have dealt completely with automated hardware (where no operator is involved). The system described in this paper is operated by humans. Some of the seminal man-machine interface work is represented by (Sheridan & Ferell, 1963); (Ferrier, 1998); (Fitts, 1954).

The system utilized for experimentation is shown in Figure 2. The platform is composed of a four-wheeled dolly, boom, motorized PTU, and camera. The dolly can be pushed and steered. The 1.2-m-long boom is linked to the dolly via a cylindrical pivot that allows the boom to sweep horizontally (pan) and vertically (tilt). Mounted on one end of the boom is a two-DOF motorized PTU and a video camera weighing 9.5 kg. The motors allow an operator to both pan and tilt the camera 360° at approximately 90°/sec. The PTU and the camera are counterbalanced by a 29.5-kg dumbbell mounted on the boom's opposite end.

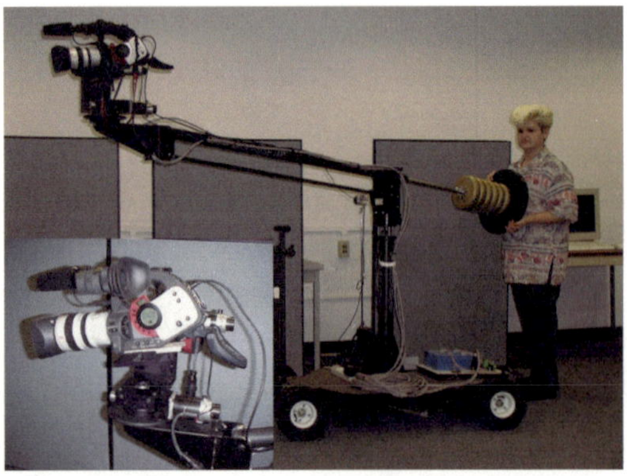

Fig. 2. The operator can boom the arm horizontally and vertically to position the camera. The pan-tilt head (*lower left inset*) provides additional DOFs.

Use of this boom-camera system normally entails one or more skilled personnel performing three different operations.

1. With a joystick, the operator servos the PTU to point the camera. A PC-104 small board computer and an ISA bus motion control card allow for accurate and relatively fast camera rotations.
2. The operator physically pushes on the counterweighted end to boom the camera horizontally and vertically. This allows one to deliver a diverse range of camera views (e.g. shots looking down at the subject), overcomes PTU joint limitations, and captures occlusion-free views.
3. The operator can push and steer the dolly in case the boom and PTU are not enough to keep the target image in the camera's desired field-of-view.

Tracking a moving object using such a tool is a particularly challenging task. Tracking performance is thus limited to how quickly the operator manipulates and coordinates multiple DOFs. Our particular interest in computer vision involves improving the camera operator's ability to track fast-moving targets. By possessing a mechanical structure, actuators, encoders, and electronic driver, this boom is a mechatronic system. *Visual-servoing* is used to control some DOFs so that the operator has fewer joints to manipulate.

This paper describes the implementation of several controllers in this human-in-the-loop system and discusses quantitatively the performance of each. The CONDENSATION algorithm is used for the image processing. As this algorithm is described in some publications (Isard & Blake, 1998), this paper will not focus on the image processing. Section 2 describes the experimental setups used. The controllers are described in Section 3. Development and validation of a boom-camera model is also presented. Section 4 describes a comparison between a well skilled operator versus a novice, both with and without the visual servoing. Section 5 presents the conclusions.

2. Experimental setups

The „artistic" side of a film shooting scenario is often very important. Because they involve humans, these scenarios are (strictly speaking) not repeatable. Therefore, to compare the behavior of different controllers, an experimental framework is needed. As such, the experiments were designed to offer the best possible answers for both scientific and artistic community.

The first experiment was people-tracking. A person was asked to walk in the laboratory. The camera attempted tracking while an operator boomed. Figure 3 (a) shows such an experiment.

Fig. 3. (a) Typical people-tracking set-up. A subject walks around and the camera attempts tracking while booming. (b) Wooden block target was mounted on the end-effector of a Mitsubishi robot arm (*background*). The boom-camera system (*foreground*) attempts to keep the target's image centered in the camera's field-of-view. (c) Novice and a well-skilled operator will manipulate the boom appropriately to move the camera along the shown path, with and without the help of the visual servoing. In addition to booming, under manual control, the operator will also have to coordinate camera's two DOFs using a joystick. Visual servoing tracking error is recorded for comparison.

Each new designed controller attempted to increase tracking performance. The second experiment was developed in an attempt to design a metric for performance. A Mitsubishi robotic arm was instructed to sinusoidally move the target back and forth [Figure 3 (b)]. While the operator boomed, the camera tracked the target. Target motion data, error, and booming data were recorded during the experiments and plotted for comparison with previous results.

At this point, it was interesting to determine whether the vision system was usable in sport broadcasting. An experiment in which the camera tried to track a ball moving between two people was set up. The experiment showed successful tracking but highlighted some challenges. This setup is described in Section 3.8.

Once the camera was considered to ensure a satisfactory tracking performance, it was interesting to determine how it can help the operator. To answer this question, another experiment was designed. Again, the Mitsubishi robotic arm was used. This time, the robot moved the target on a trajectory corresponding to the number "8". A novice and an experienced operator boomed along a predefined path and attempted tracking the robot end-effector with and without vision. Figure 3(c) shows the way the operator should boom. The visual servoing tracking error was recorded and plotted. This experiment is described in Section 4.

3. Controllers description

This section presents the hypotheses, describes the controllers in detail, and discusses the experiments and results during this research.

3.1 Proportional controller

To establish a base level, the first of our hypotheses was launched. It states that by *using a very simple controller (proportional) and a very simple image processing technique (color tracking), the camera is able to track a moving target when booming*.
The proportional controller was implemented. The current target position in the image plane is compared with the desired position and an error signal is generated. This error signal will determine the speed of the camera in its attempt to bring the target in focus. The controller gain K_x was set to 100. People-tracking experiment was attempted using this controller [Figure 3(a)]. A person wearing a red coat was asked to walk in the laboratory. The color-tracker board was trained for red. The task was to keep the red coat in the camera's field of view while an operator boomed. In this experiment, the camera-target distance was about 5 m.
To assess the controller performance quantitatively a toy-truck was to be tracked. An artificial white background was used to help the vision system to detect the target. In this experiment, the camera-target distance was 3 m. The toy moved back and forth while the camera attempted tracking. Camera motion data, booming data, and tracking error were recorded. The plots can be seen in Figure 4. Figure 4(a) shows the pan motor encoder indication, (b) shows the error (in pixels), and (c) shows the booming angle (in degrees). It can be seen that as the operator is booming and the target is moving, the controller performs a visually servoed counterrotation. The system was able to track the moving target even when using a very simple controller. Still, as one expects, there were two challenges: system stability and tracking performance.

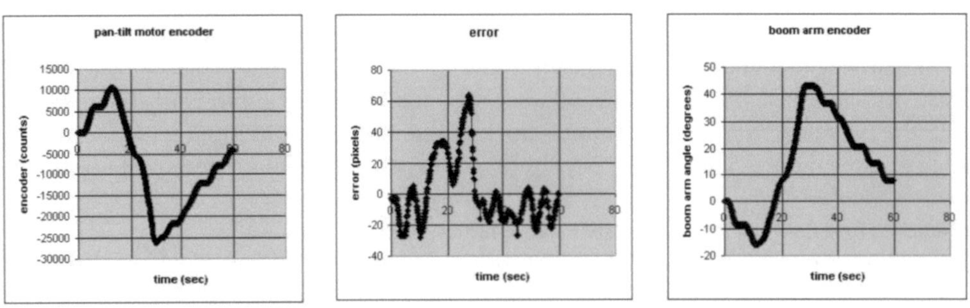

Fig. 4. $K_x = 100$ (a) PTU motor encoder. (b) Pixel error. (c) Boom-arm encoder.

The experiments have demonstrated that the key design parameter, when visually servoing redundant DOF systems, is stability, especially when the target and the boom move 180° out

of phase. If boom motion data is not included, camera pose cannot be determined explicitly because there are redundant DOFs. As a result, the system could track a slow-moving target rather well, but would be unstable when the target or boom moves quickly.
The second issue was the tracking performance. With the proportional controller, the operator boomed very slowly (less than 1°/sec). The target also moved slowly (about 10 cm/s). Any attempt to increase the booming or target speed resulted in the tracking failure. Both the experiments proved the first hypothesis. It is important to underline that the vision had no information about booming. Introducing booming information could improve tracking performance as well as stability.

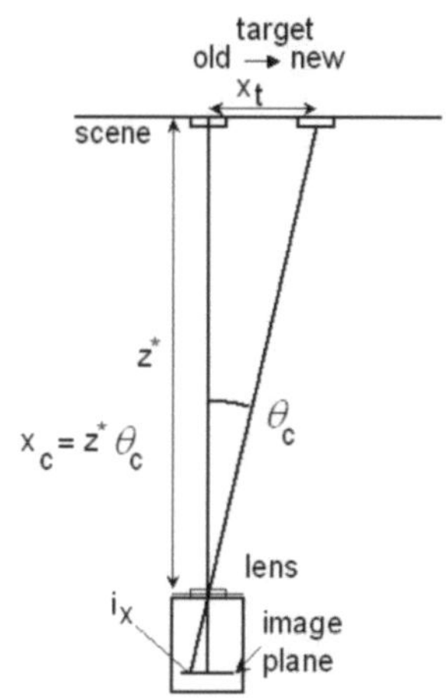

Fig. 5. Schematic of camera scene

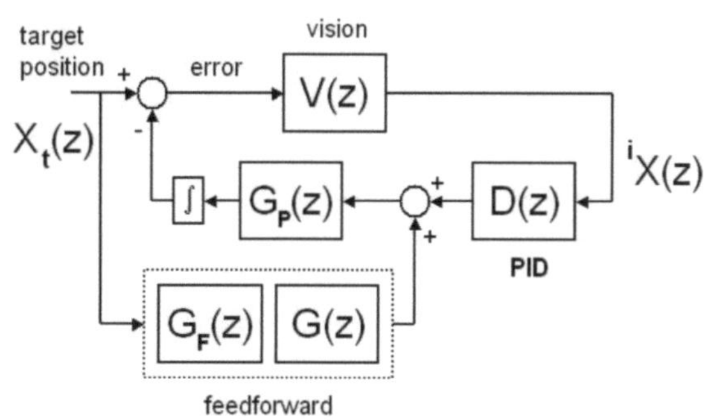

Fig. 6. Feedforward controller with a feedback compensation.

3.2 Feedforward controller

The second hypothesis was that *by using a feedforward control technique, we can improve both the performance and the stability*. A feedforward controller was designed to validate the second hypothesis. This controller provides the target motion estimation (Corke & Good, 1996). Figure 6 depicts a block diagram with a transfer function

$$\frac{{}^iX(z)}{X_t(z)} = \frac{V(z)(1 - G_p(z)D_F(z))}{1 + V(z)G_p(z)D(z)} \tag{1}$$

where ${}^{i}X(z)$ is the position of the target in the image, $X_t(z)$ is the target position, $V(z)$ and $G_p(z)$ are the transfer functions for the vision system and PTU, respectively. The previous and actual positions of the target in the image plane are used to predict its position and velocity one step ahead. Based on this, the feedforward controller will compute the camera velocity for the next step. $D_F(z)=G_F(z)G(z)$ represents the transfer function of the filter combined with the feedforward controller. $D(z)$ is the transfer function for the feedback controller. If $D_F(z)=G_p^{-1}(z)$, the tracking error will be zero, but this requires knowledge of the target position that is not directly measurable. Consequently, the target position and velocity are estimated. For a horizontally translating target, its centroid in the image plane is given by the relative angle between the camera and the target

$$ {}^{i}X(z) = K_{lens}(X_t(z) - X_r(z)) \tag{2} $$

where ${}^{i}X(z)$ and $X_t(z)$ are the target position in the image plane and world frame, respectively. $X_r(z)$ is the position of the point that is in the camera's focus (due to the booming and camera rotation) and K_{lens} is the lens zoom value. The target position prediction can be obtained from the boom and the PTU, as seen in Figure 5. Rearranging this equation yields

$$ \hat{X}_t(z) = \frac{{}^{i}\tilde{X}(z)}{K_{lens}} + X_r(z) \tag{3} $$

where $\hat{X}_t$ is the predicted target position.

3.3 The $\alpha-\beta-\gamma$ filter

Predicting the target velocity requires a tracking filter. Oftentimes, a Kalman filter is used, but is computationally expensive. Since Kalman gains often converge to constants, a simpler $\alpha-\beta-\gamma$ tracking filter can be employed that tracks both position and velocity without steady-state errors (Kalata & Murphy, 1997); (Tenne & Singh, 2000). Tracking involves a two step process. The first step is to predict the target position and velocity

$$ x_p(k+1) = x_s(k) + Tv_s(k) + T^2 a_s(k)/2 \tag{4} $$

$$ v_p(k+1) = v_s(k) + Ta_s(k) \tag{5} $$

where T is the sample time and $x_p(k+1)$ and $v_p(k+1)$ are the predictions for the position and velocity at iteration $k+1$, respectively. The variables $x_s(k)$, $v_s(k)$, and $a_s(k)$ are the corrected (smoothed) values of iteration k for position, velocity, and acceleration, respectively. The second step is to make corrections

$$ x_s(k) = x_p(k) + \alpha(x_o(k) - x_p(k)) \tag{6} $$

$$ v_s(k) = v_p(k) + (\beta/T)(x_o(k) - x_p(k)) \tag{7} $$

$$ a_s(k) = a_p(k-1) + (\gamma/2T^2)(x_o(k) - x_p(k)) \tag{8} $$

where $x_o(k)$ is the observed (sampled) position at iteration k. The appropriate selection of gains α, β, and γ will determine the performance and stability of the filter (Tenne & Singh 2000). The $\alpha-\beta-\gamma$ filter was implemented to predict the target velocity in the image plane with gains set at $\alpha=0.75$, $\beta=0.8$, and $\gamma=0.25$. This velocity was, then, used in the feedforward algorithm, as shown in Figure 7. Image processing in the camera system can be modeled as a $1/z$ unit delay that affects the camera position x_r and estimates of the target position. In Figure 7, the block $G_F(z)$ represents the transfer function of the $\alpha-\beta-\gamma$

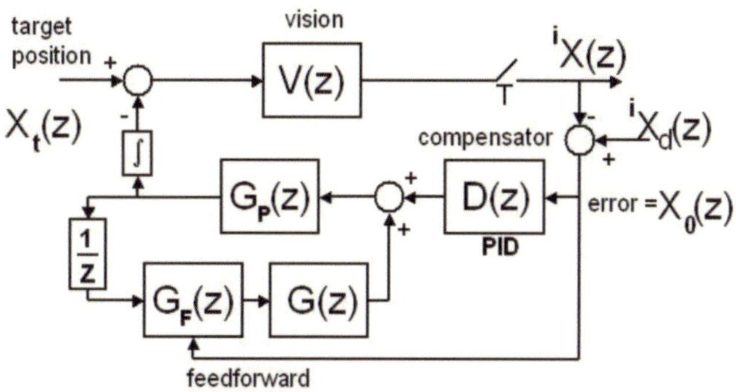

Fig. 7. Feedforward controller with a feedback compensation as it was implemented

Fig. 8. Three sequential images from videotaping the feedforward controller experiment. Camera field-of-view shows target is tracked *top row*. Boom manually controlled *middle row*. Working program *bottom row*.

filter, with the observed position as the input and the predicted velocity as the output. $X_d(z)$ represents the target's desired position in the image plane and its value is 320 pixels. $X_o(z)$ represents the position error in the image plane (in pixels).

The constant K_{lens} converts pixels in the image plane to meters. K_{lens} was assigned a constant value, and it assumes a pinhole camera model that maps the image plane and world coordinates. This constant was experimentally determined by comparing the known lengths in world coordinates to their projections in the camera's image plane. With the system equipped with the feedforward controller, a couple of experiments were performed. Again, the first was the people-tracking experiment. A subject was asked to walk back and forth in the laboratory environment. The operator boomed while the camera tracked the subject. Sequential images from the experiment can be seen in Figure 8. The first row shows the boom camera view. It can be seen that the system is not in danger of losing the target. The second row shows the operator booming while the third row shows the program working. It can be seen that the target is well detected.

To quantitatively assess the performance, the Mitsubishi robotic arm was instructed to move the target sinusoidally. The camera was instructed to track this target using the proportional as well as the feedforward controller. An operator panned the boom at the same time. Data regarding Mitsubishi motion, booming motion, and tracking error were recorded. The performance is assessed by comparing the tracking error. The setup can be seen in Figure 3(b).

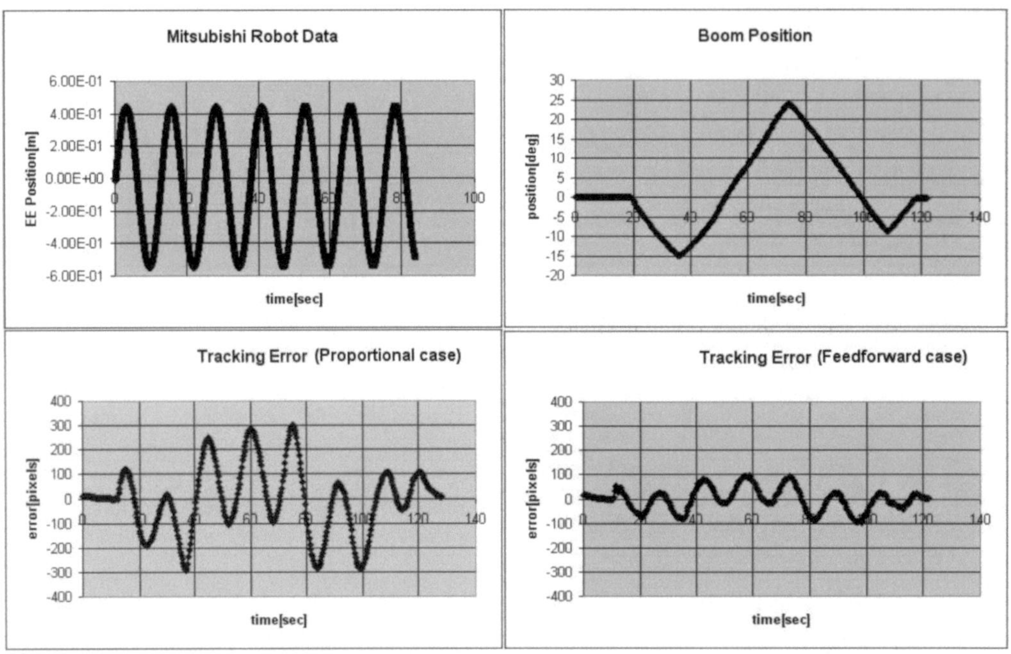

Fig. 9. Tracking errors comparing feedforward and proportional control in *human-in-the-loop visual servoing.(top row)* Target sinusoidal motion and booming. It can be seen that the operator moved the boom real slow (about 1°/sec). *(bottom row)* Tracking error using a proportional control (*left-hand side*) and a feedforward control (*right-hand side*). The image dimensions are 640×480pixels.

The experiment was set up in the laboratory. The camera-target distance was 3.15 m. The target dimensions were 8.9 × 8.25 cm^2. The robotic arm moved the target sinusoidally with a frequency of about 0.08 Hz and a magnitude of 0.5 m. CONDENSATION algorithm was employed for the target detection. As this algorithm is noisy, the target image should be kept small. The target dimensions in the image plane were 34×32pixels. While both the controllers attempted to track, the boom was manually moved from -15° to +25°. The plots can be seen in Figure 9. In the top row, the target motion and the booming plot (both versus time) can be seen. The operator moved the boom really slow (approximately 1°/sec). This booming rate was used because of the proportional controller. The tracking errors are shown in the bottom row. The bottom left image shows the error when using the proportional controller for tracking. The bottom right image shows the error when using the feedforward controller. The peak-to-peak error was about 100 pixels with the feedforward controller, while the proportional controller yielded an error of more then 300 pixels. By comparing the error in the same conditions, the conclusion was that the feedforward controller is „much better" then the proportional controller. Still, considering that the focal length was about 1200 pixels and given the camera-target distance of 3.15 m, 100 pixels represented about 35 cm of error. This value was considered to be too big.

3.4 Symbolic model formulation and validation

At this point, a model was desired for the boom-camera system. Simulation of new controllers would be much easier once the model was available. With satisfactory simulation results, a suitable controller can be implemented for experiments.

Both the nonlinear mathematical and simulation models of the boom were developed using *Mathematica* and *Tsi ProPac* (Kwatny & Blankenship, 1995); (Kwatny & Blankenship, 2000). The former is in Poincaré equations enabling one to evaluate the properties of the boom and to design either a linear or a nonlinear controller. The latter is in the form of a C-code that can be compiled as an S-function in SIMULINK. Together, these models of the highly involved boom dynamics facilitate the design and testing of the controller before its actual implementation. The boom, shown in Figure 10, comprises of seven bodies and eight joints.

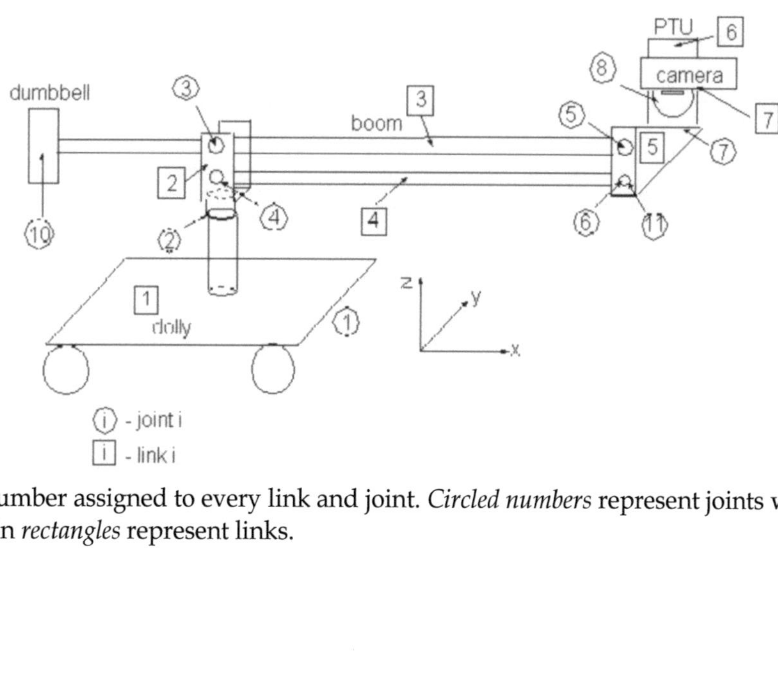

Fig. 10. Number assigned to every link and joint. *Circled numbers* represent joints while numbers in *rectangles* represent links.

Joint #	RB	JB	x	y	R_x	R_y	R_z
1		1	x	y			
2	1	2					ψ_b
3	2	3				θ_{bt1}	
4	2	4				θ_{bb1}	
5	3	5				θ_{bt2}	
6	4	5				θ_{bb2}	
7	5	6					ψ_c
8	6	7				θ_c	

Table 1. Types of motion for links.

Object	Mass $[kg]$	Moment of inertia $[kg \cdot m^2]$
Dolly (link 1)	25	$I_{xx} = 2.48$ $I_{yy} = 0.97$ $I_{zz} = 3.465$
Link 2	0.6254	$I_{xx} = 0.000907$ $I_{yy} = 0.000907$ $I_{zz} = 0.00181$
Boom (link 3)	29.5	$I_{xx} = 0$ $I_{yy} = 16.904$ $I_{zz} = 16.904$
Link 4	0.879	$I_{xx} = 0$ $I_{yy} = 0.02379$ $I_{zz} = 0.02379$
Link 5	3.624	$I_{xx} = 0.08204$ $I_{yy} = 0.00119$ $I_{zz} = 0.00701$
PTU (link 6)	12.684	$I_{xx} = 0.276$ $I_{yy} = 0.234$ $I_{zz} = 0.0690$
Camera (link 7)	0.185	$I_{xx} = 0$ $I_{yy} = 1.33 \cdot 10^{-5}$ $I_{zz} = 1.33 \cdot 10^{-5}$

Table 2. Boom links, masses, and moments of inertia.

The bodies and joints are denoted by boxes and circles, respectively. The DOFs of various joints are detailed in Table 1, while the physical data are given in Table 2. They give the position or Euler angles of the *joint body* (JB) with respect to the *reference body* (RB). At the origin, which corresponds to a stable equilibrium, the boom and the camera are perfectly aligned. One characteristic of the boom is that it always keeps the camera's base parallel to the floor. This is because bodies 3 and 4 are part of a four-bar linkage. There are two constraints for the system which can be seen in equation 9

$$\begin{aligned} \theta_{bb1} - \theta_{bt1} &= 0 \\ \theta_{bt1} + \theta_{bt2} &= 0 \end{aligned} \tag{9}$$

The inputs acting on the system are the torques Q_1 (about y) and Q_2 (about z) exerted by the operator, and the torques Q_3 and Q_4 applied by the pan and tilt motors of the camera, that is, u={ Q_1, Q_2, Q_3, Q_4}. The dumbbell at the end of body 3 is pushed to facilitate the target tracking with the camera. In this analysis, it is assumed that the operator does not move the cart, although it is straightforward to incorporate that as well. The pan and tilt motors correspond to the rotations ψ_c and θ_c, respectively.

The model can be obtained in the form of Poincaré equations [see (Kwatny & Blankenship, 1995) and (Kwatny & Blankenship, 2000) for details].

$$\begin{aligned} \dot{q} &= V(q)p \\ M(q)\dot{p} + C(q)p + Q(p,q,u) &= 0 \end{aligned} \tag{10}$$

The generalized coordinate vector q (see Table 1 for notation) is given by

$$q = [x, y, \psi_b, \theta_{bt1}, \theta_{bt2}, \theta_{bb1}, \theta_{bb2}, \psi_c, \theta_c]^T \tag{11}$$

Vector p is the 7×1 vector of quasi-velocities given by

$$p = \left[\Omega_{yc}, \Omega_{zc}, \Omega_{bb2}, \Omega_{bt2}, \Omega_{zb}, v_y, v_x\right]^T \tag{12}$$

They are the quasi-velocities associated with joints 8, 7, 6, 5, 2, and a double-joint 1, respectively. The first set of equations are the kinematics and the second are the dynamics of the system.

3.5 Model validation

The simulation model is generated as a C-file that can be compiled using any standard C-compiler. The MATLAB function *mex* is used to compile it as a dll file, which defines an S-function in SIMULINK. To ascertain the fidelity of the model, the experimental results in (Stanciu & Oh, 2004) were simulated in SIMULINK. The experimental setup is depicted in Figure 3(b). The booming angles, the target motion, and the errors are shown in Figures 11 and 12, respectively. In spite of the fact that the dynamics of the wheels and the friction in the joints are neglected, the experimental and simulated results show fairly good agreement.

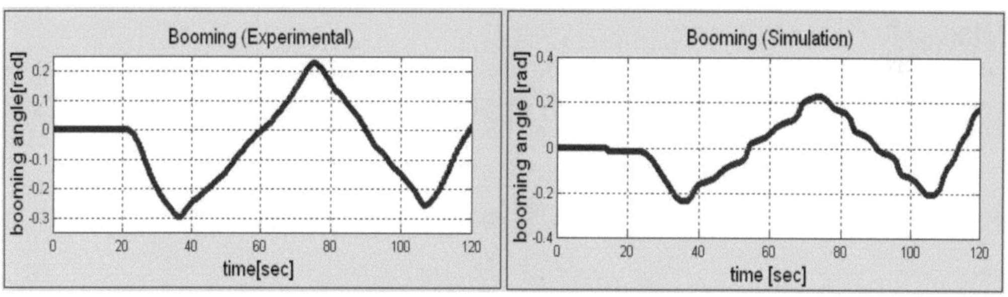

Fig. 11. Booming. Experiments (*left*) and simulation (*right*).

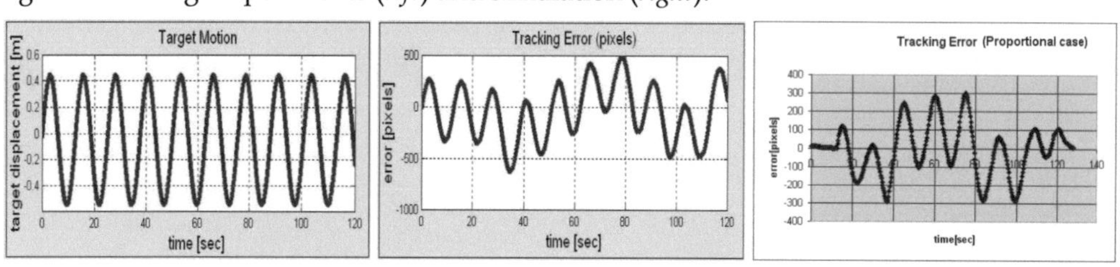

Fig. 12. Target motion (*left*). Simulation and experimental errors in pixels (central and right).

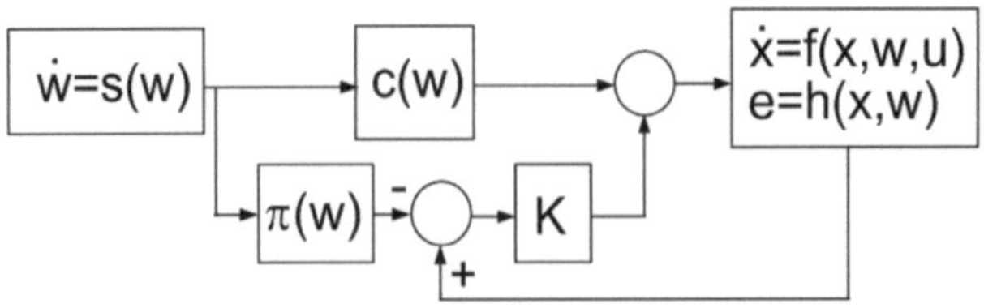

Fig. 13. Output tracking regulation controller as it was implemented.

3.6 Output Tracking Regulation Controller (OTR)

The target position in the image plane is a time-dependent function. By applying the Fourier theory, such a function can be expressed as a sum of sinusoids with decaying magnitudes and increasing frequencies. If the controller can be fine-tuned to ensure lower frequency sinusoids tracking, then the tracking error will be acceptable. The last of our hypotheses was that *adding such a controller to our system will improve the performance by reducing the error to* ±50 *pixels (50%) in case of the Mitsubishi Robot experiment.*

This paper investigated the effectiveness and advantages of the controller implemented as a regulator with disturbance rejection properties. This approach guarantees regulation of the desired variables, while simultaneously stabilizing the system and rejecting the exogenous disturbances. As a first step, a linear controller was designed to regulate only the pan motion. Its structure can be seen in Figure 13. The linearized equations are recast as

$$\begin{aligned} \dot{x} &= Ax + Pw + Bu \\ \dot{w} &= Sx \\ e &= Cx + Qw \end{aligned} \tag{13}$$

The regulator problem is solvable if and only if Π and Γ satisfy the linear matrix equations 14 [(Kwatny & Kalnitsky, 1978); (Isidori 1995)]:

$$\begin{aligned} \Pi S &= A\Pi + P + B\Gamma \\ 0 &= C\Pi + Q \end{aligned} \tag{14}$$

A regulating control can, then, be constructed as

$$u = \Gamma w + K(x - \Pi w) \tag{15}$$

where K is chosen so that the matrix $A + BK$ has the desired eigenvalues. These eigenvalues determine the quality of the response. The PTU motor model has the transfer function

$$\frac{\theta(s)}{V_a(s)} = \frac{0.01175}{1.3s^2 + 32s} \tag{16}$$

where the output is the camera angle. In this case, the state space description of the system is given by matrices A, B, and C

$$A = \begin{bmatrix} -24.61 & 0 \\ 1 & 0 \end{bmatrix}$$
$$B = \begin{bmatrix} 0.0088 \\ 0 \end{bmatrix} \quad (17)$$
$$C = \begin{bmatrix} 0 & 1 \end{bmatrix}$$

From equations (14)

$$\Pi = \begin{bmatrix} 1 & 0 \\ 0 & 1 \end{bmatrix}$$
$$\Gamma = \begin{bmatrix} -113.6 & 2796.6 \end{bmatrix} \quad (18)$$

The matrix K was

$$K = \begin{bmatrix} -10000 & -380 \end{bmatrix} \quad (19)$$

3.7 Simulation and experiments using output tracking regulation controller

Prior to the implementation experiment, a new controller was simulated using MATLAB SIMULINK. Sinusoidal reference signals corresponding to 1, 5, and 10 rad/sec were applied to the controller (in simulation). Both the reference and the output of the system were plotted on the same axes frame. The plots corresponding to the 5 rad/sec input can be seen in Figure 14. After the implementation, several experiments were performed using this controller. First, the controller was tested with the Mitsubishi robotic arm for a comparison of the performance of the feedforward and proportional controllers. In the second experiment, the system attempted to track a ball kicked by two players.

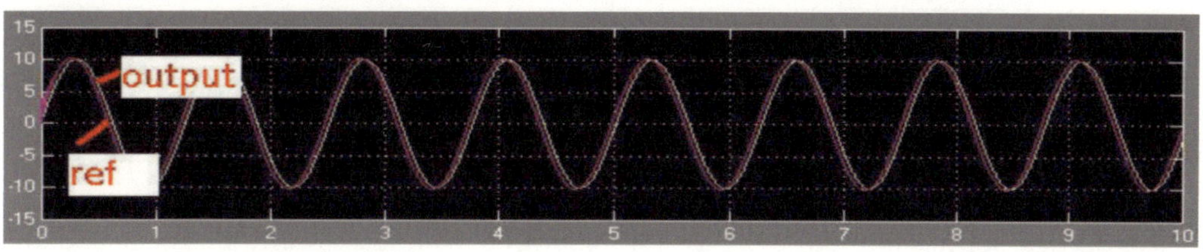

Fig. 14. Reference (5 rad/sec) as well as the output of the PTU using the new controller (the horizontal axis represents time in seconds).

In the first experiment, the robotic arm was instructed to sinusoidally move the target with the same frequency and magnitude as in the case of the feedforward controller. The camera tracked the target while the operator boomed. The booming data and the tracking error were recorded. The plots can be seen in Figure 15. In this figure, the top left plot represents the target motion while the top right plot shows the operator booming. It can be seen that the booming takes place with a frequency of about 3°/sec (when comparing the proportional and the feedforward controllers, the booming speed was about 1°/sec). The bottom left plot is the horizontal error when using the OTR controller (provided for comparison). It can be seen that when the OTR controller is used, the error becomes ±50 pixels (half of the value obtained using only the feedforward controller).

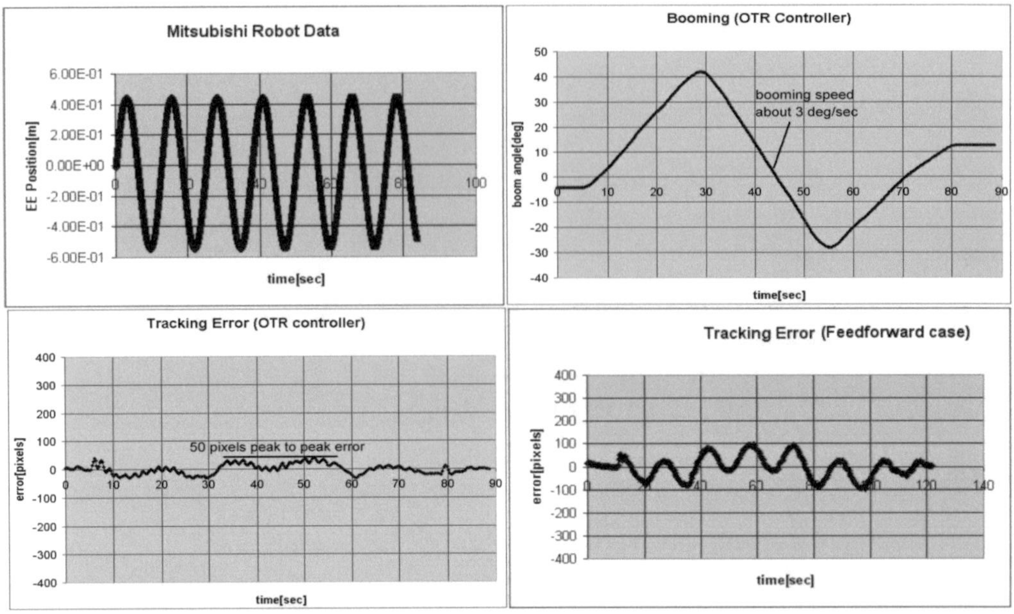

Fig. 15. Mitsubishi experiment using the OTR controller. The first figure shows the moving target. The second figure shows the boom motion. The third figure shows the tracking error in case of the output tracking controller. The fourth figure shows the error using the feedforward controller. It can be seen that by using the OTR controller, the error is less then ±50 pixels. This value reflects a gain in performance of 50%.

3.8 Ball-tracking experiment

Since the tracking error reduced when the robotic arm was used, it was interesting to see its behavior in a more natural environment. This time the task was to track a ball moving between two players. The experiment was set up in the laboratory and videotaped using three cameras. Sequential pictures can be seen in Figure 16. The top row shows the operator booming as the camera tracks the ball. The bottom row shows the boom camera point of view. It can be seen that the target is precisely detected and tracked. Despite its „not so scientific nature“ (no data was recorded), this experiment highlighted one challenge. If the ball is kicked softly, the image processing algorithm will successfully detect it and the camera is able to track it. If the ball is kicked harder, the camera fails to track it. This means that at a frequency of 3-4 Hz (the total time to process a frame and compute the controller outputs was around 340 ms), the target acceleration is limited to small values. This particular challenge was not revealed by experiments involving the robotic arm.

4. Human versus human-vision control: a comparison

It was interesting to determine if and how this system is able to help the operator. To assess the increase in performance due to the vision system, an experiment was set up. Again, the Mitsubishi robot was used. Its end-effector moved the target on a trajectory corresponding to a figure „8“ for 60 sec. An experienced operator and a beginner were asked to handle the boom with and without the help of vision. When vision was not used, the operator manually controlled the camera using a joystick.

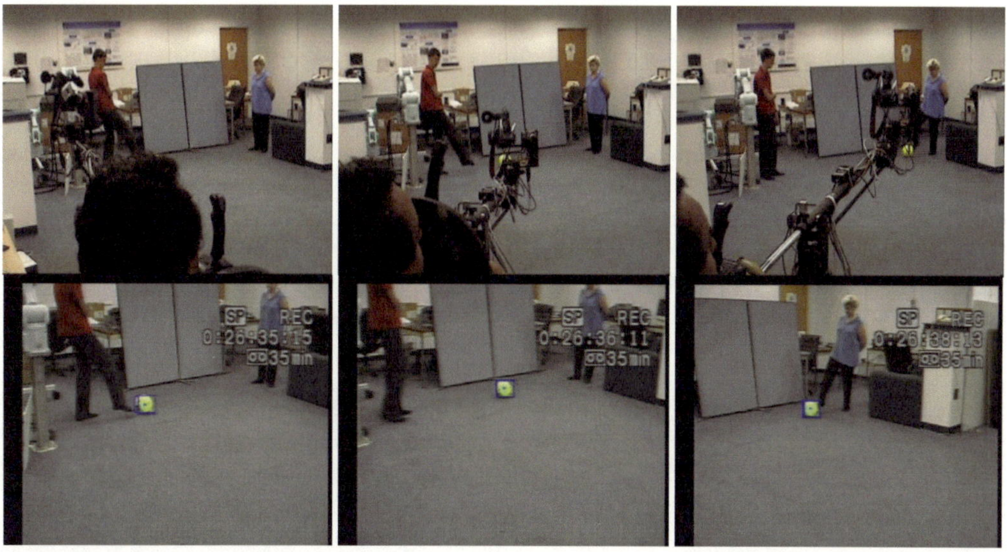

Fig. 16. Ball-tracking experiment. Operator booming and camera point of view (*top row*). Program working (*bottom row*).

A booming path was set up in an attempt to increase the experiment repeatability [shown in Figure 3(c)]. Each operator boomed two times: first, when using the vision system, and second, when manually manipulating the camera using a joystick. The objective was to keep the target in the camera's field-of-view while both the target and boom move. Several positions of interest were marked along the booming path using numbers [see Figure 3(c)]. Tracking error was recorded when using vision. Under the manual manipulation experiment, both the operators lost the target. When the target was outside the image plane, the image processing algorithm focused on other objects in the image. Because of this, the tracking error had no relevance during manual manipulation.
Sequential images from the experiment can be seen in Figures 17-20. The images are taken when the camera was in one of the positions marked in Figure 3(c).

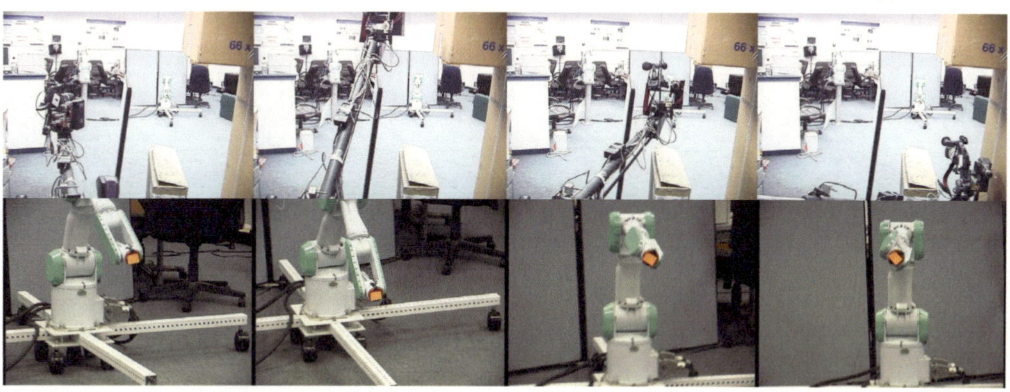

Fig. 17. Unexperienced operator with the vision system.

In the case of using the vision system, the target was never lost (Figures 17 and 19). Moreover, the output regulation controller (which is implemented for the pan motion) maintains the target very close to the image center.

In the case of manual tracking (Figures 18 and 20), the operator has to manipulate the boom as well as the camera. It can be seen that both the operators have moments when the target is lost. In case of an unexperienced operator without vision, the booming took longer than the motion of the robotic arm simply because there are more DOFs to be controlled simultaneously. The unexperienced operator lost the target eight times. The experienced operator was able to finish booming within 60 sec, but he lost the target five times. Because the program focuses on something else in the absence of the target, the data regarding the tracking error is not relevant when the target was lost. The target was never lost when using vision. The absolute value of the error in both the cases is shown in Figure 21. One can see that the values are in the same range. This means that visual servoing helps the novice operator to obtain performance similar to that of the expert.

Fig. 18. Unexperienced operator without the vision system. The target was lost eight times. The pictures were taken when the camera was in positions of interest shown in Fig. 3(c) and the top row of Fig 17. Because the target was lost, the tracking error curve has no relevance.

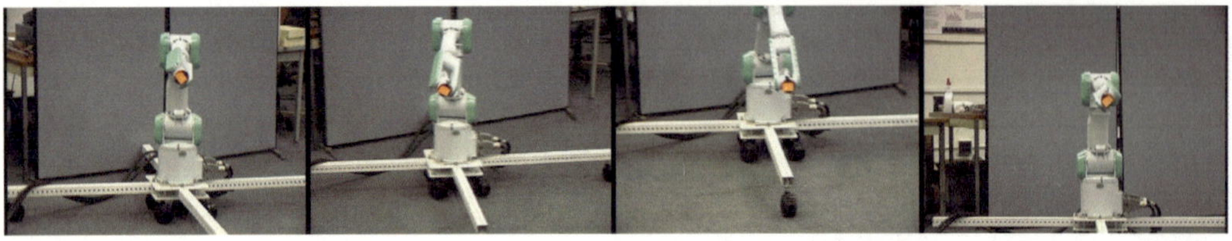

Fig. 19. Experienced operator with the vision system. Again, target is never lost. The pictures were taken when the camera was in positions of interest shown in Fig. 3(c) and the top row of Fig. 17.

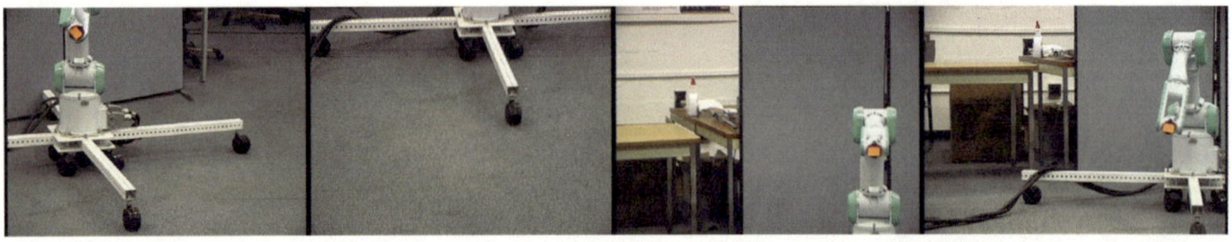

Fig. 20. Experienced operator without the vision system. The target was lost five times. Because the target was lost, the tracking error curve has no relevance. The pictures were taken when the camera was in positions of interest shown in Fig. 3(c) and the top row of Fig. 17.

5. Conclusion and future work

This paper integrates visual-servoing for augmenting the tracking performance of camera teleoperators. By reducing the number of DOFs that need to be manually manipulated, the operator can concentrate on coarse camera motion. Using a broadcast boom system as an experimental platform, the dynamics of the boom PTU were derived and validated experimentally. A new controller was added to the feedforward scheme and tested experimentally. The performance of the new control law was assessed by comparing the use of the vision system versus manual tracking for both an experienced and an unexperienced operator. The addition of the OTR controller to the feedforward scheme yielded lower errors. The use of the vision system helps the operator (the target was precisely detected and tracked). This suggests that by using the vision system, even an unexperienced operator can achieve a performance similar to that of a skilled operator. Also, there are situations when vision is helpful for a skilled operator. Still, there are situations when the target detection and tracking fail. A mechanism to detect such situations and alert the operator is desirable. When such situations occur, the camera can be programmed to automatically move to a particular position. The ball-tracking experiment proves to be successful if the ball is hit softly. When the ball is hit harder, the image processing fails to detect it, and tracking fails. However, there is no proof that controllers would be able to track a harder-hit ball if image processing did not fail.

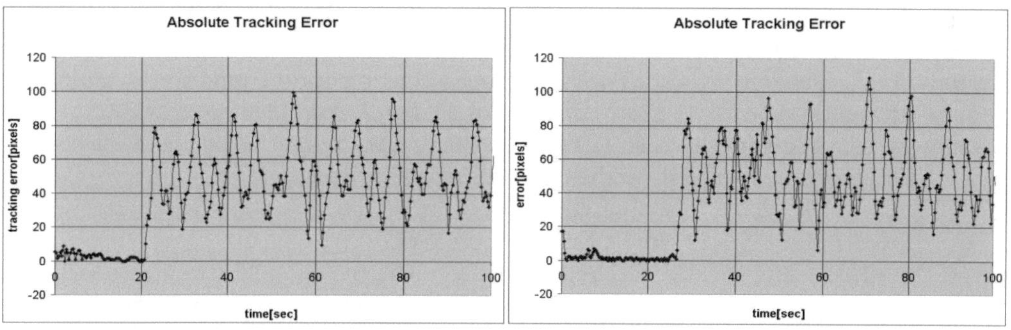

Fig. 21. Tracking error. Experienced operator with vision (*left-hand side*). Unexperienced operator using vision (*right-hand side*). Booming path was restricted. It can be seen that there are no significant diffrences between these two plots.

Another case that is not investigated in this paper is occlusion. Such experiments were not performed. They should be studied in future work. Because the focus of this research was the control part, the case of appearance of similar targets in the image plane was not studied. The effect of the image noise, when the camera moves quickly was also not studied. Future work will also have to focus on increasing tracking performance. If this tracking system is to be used in sports broadcasting, it will have to be able to track objects moving with higher acceleration. The sampling time (which now corresponds to 3-4 Hz) will have to decrease (perhaps one way to achieve this is to use a faster computer). When tracking sports events (football, soccer, etc.), when the target moves with high accelerations and its dimensions vary in the image, a target estimation mechanism will be desirable. Such a mechanism would record ball positions and estimate its trajectory. Once the estimation is done, this mechanism would command the camera to move to the estimated „landing" position and try to re-acquire the ball. Combining this mechanism with zooming in and out would allow tracking of faster objects.

6. References

Corke P.I.; Good M.C. (1996). Dynamic effects in visual closed-loop systems. *IEEE Trans. Robot. Autom.* Vol. 12, No. 5, Oct 2001, pp. 671-683,

Ferrier N. (1998). Achieving a Fitts Law relationship for visual guided reaching. *Proceedings of Int. Conf. Comput. Vis. (ICCV)* pp. 903-910 Bombay, India, Jan. 1998,

Fitts P. M. (1954). The information capacity of the human motor system in controlling the amplitude of movement. *J. Exp. Psychology,* Vol. 47, No. 6, 1954, pp. 381-391,

Hutchinson S., Hager G.D., Corke P.I. (1996). A tutorial on visual servo control. *IEEE Trans. Robot. Autom.* pp. 651-670, Vol.12, No. 5, Oct. 1996,

Hill J., Park W.T. (1973). Real time control of a robot by visual feedback in assembling tasks. *Pattern Recognit* Vol. 5 pp. 99-108, 1973,

Isard M., Blake A. (1998). CONDENSATION - Conditional density propagation for visual tracking. *Int. J. Comput. Vis.,* Vol. 29, No. 1, pp. 5-28, 1998,

Isidori A. (1995) *Nonlinear Control Systems* 3rd ed. Springer-Verlag, 3-540-19916-0, New York,

Kalata P.R., Murphy K.M. (1997). $\alpha - \beta$ target tracking with track rate variations. *Proceedings of 29th Southeastern Symp. Syst. Theory,* pp. 70-74, Mar. 1997,

Kwatny H.G., Kalnitsky K.C. (1978). On alternative methodologies for the design of robust linear multivariable regulators. *IEEE Trans. Autom. Control,* Vol. AC-23, No. 5, Oct. 1978, pp. 930-933,

Kwatny H. G., Blankenship G.L. (1995). Symbolic construction of models for multibody dynamics. *IEEE Trans. Robot. Autom.,* Vol. 11, No. 2, Apr. 1995 pp. 271-281,

Kwatny H. G., Blankenship G.L. (2000). *Nonlinear Control and Analytical Mechanics: A Computational Approach,* Birkhauser, 0-8176-4147-5, Boston, MA,

Oh P. Y., Allen P. K. (2001). Visual servoing by partitioning degrees of freedom. *IEEE Trans. Robot. Autom.* Vol. 17 No. 1, , Feb. 2001, pp. 1-17,

Oh P. Y. (2002). Biologically inspired visual-servoing using a macro/micro actuator approach. *Int. Conf. Imaging Sci., Syst. Technol (CISST),* Jun 2002 Las Vegas CA,

Papanikolopoulos N.P., Khosla P.K., Kanade T. (1993). Visual tracking of a moving target by a camera mounted on a robot: A combination of vision and control. *IEEE Trans. Robot. Autom.,* Vol. 9, No. 1, Feb. 1993, pp. 14-35,

Sanderson A.C., Weiss L.E. (1980). Image-based visual servo control using relational graph error signals. *Proceedings of IEEE Int. Conf. Robot. Autom.,* 1980, pp. 1074-1077,

Sheridan T.B., Ferell W.R. (1963). Remote manipulative control with transmission delay. *IEEE Trans. Human Factors in Electronics* Vol. HFE-4, 1963, pp. 25-29,

Stanciu R., Oh P.Y. (2002). Designing visually servoed tracking to augment camera teleoperators. *Proceedings of IEEE Int. Conf. Intell. Robots Syst. (IROS),* Vol. 1, Lausanne, Switzerland, 2002, pp. 342-347,

Stanciu, R., Oh, P.Y. (2003). Human-in-the-loop visually servoed tracking. *Proceedings of Int. Conf. Comput. Commun. Control Technol. (CCCT),* Vol. 5, pp. 318-323, Orlando, FL, Jul. 2003,

Stanciu R., P.Y. Oh P.Y. (2004), Feedforward control for human-in-the-loop camera systems. *Proceedings of Int. Conf. Robot. and Autom. (ICRA),* Vol. 1, pp. 1-6, New Orleans, LA, Apr. 2004,

Tenne D., Singh T. (2000). Optimal design of $\alpha - \beta - (\gamma)$ filters. *Proceedings of Am. Control Conf.,* Vol. 6, pp. 4348-4352, Chicago, Illinois, Jun. *2000.*

5

Vision-Based Control of the Mechatronic System

Rong-Fong Fung[1] and Kun-Yung Chen[2]
[1]*Department of Mechanical and Automation Engineering*
[2]*Institute of Engineering Science and Technology*
National Kaohsiung First University of Science and Technology
1 University Road, Yenchau, Kaohsiung County 824,
Taiwan

1. Introduction

The mechatronic system is employed widely in the industry, transportation, aviation and military. The system consists of an electrical actuator and a mechanism, and commonly is effective in industry territory. The toggle mechanism has many applications where overcome a large resistance with a small driving force is necessary; for examples, clutches, rock crushers, truck tailgates, vacuum circuit breakers, pneumatic riveters, punching machines, forging machines and injection modeling machines, etc. The motion controls of the motor-toggle mechanism have been studied (Lin *et al.*, 1997; Fung & Yang, 2001; Fung *et al.*, 2001). (Lin *et al.1997)* proposed a fuzzy logic controller, which was based on the concept of hitting condition without using the complex mathematical model for a motor-mechanism system. The fuzzy neural network controller (Wai *et al.*, 2001; Wai, 2003) was applied to control a motor-toggle servomechanism. The numerical results via the inverse dynamics control and variable structure control (VSC) were compared for an electrohydraulic actuated toggle mechanism (Fung & Yang, 2001). The VSC (Fung *et al.*, 2001) was employed to a toggle mechanism, which was driven by a linear synchronous motor and the joint coulomb friction was considered. In the previous studies, the motion controller for the toggle mechanism had been performed extensively. But the controllers are still difficult to realize if the linear scales can not be installed in the toggle mechanism for real feedbacks of positions and speeds.

In the adaptive control territory, (Li *et al.* 2004) proposed a hybrid control scheme for the flexible structures to obtain both dynamic and static characteristics. A nonlinear strategy is proposed by (Beji & Bestaoui, 2005) to ensure the vehicle control, in which the proof of main results is based on the Lyapunov concept. In these studies, the linear scale or encoder was employed as the sensor to feedback the positions and speeds. If the sensor is difficult to install, the non-contact measure vision-based is necessary and effective to apply in the mechatronic system.

In such motor-mechanism coupled systems, the non-contact machine vision exhibits its merits to measure the output responses of the machine. In previous references (Petrovic & Brezak, 2002; Yong et al., 2001), the machine vision was implemented with the PI and PD controllers, but didn't concern about the robustness of the vision system associated with controllers. (Park & Lee, 2002) presented the visual servo control for a ball on a plate and tracked its desired trajectory by the SMC. But there was no comparison with any other controller, and the mathematical equations of motion must be exactly obtained first, then the SMC can be implemented. (Petrovic & Brezak, 2002) applied the vision systems to motion control, in which the hard real-time constrains was put on image processing and was suitable for real-time angle measurement. In the autonomous vehicle (Yong *et al.*, 2001), the reference lane was continually detected by machine vision system in order to cope with the steering delay and the side-slip of vehicle, and the PI controller was employed for the yaw rate feedback. (Nasisi & Carelli, 2003) designed the adaptive controllers for the robot's positioning and tracking by use of direct visual feedback with camera-in-hand configurations. In these previous studies, they did not either discuss about the robustness of the vision system associated with the controllers or investigate robustness performances of the controllers for robot systems in experimental realization.

The control techniques are essential to provide a stable and robust performance for a wide range of applications, e.g. robot control, process control, etc., and most of the applications are inherently nonlinear. Moreover, there exist relatively little general theories for the adaptive controls (Astrom & Wittenmark, 1995; Slotine & Li, 1991) of nonlinear systems. As the application of a motor-toggle mechanism has similar control problems to the robotic systems, the adaptive control technique developed by (Slotine and Li, 1988, 1989), which exploited the conservation of energy formulation to design control laws for the fixed position control problem, is adopted to control the motor-toggle mechanism in this chapter. The techniques made use of matrix properties of a skew-symmetric system so that the measurements of acceleration signals and the computations of inverse of the inertia matrix are not necessary. Moreover, an inertia-related Lyapunov function containing a quadratic form of a linear combination of speed- and position-error states will be formulated. Furthermore, the SMC, PD-type FLC (Rahbari & Silva, 2000) and PI-type FLC (Aracil & Gordillo, 2004) are proposed to positioning controls, and their performances by machine vision are compared between numerical simulations and experimental experiments.

In this chapter, the machine vision system is used as the sensor to measure the output state of the motor-toggle mechanism in real operational conditions. The shape-pattern and color-pattern (Hashimoto & Tomiie, 1999) on the link and slider are applied as the output objects to measure by the machine vision system. The main advantage of a vision-based measuring system is its non-contact measurement principle, which is important in cases when the contact measurements are difficult to implement.

In the theoretical analysis, Hamilton's principle, Lagrange multiplier, geometric constraints and partitioning method are employed to derive the dynamic equations. In order to control the motor-mechanism system with robust characteristics, the SMC is designed to control the slider position. However, the general problem encountered in the design of a SMC system is that the bound of uncertainties and the exact mathematical mode of the motor-mechanism system are difficult to obtain in practical applications. In order to overcome the difficulties, the PI-type FLC, which is based on the concept of hitting conditions and without using the complex mathematical model of the motor-mechanism system, is successfully proposed by machine vision numerically and experimentally.

This chapter is organized as follows. After an introduction in Section 1, a mathematical modeling is in Section 2. Section 3 shows the design of the vision-based controller. Section 4 is the numerical simulations. The machine-vision experiments are in Section 5. Finally, experimental results and conclusions are shown in Section 6 and 7, respectively.

2. Mathematical modeling of the mechatronic system

In this chapter, the motor-toggle mechanism is a representative mechatronic system and consists of a servo motor and a toggle mechanism. The electric power is transferred to mechanical power by the motor. This is the basic goal of the mechatronic system.

2.1 Mathematical model of the motor-toggle mechanism

The toggle mechanism driven by a PMSM is shown in Fig. 1(a) and its experimental equipment is shown in Fig. 1(b). The screw is a media that makes the small torque τ to convert into the large force F_C acting on the slider C. The conversion relationship is

$$\tau = \frac{F_C l_d}{2\pi n}, \tag{1}$$

where l_d is the lead of screw, n is the gear ratio number. (Huang et al., 2008) have shown the holomonic constraint equation for the toggle mechanism as follows:

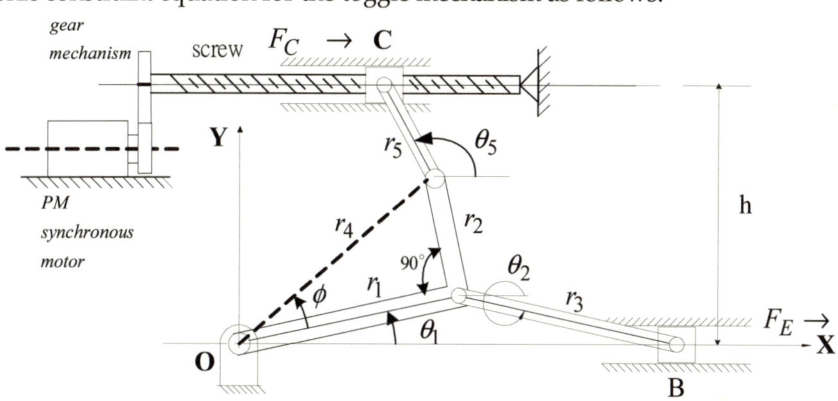

(a)

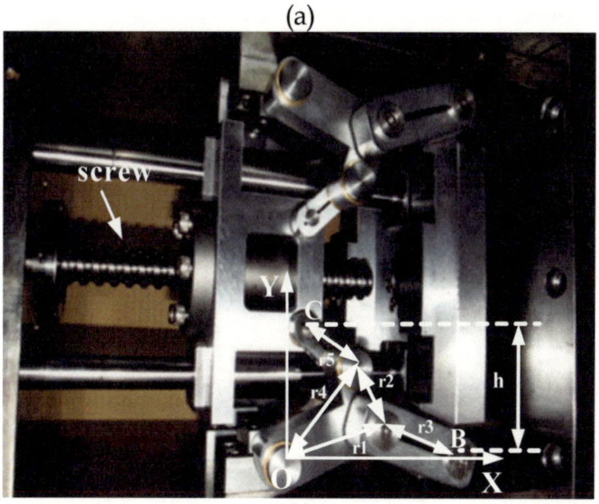

(b)

Fig. 1. The toggle mechanism driven by a PMSM. (a) The physical model. (b) The experimental equipment.

$$\Phi(\theta)=\begin{bmatrix} r_3\sin\theta_2+r_1\sin\theta_1 \\ r_5\sin(\pi-\theta_5)+r_4\sin(\theta_2+\phi)-h \end{bmatrix}=0, \tag{2}$$

where $\theta=[\theta_5\ \theta_2\ \theta_1]^T$ is the vector of generalized coordinates. Similar to the previous study (Chuang et al. 2008) one obtains Euler-Lagrange equation of motion, accounting for both the applied and constraint forces, as

$$\mathbf{M}(\theta)\ddot{\theta}+\mathbf{N}(\theta,\dot{\theta})-\mathbf{B}\mathrm{U}-\mathbf{D}+\Phi_\theta^T\lambda=0, \tag{3}$$

and the details of $\mathbf{M}$, $\mathbf{N}$, $\mathbf{B}$, U and $\mathbf{D}$ can also be found in (Chuang et al. 2008) . Taking the first and second derivatives of the constraint position Equation (2), we obtain

$$\Phi_\theta\dot{\theta}=\begin{bmatrix} r_3\dot{\theta}_2\cos\theta_2+r_1\dot{\theta}_1\cos\theta_1 \\ r_5\dot{\theta}_5\cos\theta_5+r_4\dot{\theta}_1\cos(\theta_1+\phi) \end{bmatrix}=0, \tag{4}$$

$$\Phi_\theta\ddot{\theta}=-(\Phi_\theta\dot{\theta})_\theta\dot{\theta}=\gamma=\begin{bmatrix} r_3\dot{\theta}_2^2\sin\theta_2+r_1\dot{\theta}_1^2\sin\theta_1 \\ r_5\dot{\theta}_5^2\sin\theta_5+r_4\dot{\theta}_1^2\sin(\theta_1+\phi) \end{bmatrix}=0. \tag{5}$$

By using these equations and Euler-Lagrange Eq. (3), we obtain the equation in the matrix form as

$$\begin{bmatrix} \mathbf{M} & \Phi_\theta^T \\ \Phi_\theta & \mathbf{0} \end{bmatrix}\begin{bmatrix} \ddot{\theta} \\ \lambda \end{bmatrix}=\begin{bmatrix} \mathbf{B}\mathrm{U}+\mathbf{D}(\theta)-\mathbf{N}(\theta,\dot{\theta}) \\ \gamma \end{bmatrix}. \tag{6}$$

This is a system of differential-algebraic equations.

2.2 Reduce formulation of the differential equations

The motion equations of the toggle mechanism are summarized in the matrix form of Eq. (6) and the constraint equation (2). The following implicit method is employed to reduce the system equations.

Equations (2) and (6) may be reordered and partitioned according to the decomposition of $\theta=[\theta_5\ \ \theta_2\ \ \theta_1]^T=[\mathbf{u}^T\ v^T]^T$. Thus, equation (6) can be written in the matrix form as:

$$\hat{M}(v)\ddot{v}+\hat{N}(v,\dot{v})=\hat{Q}U+\hat{D}. \tag{7}$$

where

$$\hat{M}=M^{vv}-\mathbf{M}^{vu}\Phi_u^{-1}\Phi_v-\Phi_v^T\left(\Phi_u^{-1}\right)^T\left[\mathbf{M}^{uv}-\mathbf{M}^{uu}\Phi_u^{-1}\Phi_v\right],$$

$$\hat{N}=\left[N^v-\Phi_v^T\left(\Phi_u^{-1}\right)^T\mathbf{N}^u\right]+\left[\mathbf{M}^{vu}\Phi_u^{-1}-\Phi_v^T\left(\Phi_u^{-1}\right)^T\mathbf{M}^{uu}\Phi_u^{-1}\right]\gamma,$$

$$\hat{Q}=B^v-\Phi_v^T\left(\Phi_u^{-1}\right)^T\mathbf{B}^u,\ \ U=\left[i_q^*\right],\ \ \hat{D}=D^v-\Phi_v^T\left(\Phi_u^{-1}\right)^T\mathbf{D}^u.$$

The elements of the vectors *u*, *v* and matrices $\mathbf{\Phi}_\mathbf{u}$, $\mathbf{\Phi}_\mathbf{v}$, $\mathbf{M}^{\mathbf{uu}}$, $\mathbf{M}^{\mathbf{uv}}$, $\mathbf{M}^{\mathbf{vu}}$, M^{vv}, $\mathbf{N}^{\mathbf{u}}$ and N^{v} are detailed in (Huang et al., 2008) . The resultant equation (7) is a differential equation with only one independent generalized coordinate v , which is the rotation angle θ_1 of link 1 in Fig. 1(a). The system becomes an initial value problem and can be integrated by using the fourth-order Runge-Kutta method.

2.3 Field-oriented PMSM

A machine model (Lee *et al.*, 2005) of a PMSM can be described in a rotating rotor and the electric torque equation for the motor dynamics is

$$\tau_e = \tau_m + B_m \omega_r + J_m \dot{\omega}_r. \tag{8}$$

where τ_m is the load torque, B_m is the damping coefficient, ω_r is the rotor speed and J_m is the moment of inertia.

With the implementation of field-oriented control, the PMSM drive system can be simplified to a control system block diagram as shown in Fig. 2, in which

$$\tau_e = K_t i_q^*, \tag{9}$$

$$K_t = \frac{3}{2} P L_{md} I_{fd}, \tag{10}$$

$$H_p(s) = \frac{1}{J_m s + B_m}, \tag{11}$$

where i_q^* is the torque current command. By substituting (9) into (8), the applied torque can be obtained as follows:

$$\tau_m = K_t i_q - J_m \dot{\omega}_r - B_m \omega_r, \tag{12}$$

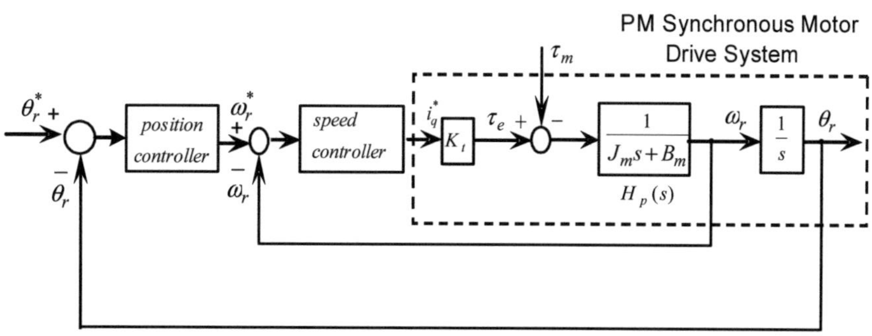

Fig. 2. A simplified control block diagram.

3. Design of the vision-based controllers

The control strategies are to use the non contact measurement CCD as the feedback sensor and design the controller to control the output status of the mechatronic system. Based on the CCD vision, we will propose the adaptive controller, slider mode controller and fuzzy controller for the mechatronic system. Because the dynamic formulation is obtained, we can perform the controllers in the mechanism modeling numerically, and realize the proposed controllers experimentally.

3.1 Design of an adaptive vision-based controller

The block diagram of the adaptive vision-based control system is shown in Fig. 3, where x_B^*, x_B and θ_1 are the slider command position, slider position and the rotation angle of link 1 of the motor-mechanism system, respectively. The slider position x_B is the desired control objective and can be manipulated from the rotation angle θ_1 by the relation $x_B = 2r_1 \cos\theta_1$, where the angle θ_1 is the experimental measured state by use of a shape pattern in the machine vision system.

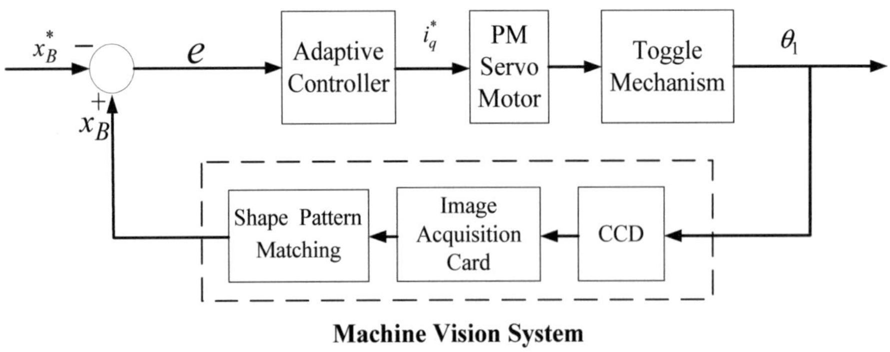

Fig. 3. Block diagram of an adaptive vision-based control system.

In order to design an adaptive control, we rewrite equation (7) as the second-order nonlinear one:

$$U(t) = f(\mathbf{X};t)\ddot{v}(t) + G(\mathbf{X};t) - d(t), \tag{13}$$

where

$$f(\mathbf{X};t) = \hat{\mathbf{Q}}^{-1}\hat{\mathbf{M}},\ G(\mathbf{X};t) = \hat{\mathbf{Q}}^{-1}\hat{\mathbf{N}},\ d(t) = \hat{\mathbf{Q}}^{-1}\hat{\mathbf{D}},$$

and $U(t)$ is the control input current i_q^*. It is assumed that the mass of slider B and the external force F_E are not exactly known. With these uncertainties, the first step in designing an adaptive vision-based controller is to select a Lyapunov function, which is a function of tracking error and the parameters' errors. An inertia-related Lyapunov function (Slotine & Li, 1988; Slotine & Li, 1989; Lin et al., 1997) containing a quadratic form of a linear combination of speed- and position-error states is chosen as follows:

$$V = \frac{1}{2}s^T f(X;t)s + \frac{1}{2}\tilde{\varphi}^T\Gamma^{-1}\tilde{\varphi}, \tag{14}$$

where

$$s = \lambda_e e + \dot{e},\ \ e = x_B - x_B^*,\ \ \Gamma = \begin{bmatrix} \gamma_1 & 0 \\ 0 & \gamma_2 \end{bmatrix},\ \ \tilde{\varphi} = \varphi - \hat{\varphi},\ \ \varphi = \begin{bmatrix} m_B \\ F_E \end{bmatrix},\ \ \hat{\varphi} = \begin{bmatrix} \hat{m}_B \\ \hat{F}_E \end{bmatrix},$$

and λ_e, γ_1 and γ_2 are positive scalar constants. The auxiliary signal s may be considered as a filtered tracking error.
Differentiating Eq. (14) with respect to time gives

$$\dot{V} = s^T f(X;t)\dot{s} + \frac{1}{2}s^T\dot{\hat{Q}}^{-1}\hat{M}s + \frac{1}{2}s^T\hat{Q}^{-1}\dot{\hat{M}}s + \tilde{\varphi}^T\Gamma^{-1}\dot{\tilde{\varphi}}, \tag{15}$$

and multiplying the variable $\dot{s}$ with Eq. (13), we have

$$\begin{aligned} f(X;t)\dot{s} &= f(X;t)(\lambda_e\dot{e} - \ddot{x}_B^* + \ddot{x}_B) \\ &= f(X;t)(\lambda_e\dot{e} - \ddot{x}_B^*) + f(X;t)\ddot{x}_B \\ &= Y(\bullet) + Z(\bullet)\varphi - 2r_1\sin\theta_1 U, \end{aligned} \tag{16}$$

Substituting Eq. (16) into Eq. (15) gives

$$\begin{aligned} \dot{V} &= s^T(Y(\bullet) + Z(\bullet)\varphi - 2r_1\sin\theta_1 U) + \frac{1}{2}s^T\dot{\hat{Q}}^{-1}\hat{M}s + \frac{1}{2}s^T\hat{Q}^{-1}\dot{\hat{M}}s + \tilde{\varphi}^T\Gamma^{-1}\dot{\tilde{\varphi}} \\ &\equiv s^T(Y'(\bullet) + Z'(\bullet)\varphi - 2r_1\sin\theta_1 U) + \tilde{\varphi}^T\Gamma^{-1}\dot{\tilde{\varphi}}, \end{aligned} \tag{17}$$

where $Y(\bullet)$, $Z(\bullet)$, $Y'(\bullet)$ and $Z'(\bullet)$ can be found in (Chuang et al., 2008). If the control input is selected as

$$U = \frac{1}{2r_1\sin\theta_1}(Y'(\bullet) + Z'(\bullet)\hat{\varphi} + K_V s), \tag{18}$$

where K_V is a positive constant. Eq. (17) becomes

$$\dot{V} = -s^T K_V s + \tilde{\varphi}^T(\Gamma^{-1}\dot{\tilde{\varphi}} + Z'(\bullet)^T s). \tag{19}$$

By selecting the adaptive update rule as

$$\dot{\tilde{\varphi}} = -\dot{\hat{\varphi}} = -\Gamma Z'(\bullet)^T s, \tag{20}$$

and substituting into Eq. (19), it then becomes

$$\dot{V} = -s^T K_V s \le 0. \tag{21}$$

As $\dot{V}$ in Eq. (21) is negative semi-definite, then V in Eq. (14) is upper-bounded. As V is upper-bounded and $f(\mathbf{X};t)$ is a positive-definite matrix, i, e, s and $\tilde{\varphi}$ are bounded.
Let function $P(t) = -\dot{V}(t) = s^T K_V s$, and integrate function $P(t)$ with respect to time

$$\int_0^t P(t)dt = V(0) - V(t). \tag{22}$$

Because $V(0)$ is bound, and $V(t)$ is non-increasing and bounded, then

$$\lim_{t\to\infty}\int_0^t P(\tau)d\tau < \infty . \tag{23}$$

Differentiate $P(t)$ with respect to time, we have

$$\dot{P}(t) = s^T K_V \dot{s} + \dot{s}^T K_V s . \tag{24}$$

Since K_V, s, and $\dot{s}$ are bounded, $P(t)$ is uniformly continuous. From the above description, Barbalat's Lemma (Narendra & Annaswamy, 1988) can be used to state that

$$\lim_{t\to\infty} P(t) = 0. \tag{25}$$

Therefore, it can be obtained as follows

$$\lim_{t\to\infty} s = 0. \tag{26}$$

As a result, the system is asymptotically stable. Moreover, the tracking error of the system will converge to zero according to $s = \lambda_e e + \dot{e}$.

3.2 Design of a sliding mode controller

Rewriting Eq. (7) as a second-order nonlinear, single-input-single-output (SISO) motor-mechanism coupled system as follows:

$$\ddot{v}(t) = f(\mathbf{X};t) + G(\mathbf{X};t)U(t) + d(t) \tag{27}$$

where

$$f(\mathbf{X};t) = -\hat{M}^{-1}\hat{N} \quad G(\mathbf{X};t) = \hat{M}^{-1}\hat{Q} \quad d(t) = \hat{M}^{-1}\hat{D}$$

and $U(t)$ is the control input v_q^*. It is assumed that the function f is not exactly known, but the extent of the imprecision Δf is bounded by a known continuous function $F(\mathbf{X};t)$. Similarly, the control gain $G(\mathbf{X};t)$ is not exactly known but having a constant sign and known bounds, *i.e.*

$$0 < G_{\min} \le G(\mathbf{X};t) \le G_{\max}. \tag{28}$$

Disturbance $d(t)$ is unknown, but is bounded by a known continuous function $D(\mathbf{X};t)$. According to the above descriptions, we have

$$\left| f - \hat{f} \right| \le F(\mathbf{X};t) \tag{29a}$$

$$\frac{1}{\alpha} \le \frac{\hat{G}(\mathbf{X};t)}{G(\mathbf{X};t)} \le \alpha \tag{29b}$$

$$|d| \le D(\mathbf{X};t) \tag{29c}$$

where $\hat{f}$ and $\hat{G}$ are nominal values of f and G, respectively, and

$$\alpha = \left(G_{\max}/G_{\min}\right)^{1/2}.$$

The control problem is to find a control law so that the state $\mathbf{X}$ can track the desired trajectory $\mathbf{X}_d$ in the presence of uncertainties.
Let the tracking error vector be

$$\mathbf{e} = \mathbf{X} - \mathbf{X_d} = \left[e\, \dot{e}\right]^T \tag{30}$$

where $\mathbf{X_d} = \left[\overset{*}{x_B}\ \dot{\overset{*}{x}}_B\right]^{\mathbf{T}}$. Define a sliding surface $s(t)$ in the state space R^2 by the scalar function $s(\mathbf{X};t) = 0$, where

$$s(\mathbf{X},\, t) = Ce + \dot{e} \quad C > 0\,. \tag{31}$$

The sliding mode controller is proposed as follows:

$$U = U_{eq} + U_n \tag{32}$$

where

$$U_{eq} = \left(\hat{G}\right)^{-1} \hat{U} \tag{33a}$$

$$U_n = -\left(\hat{G}\right)^{-1} K \operatorname{sgn}(s) \tag{33b}$$

and

$$\hat{U} = -\hat{f} - \ddot{\overset{*}{v}}_d + d(t) + C\dot{e} \tag{34}$$

$$K \geq \alpha(F + D + \eta) + (\alpha - 1)\left|\hat{U}\right|,\ \operatorname{sgn}(s) = \begin{cases} 1 & if \quad s > 0 \\ 0 & if \quad s = 0 \\ -1 & if \quad s < 0 \end{cases} \tag{35}$$

where η is a positive constant. The detailed derivations of the sliding mode controller are similar to the work of (Slotine & Li, 1992). Some discussions about the sliding mode control could refer to the References (Gao & Hung, 1993; Hung et al., 1993).
To alleviate the chattering phenomenon, we adopt the quasi-linear mode controller (Slotine & Li, 1992), which replaces the discontinuous control laws of Eq. (33b) by a continuous one and insides a boundary layer around the switching surface. That is, U_n is replaced by

$$U_n = -\left(\hat{G}\right)^{-1} K sat\left(\frac{s}{\varepsilon}\right), \tag{36}$$

where $\varepsilon > 0$ is the width of boundary, and the function of $sat\left(\dfrac{s}{\varepsilon}\right)$ is defined as

$$sat\left(\frac{s}{\varepsilon}\right)=\begin{cases}1 & if \;\; s>\varepsilon \\ \dfrac{s}{\varepsilon} & if \;\; -\varepsilon\le s\le\varepsilon \\ -1 & if \;\; s<-\varepsilon\end{cases} \tag{37}$$

This leads to tracking within a guaranteed precision ε while allowing the alleviation of the chattering phenomenon. The block diagram of the SMC by use of a machine vision system is shown in Fig. 4, where the tracking error is $e=x_B-x_B^*$ and the output displacements of slider B are measured by a machine vision system, which includes the CCD, image acquisition card and color pattern matching.

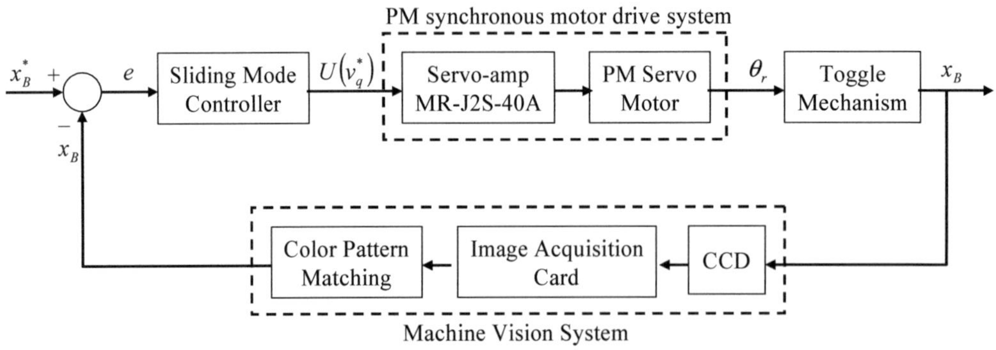

Fig. 4. Block diagram of the sliding mode control by the machine vision system.

3.3 Design of a fuzzy logic controller

In the real situations, the general problem encountered in designing a controller is that the bounds of the uncertainties and exact mathematical models of the motor-toggle mechanism system are difficult to obtain for practical applications. Moreover, the parameters can not be obtained directly and the output responses of slider B must be able to measure. In this chapter, the PD-type FLC (Rahbari & Silva, 2000) and PI-type FLC (Aracil & Gordillo, 2004), which are without using complex mathematical model, are proposed to overcome the difficulties of uncertainties and un-modeling.

3.3.1 The PD-type fuzzy logic controller

The control problem is to find the PD-type FLC law such that the output displacement x_B can track the desired trajectories x_B^* in the presence of uncertainties. Let the tracking error be

$$e=x_B-x_B^* \tag{38}$$

As shown in Fig. 5, the signals of e and $\dot{e}$ are selected as the inputs for the proposed PD-type FLC.

The control output of the PD-type FLC is u, which denotes the change of controller outputs. The signals of e and $\dot{e}$ could be respectively transferred to their corresponding universes of discourse by multiplying scaling factors k_1 and k_2, namely,

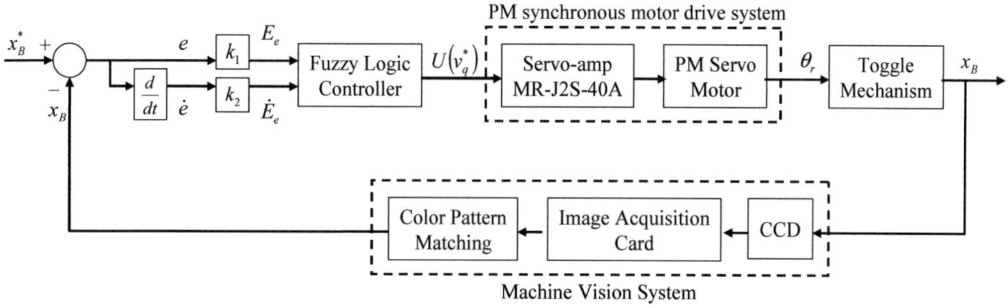

Fig. 5. Block diagram of a PD-type FLC for the motor-toggle mechanism.

$$E_e = e * k_1 , \quad E_e = e * k_2 \tag{39}$$

Since the output u of the FLC is in its corresponding universe of discourse, the u could be transferred, by multiplying a scaling factor G_u, to an actual input of the plant, namely,

$$U = u * G_u \tag{40}$$

Because the data manipulation in the PD-type FLC is based on the fuzzy set theory, the associated fuzzy sets involved in the linguistic control rules are defined as follows:
N: Negative *Z*: Zero *P*: Positive
NB: Negative Big *NM*: Negative Medium *NS*: Negative Small
ZE: Zero *PS*: Positive Small *PM*: Positive Medium *PB*: Positive Big
and their universe of discourse are all assigned to be [-10, 10] for a real experimental motor. The membership functions for these fuzzy sets corresponding to E_e, $\dot{E}_e$ and u are defined in Fig. 6.
In the following, the rule bases of the proposed PD-type FLC are systematically constructed on the basis of a Lyapnuov function L_f:

$$L_f = \frac{1}{2} e^2 \geq 0 \text{ and } \dot{L}_f = e\dot{e} \tag{41}$$

According to Lyapnuov stable theory (Cheng & Tzou, 2004), if the system is stable, the conditions $L_f = \frac{1}{2} e^2 \geq 0$ and $e\dot{e} < 0$ are necessary. Therefore, according to Eq. (41), if $e<0$, increasing u will result in decreasing $e\dot{e}$; if $e>0$, decreasing u will result in decreasing $e\dot{e}$. Hence, the control input u can be designed in an attempt to satisfy the condition $e\dot{e} < 0$. The resulting fuzzy control rules are shown in the following:

Rule 1: If E_e is *P* and $\dot{E}_e$ is *P* Then u is *NB*

Rule 2: If E_e is *P* and $\dot{E}_e$ is *Z* Then u is *NM*

Rule 3: If E_e is *P* and $\dot{E}_e$ is *N* Then u is *NM*

Rule 4: If E_e is *Z* and $\dot{E}_e$ is *P* Then u is *NS*

Rule 5: If E_e is *Z* and $\dot{E}_e$ is *Z* Then u is *ZE*

Rule 6: If E_e is Z and $\dot{E}_e$ is N Then u is PS

Rule 7: If E_e is N and $\dot{E}_e$ is P Then u is PM

Rule 8: If E_e is N and $\dot{E}_e$ is Z Then u is PM

Rule 9: If E_e is N and $\dot{E}_e$ is N Then u is PB.

By using the centre-of-area (COA) method, the output can be obtained as

$$u = \frac{\sum u_k \left[(u_{A_i}(e))(u_{B_j}(\dot{e})) \right]}{\sum ((u_{A_i}(e))(u_{B_j}(\dot{e})))} . \tag{42}$$

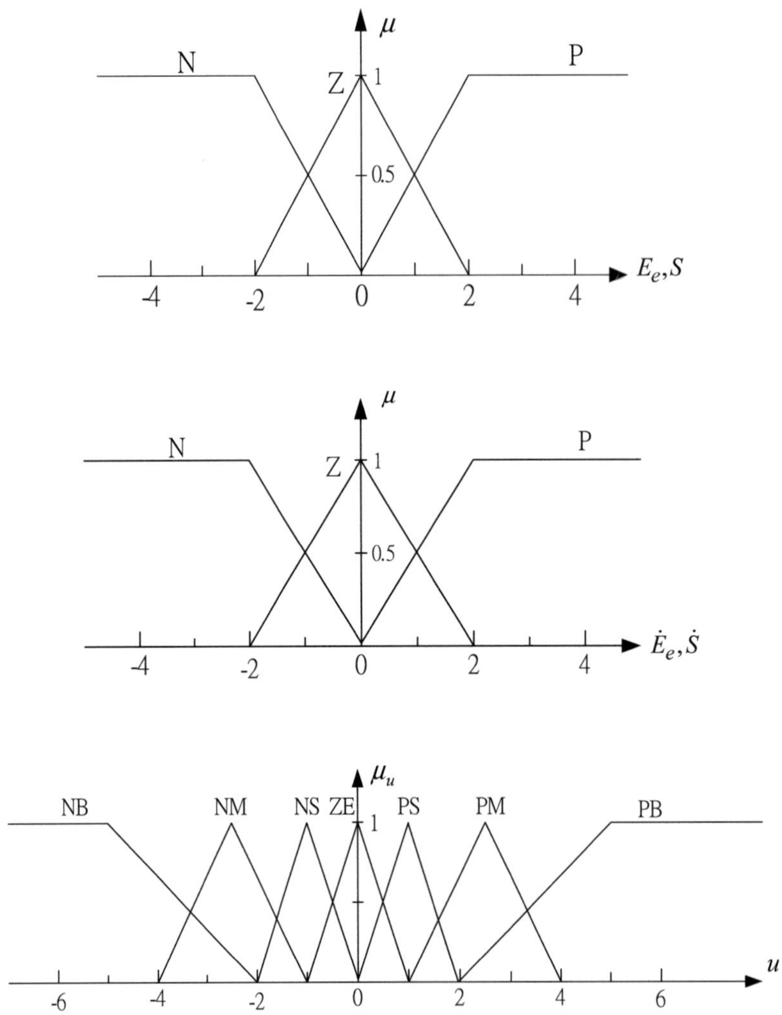

Fig. 6. Membership functions of E_e, S, $\dot{E}_e$, $\dot{S}$ and u.

3.3.2 The PI-type fuzzy logic controller

In this section, the proposed PI-type FLC is designed based on the concept of hitting switch conditions. As shown in Fig. 7, the switching functions are selected as the inputs. In practical implementation, it can be approximated by

$$\dot{s}(kT) = (s(k-1)T) / T , \quad (43)$$

where k is the number of iteration and T is the sampling period.

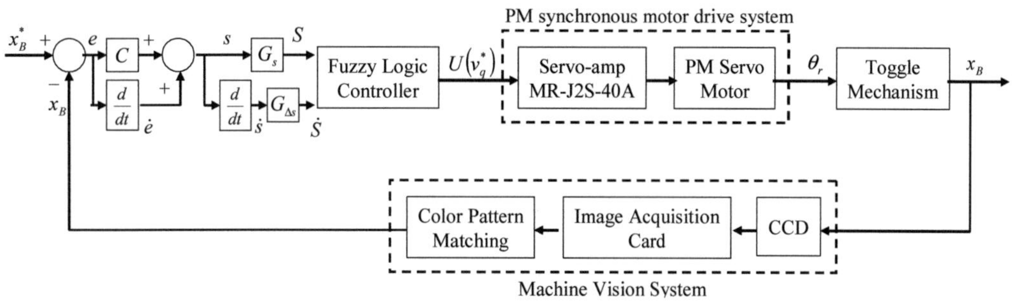

Fig. 7. Block diagram of a PI-type FLC for the motor-toggle mechanism.

The control input of the PI-type FLC is u which denotes the change of the controller outputs. The s and $\dot{s}$ signals could be transferred to their corresponding universes of discourse by multiplying scaling factors G_s and $G_{\Delta s}$ respectively, namely,

$$S = s \cdot G_s, \quad \dot{S} = \dot{s} \cdot G_{\Delta s}. \quad (44)$$

Since the output u of the PI-type FLC is in its corresponding universe of discourse, the u could be transferred, by multiplying a scaling factor $G_{\Delta u}$, to an actual input of the plant, namely,

$$\Delta U = u \cdot G_{\Delta u}. \quad (45)$$

Because the data manipulation in a PI-type FLC is based on fuzzy set theory, the associated fuzzy sets involved in the linguistic control rules are defined as the same as the previous section and their universe of discourse are all assigned the same as the previous section. The membership functions for these fuzzy sets corresponding to S, $\dot{S}$ and u. are also defined in Fig. 6.

In the following, the rule base of the proposed PI-type FLC are systematically constructed on the basis of hitting switching conditions of the SMC. Multiplying of Eq. (43) by s then we have

$$\dot{s}s = fs + GUs + ds - \ddot{v}s + C\dot{e}s. \quad (46)$$

It is similar to the PD-type FLC, Lyapunov function for the PI-type FLC is assigned as $L_f = \frac{1}{2}s^2 \geq 0$. The controller is designed to satisfy the condition $s\dot{s}$ <0, and the whole control system is stable. According to Eq. (44), if s < 0, increasing u will result in decreasing; if s > 0, decreasing u will result in decreasing $s\dot{s}$. Hence, the control input u can be designed in an attempt to satisfy the hitting condition $s\dot{s}$ <0.

The resulting PI-type FLC rules are shown as follows:

Rule 1: If S is P and $\dot{S}$ is P Then u is NB

Rule 2: If S is P and $\dot{S}$ is Z Then u is NM

Rule 3: If S is P and $\dot{S}$ is N Then u is NM

Rule 4: If S is Z and $\dot{S}$ is P Then u is NS

Rule 5: If S is Z and $\dot{S}$ is Z Then u is ZE

Rule 6: If S is Z and $\dot{S}$ is N Then u is PS

Rule 7: If S is N and $\dot{S}$ is P Then u is PM

Rule 8: If S is N and $\dot{S}$ is Z Then u is PM

Rule 9: If S is N and $\dot{S}$ is N Then u is PB.

By using the centre-of-area (COA) method, the output can be obtained as

$$u = \frac{\sum u_k \left[(u_{F_i}(e))(u_{G_j}(\dot{e})) \right]}{\sum ((u_{F_i}(e))(u_{G_j}(\dot{e})))}. \tag{47}$$

In this chapter, the mean of maximum (MOM) of defuzzifier is adopted in both the PD-type and PI-type FLCs.

4. Numerical simulations

For numerical simulations, the parameters of the mechatronic system of the motor-toggle mechanism are chosen as follows:

$m_B = 4.12$ Kg, $m_C = 5.58$ Kg, $m_2 = 1.82$ Kg, $m_3 = 1.61$ Kg, $m_5 = 0.95$ Kg, $\mu = 0.17$, $r_1 = 0.06$ m, $r_2 = 0.032$ m, $r_3 = 0.06$ m, $r_4 = 0.068$ m, $r_5 = 0.03$ m, $h = 0.068$ m, $\phi = 0.4899$ rad, $K_t = 0.5652$ Nm/A, $J_m = 6.7 \times 10^{-5}$ Nms^2, and $B_m = 1.12 \times 10^{-2} Nms/rad$.

The above known parameters are to substitute into Eq. (7), and the system becomes an initial value problem and can be integrated by using the fourth-order Runge-Kutta method with time step $\Delta t = 0.001$ sec and tolerance error 10^{-9}. The control objective is to control the position of slider B to move from the left side to the right side. The initial position is 0.06 m, the desired position is 0.1 m, and the controlled stroke of the slider B is equal to $\Delta x_B = 0.04$ m.

4.1 Numerical simulations of adaptive controller

For numerical simulations, the external disturbance force F_E will be added to test the robustness of the adaptive controller. The gains of the adaptive control law (18) are given as follows: $\lambda_e = 10$, $K_V = 194$, $\gamma_1 = 248$ *and* $\gamma_2 = 173$. They are chosen to achieve the best transient performance in the limitation of control effort and the requirement of stability. In the real system, the angles of θ_1, θ_2, and θ_5 are limited in the following ranges: $23^\circ < \theta_1 < 63^\circ$, $325^\circ < \theta_2 < 350^\circ$, and $145^\circ < \theta_5 < 170^\circ$. Therefore, the invertible property of $\hat{\mathbf{Q}}^{-1}$ can be guaranteed and the system function $f(\mathbf{X};t) = \hat{\mathbf{Q}}^{-1}\hat{\mathbf{M}} > 0$ can be proved.

The dynamic responses of slider B and the control efforts i_q^* with and without external disturbance force are compared in Figs. 8(a) and 8(b), respectively. The dotted lines are the desired positions, the dash lines are the transient responses of numerical simulations with external disturbance force $F_E = 0$ Nt and the solid lines are for $F_E = -100$ Nt. The negative sign in the external disturbance force indicates the action direction is opposite to the X-direction in Fig. 1(a). In Fig. 8(a), the transient responses are almost the same and are stable after 0.5 sec and the steady-state error is about 1×10^{-5} m. Since the transient responses are almost the same in the presence of uncertainties, it shows the proposed adaptive control is robust. In Fig. 8(b), the maximum control effort $i_q^* = 0.218A$ for $F_E = 0$ Nt is smaller than that $i_q^* = 0.710A$ for $F_E = -100$ Nt.

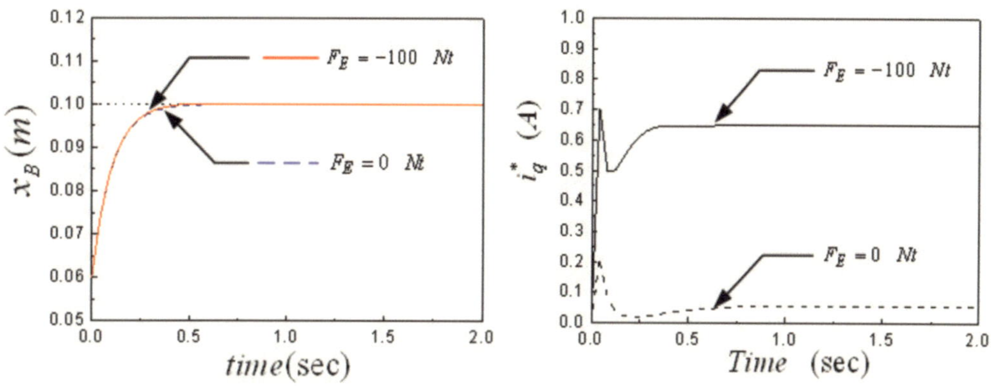

(a) (b)

Fig. 8. The numerical simulations of a motor-toggle mechanism by an adaptive controller with and without external disturbance forces. (a) The dynamic responses of the slider B. (b) The control efforts i_q^*.

4.2 Numerical simulations of sliding mode controller

The nominal case is the system without external disturbance force, i.e., $F_E = 0$ Nt and the gains of the SMC are given as C=5 and $\varepsilon = 0.3$. The dynamic responses of slider B for the nominal case are shown in Fig. 9 (a), and it is seen that the response is stable after 1 sec, and the numerical error is about 0.01mm. The trajectories in the phase plane $(e, \dot{e})$ are shown in Fig. 9 (b), where the representative point lies on the designed sliding surface after it hits the switching hyperplane.

Another case with external disturbance force F_E = 100 Nt is also considered and the simulation results are shown in Figs. 10(a) and 10(b) for its dynamic responses and trajectories, respectively. It is found that the smooth step-command tracking responses are also obtained well and the SMC is robust to the presence of uncertainties.

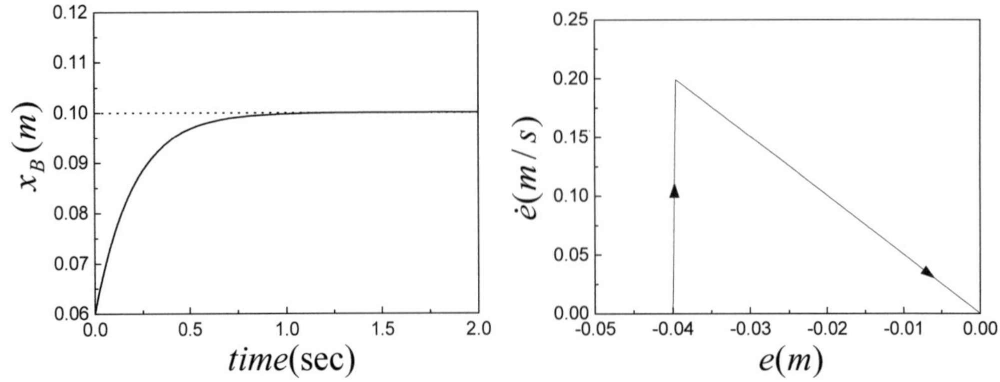

(a) (b)

Fig. 9. (a) The dynamic responses of slider B by the SMC with $F_E = 0 \quad Nt$;
(b) The trajectories in the phase plane by the SMC with $F_E = 0 \quad Nt$.

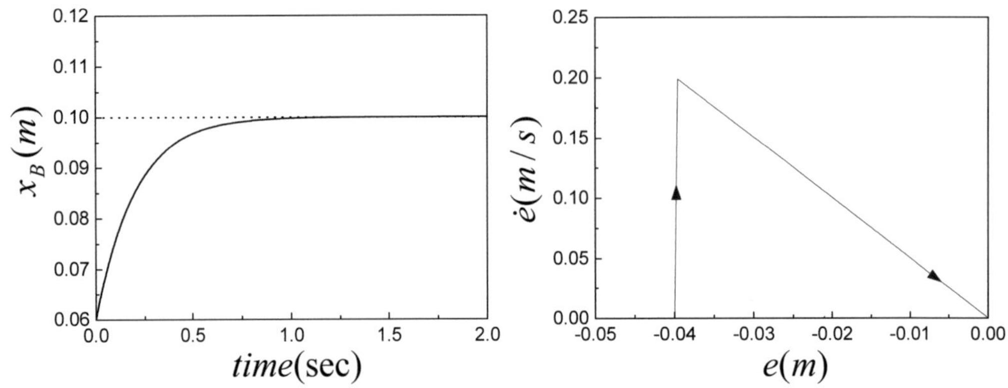

(a) (b)

Fig. 10. (a) The dynamic responses of slider B by the SMC with $F_E = 100 \quad Nt$;
(b)The trajectories in the phase plane by the SMC with $F_E = 100 \quad Nt$.

4.3 Numerical simulations of the PD-type fuzzy logic controller

Here, the PD-type FLC is applied to control the motor-toggle mechanism system numerically. In order to minimize the hitting time and track stable, the scaling factors are determined by observing numerical simulations and are selected as $k_1 = 1082$, $k_2 = 849$ and $G_u = 0.5$. Simulation results of the nominal case without external disturbance force are shown in Fig. 11(a), where the dynamic responses are stable after 1.25 sec, and the error between the desired position and numerical response of slider B is about 0.3 mm. Figure 11(b) illustrates the dynamic responses of the case with external disturbance force F_E = 100 *Nt*. It is seen that the responses are stable after 1.25 sec and the error is about 0.5 mm.

In conclusion, the responses of the PD-type FLC for a motor-toggle mechanism exhibit overshoot phenomenon, and the affection of external disturbance forces to the system is influenced. Therefore, the performance of the proposed PD-type FLC is not robust.

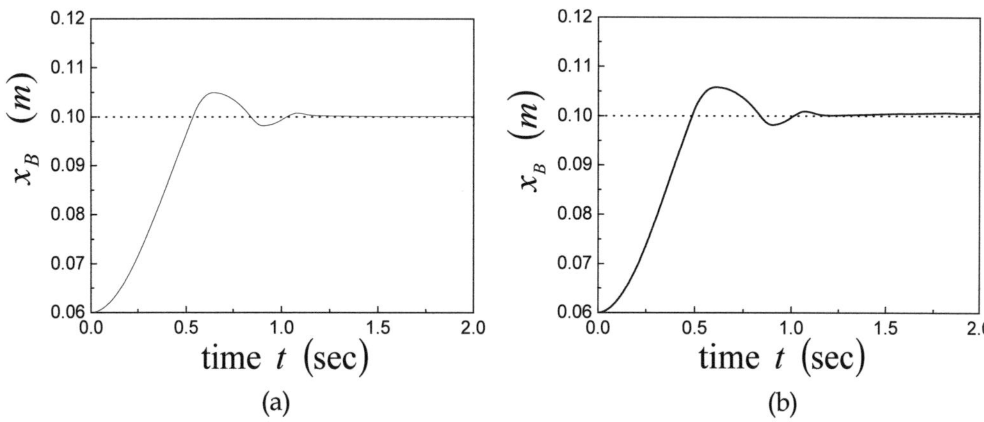

Fig. 11. The dynamic responses of slider B by the PD-type FLC: (a) $F_E = 0 \ \ Nt$; (b) $F_E = 100 \ \ Nt$.

4.4 Numerical simulations of the PI-type fuzzy logic controller

In this section, the PI-type FLC is applied to control the system numerically and compares with the PD-type FLC. The scaling factors are also determined by observing the numerical simulations, and are selected as:

$$\begin{cases} G_s = 374 \\ G_{\Delta s} = 54 \\ G_u = 0.05 \times |s|^{0.25} \end{cases} \quad if \ \ s \geq -0.01, \tag{48}$$

$$\begin{cases} G_s = 534 \\ G_{\Delta s} = 84 \\ G_u = 0.08 \times |s|^{0.25} \end{cases} \quad if \ \ s < -0.01. \tag{49}$$

First, the simulation results of the nominal case without external disturbance force are given in Fig. 12(a), where the responses of slider B are stable after 1 sec and the error between the desired position and numerical response is about 0.3 mm. It is noted that the control input is adjusted by the fuzzy inference mechanism, which is based on the concept of hitting conditions regardless of the exact mathematical model. Figure 12(b) illustrates the trajectories in the phase plane. It is seen that the representative point lies on the designed sliding surface $s = 0$ after it hits the switching hyperplane, and the smooth step-command tracking responses are obtained for x_B. Figures 13 (a) and (b) respectively show the trajectories in the phase plane for the system without and with external disturbance force $F_E = 100 \ \ Nt$.

In conclusion, the dynamic responses utilizing the PI-type FLC to a motor-toggle mechanism system has no overshoot phenomenon and are stable fast with external force. Furthermore, the PI-type fuzzy controlled motor-toggle mechanism system is robust with respect to the external disturbance forces.

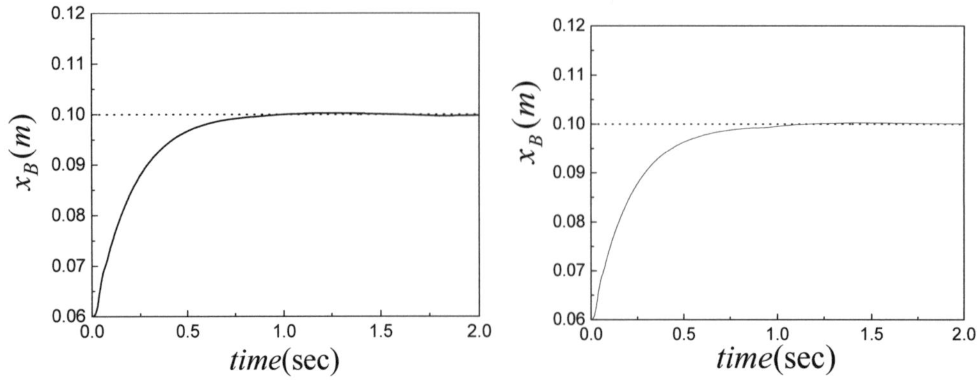

(a) (b)

Fig. 12. The dynamic responses of slider B by the PI-type FLC: (a) $F_E = 0 \quad Nt$; (b) $F_E = 100 \quad Nt$

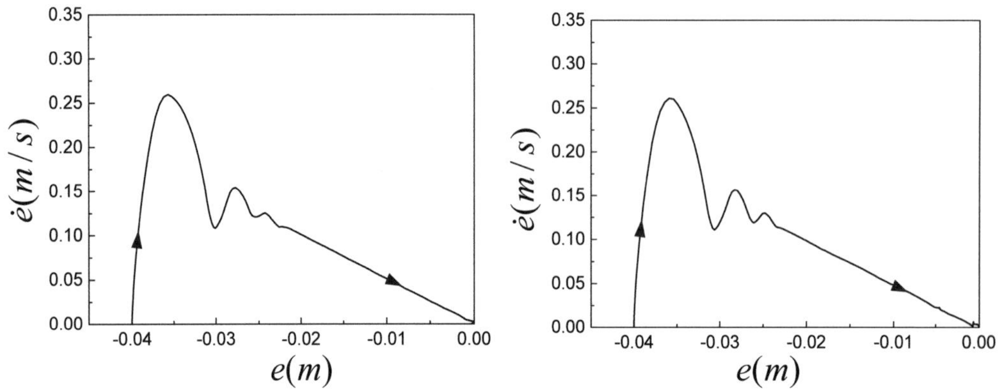

(a) (b)

Fig. 13. The trajectories in the phase plane by the PI-type FLC: (a) $F_E = 0 \quad Nt$; (b) $F_E = 100 \quad Nt$

5. Experiments

In the real operations of an experimental system, the main merit of this study is that the machine vision system of a digital CCD camera is employed as an unconstrained feedback sensor. In Fig. 14(a), the slider position can be measured by non-contacted equipments and a color pattern is pasted to measure and vision-based control. In Fig. 14(b), the state angle θ_1 of the motor-toggle mechanism system is difficult to be measured by an installed encoder and will be experimentally measured by a shape pattern of the machine vision system, which is needed only to paste a pattern on where want to be measured and can be controlled to the desired position by a digital CCD camera.

5.1 Visual control system

The machine vision servoing system takes a color-pattern and a shape-pattern is pasted up on the link and is shown in Fig. 14. It has the advantage in distinguishing the link from its

(a) The color pattern (b) The shape pattern

Fig. 14. The control image frame with (a) the color pattern and (b) the shape pattern.

near monotonic surrounding fast. The directional shape pattern is easy identified with measuring the angle θ_1 of the motor-mechanism system. In this vision system, one full-frame of image consists of 752×582 pixels. Searching the whole video data of a full-frame for the shape pattern usually takes quite long time, and degrades the performance of the visual servoing system. Thus, based on the range of the angle θ_1, the image frame is adjusted by a CCD camera to contain the controlled degree of the angle θ_1. Before using the machine vision system, it is very important to do a calibration between one pixel and a real-word unit such as millimeter (mm). Therefore, a standard calibration grid is shown in Fig.15. The real distance in the standard calibration grid of one block point center to another one point center is measured 7 mm in both the X- and Y-directions. The result of calibration is that one pixel is 0.2 mm in the real world. According to this the color pattern image coordinate can be transformed into a real-world unit in designing a control algorithm.

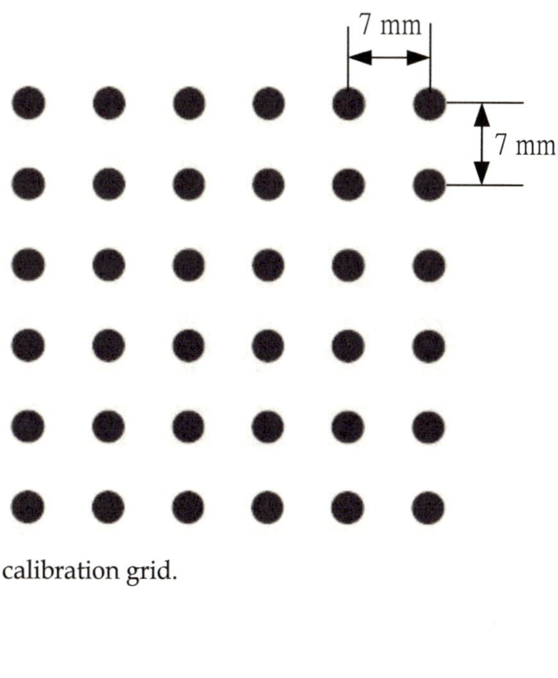

Fig. 15. The standard calibration grid.

After the experimental calibration, pattern matching is the first step for implementing the machine vision system. The servoing control algorithm is implemented by LabVIEW and the image acquisition card is implemented by PCI 1405. In the controlled image frame, the shape pattern is selected as the region of interest (ROI) to save in a disk for searching. Finally, the shape pattern and color pattern matching algorithm is realized by LabVIEW and the position of slider B is controlled by the visual controller. In this study, the image processing time is 0.2 sec by using the CCD camera to feedback the slider position.

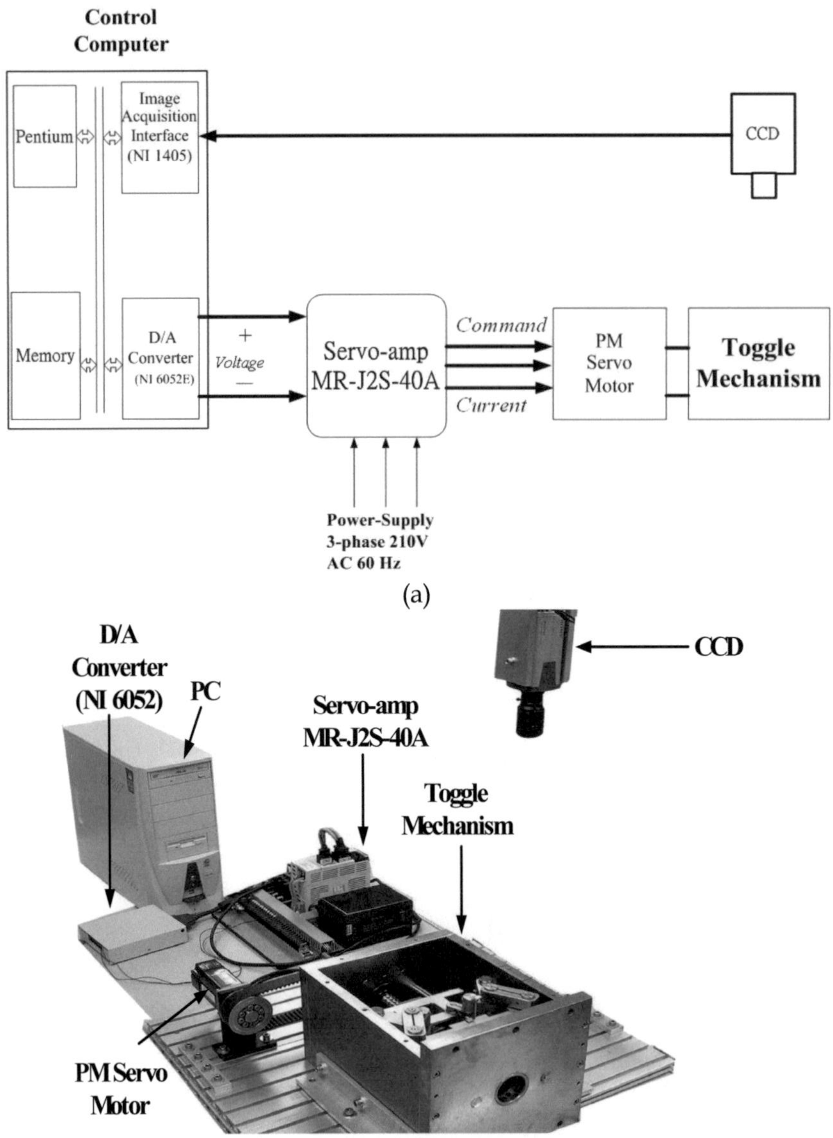

Fig. 16. The visual control system. (a) The computer control block diagram. (b) The experimental equipments.

5.2 Experimental setup

The visual control block diagram of the motor-toggle mechanism is shown in Fig. 16(a) and its experimental equipments are shown in Fig. 16(b). The control algorithm is implemented by using a Pentium computer and the control software is LabVIEW. The PMSM is implemented by the MITSUBISHI HC-KFS43 series. The specifications are described as follows: rated output 400 (W), rated torque 1.3 (Nm), rated rotation speed 3000 (rpm) and rated current 2.3 (A). The servo is implemented by the MITSUBISHI MR-J2S-40A1. The control system is a sine-wave PWM control, which is a current control system. The digital CCD camera is implemented by the SONY SSC-DC393 series. The specifications are imaging device 1/3-type interline transfer, picture elements 752(horizontal) ×582(vertical), and Lent CS-mount.

6. Experimental results

6.1 The vision-based adaptive controller

The adaptive vision-based control for the motor-mechanism system is performed by comparing the external disturbance force $F_E = 0\ Nt$ with $F_E = -10\ Nt$. The experimental results of the measured angle θ_1 via the machine vision system, the transient responses of slider B via the manipulating relation $x_B = 2r_1 \cos\theta_1$ and the control efforts i_q^* are shown in Figs. 17-18, respectively. It is seen that the experimentally measured angle θ_1 in Fig. 17 and the transient responses of slider B in Fig.18 are almost the same for the system with and without F_E, and are stable after 0.75 sec. However, the control efforts i_q^* are quite different. The maximum control effort $i_q^* = 0.28A$ for $F_E = 0\ Nt$ is much smaller than that $i_q^* = 0.75A$ for $F_E = -10\ Nt$. The maximum control efforts are near to those of numerical simulations.
Moreover, in order to demonstrate the robust control performance of the adaptive vision-based controller, the experiments are performed by suddenly adding an extra mass 10 kg on slider B at 0.6 sec, and suddenly adding 10 $Nt$of the external force at 0.6 sec. Figure 19(a) show the good performance of regulation problems, and Figure 19(b) show the control input efforts, where the jumps occurs when the extra mass and external force are suddenly added.

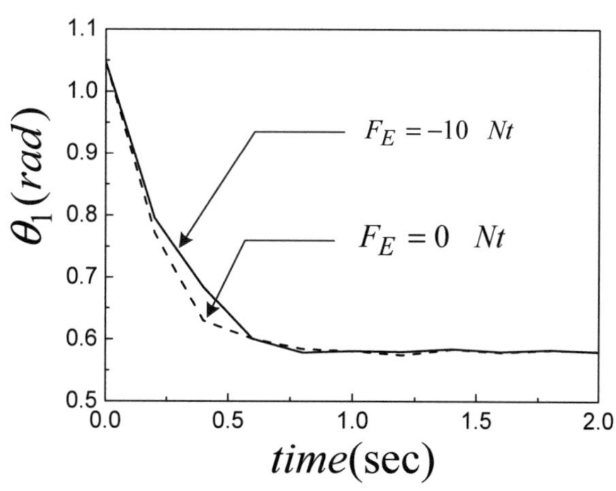

Fig. 17. The experimentally measured angle θ_1 with and without external disturbance forces.

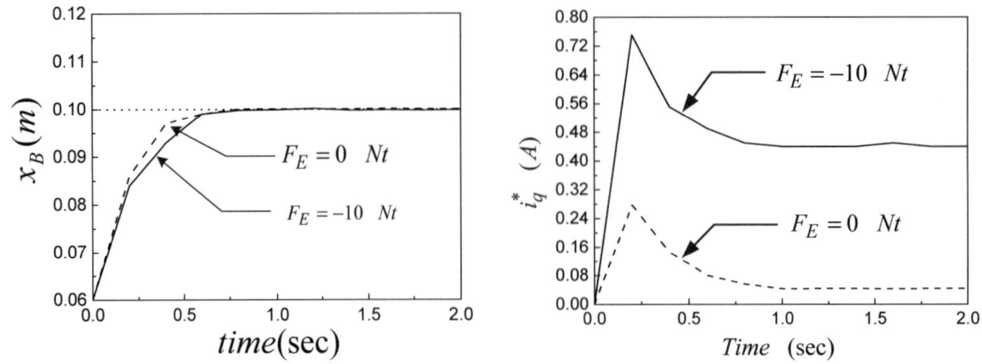

(a) (b)

Fig. 18. (a) The experimentally dynamic responses of slider B with and without external disturbance forces; (b) The experimental control efforts i_q^*

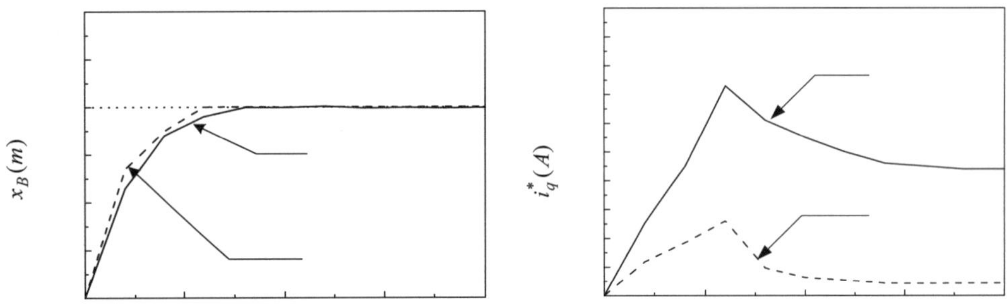

(a) (b)

Fig. 19. (a)The experimentally dynamic responses of slider B with the time-varying external force and mass variation at slider B. (b) Control input efforts with the time-varying external force and mass variation at slider B

6.2 The vision-based sliding mode and fuzzy logic controller

The experiments are performed by suddenly adding an extra mass 10 kg on slider B at 0.6 sec, and suddenly adding 10 Nt force of the external force at 0.6 sec. The initial state is $x_B(0) = 0.06\ m$ while the desired position is $x_B^* = 0.1\ m$. The SMC, PD-type FLC and PI-type FLC are performed for the cases with external disturbance forces $F_E = 0\ Nt$ and $F_E = 10\ Nt$. Some experimental results are provided to demonstrate the effectiveness of the proposed controllers by the machine vision system. First, the SMC is applied to control the motor-toggle mechanism system and the experimentally controlled responses of slider B without and with external disturbance forces are shown in Fig. 20. It is seen that the experimental responses of slider B are all stable after 1 sec and the errors between the desired position and experimental one are about 1 mm. The results show that the smooth step-command responses are obtained for the slider B due to the robust control characteristics of the SMC.

Furthermore, the results via the PD-type and PI-type FLC are compared without and with external disturbance force in Figs. 21(a) and 21(b), respectively. It is seen that the PI-type FLC performance is always superior to PD-type FLC for the system without and with the external disturbance forces. Finally, the control current inputs of the PD-type and PI-type FLCs with and without external disturbance forces are respectively shown in Figs 22(a) and 22(b).

In conclusions of the experiments, the general problems encountered in designing controllers are that the bounds of uncertainties and the exact mathematical models of a motor-mechanism system are difficult to obtain in advance for practical applications. Moreover, the parameters of the motor-mechanism system can not be obtained directly and the output responses must be measured without constraint. From the experimental results, the PI-type FLC owns more robust control characteristics for the motor-mechanism system by using machine vision.

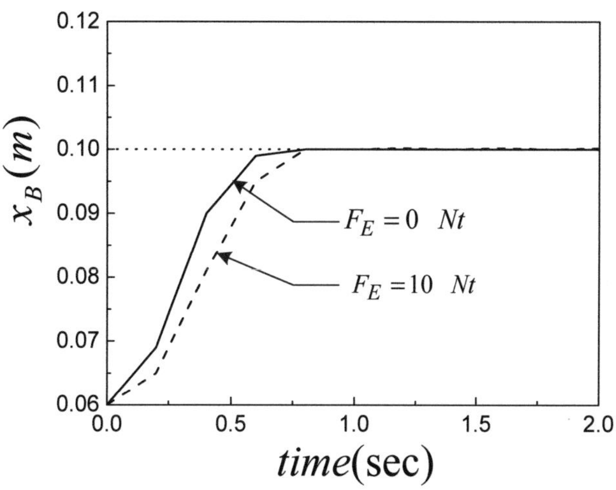

Fig. 20. The experimentally dynamic responses of slider B by the SMC.

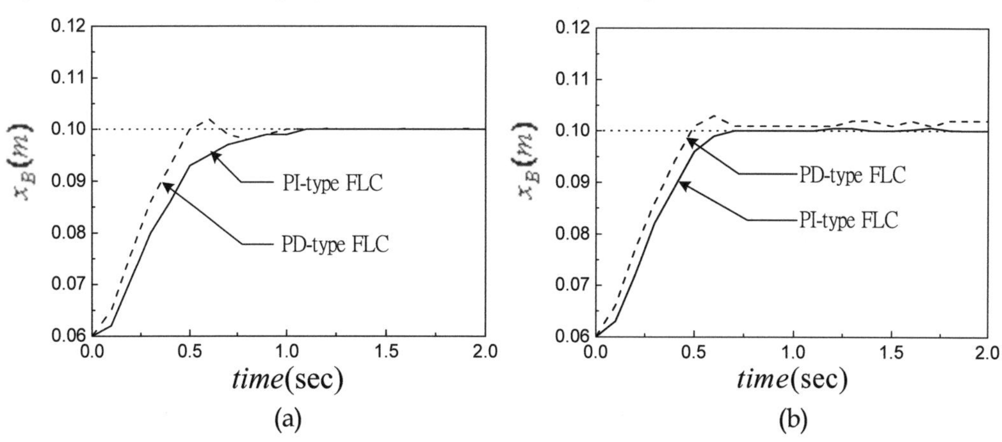

Fig. 21. The experimentally dynamic responses by the PI-type and PD-type FLCs (a) with $F_E = 0\,Nt$; (b) with $F_E = 10\ \ Nt$.

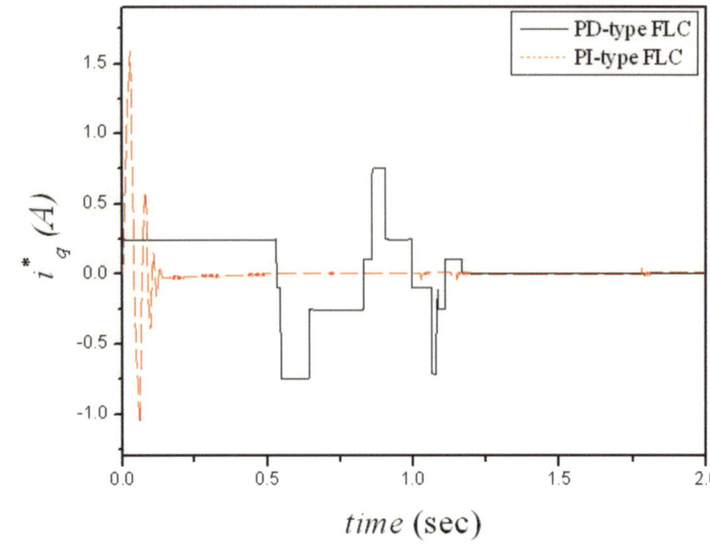

(a)

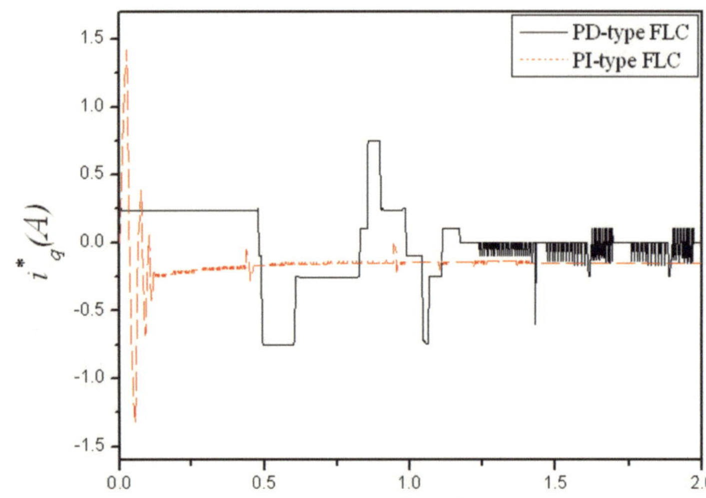

(b)

Fig. 22. The control currents of the PD-type and PI-type FLCs with and without external disturbance forces: (a) The disturbance forces $F_E = 0 \ \ Nt$. (b) The disturbance forces $F_E = 10 \ \ Nt$.

7. Conclusions

In this chapter, we successfully demonstrate the applications of the proposed adaptive, sliding mode and fuzzy logic vision-based controller to position control of the motor-toggle mechanism system, which is made up by the toggle mechanism driven by a field-oriented

PMSM. In order to overcome the general difficulties of non-contact measuring and external-force uncertainties of the system, the shape pattern and color pattern are designed to measure the rotating angle and slider position, respectively. Finally, the numerical simulations and experimental results are provided to demonstrate the robust control performance of the proposed vision-based controllers.

The main contributions of this study are summarized as:

1. We developed a complete mathematical model for the mechatronic system, which is made up by the toggle mechanism driven by a PMSM.
2. We successfully employed the controllers by machine vision to control the slider position of a complex motor-mechanism coupled system with a simple rule base instead of its complex mathematical model. Moreover, the robust control performance of the mechatronic system is presented with external disturbance forces numerically and experimentally.
3. The color-pattern and shape-pattern matching method of the machine vision are implemented successfully for the mechatronic system. It is shown that the applications of machine vision for industrial equipments are convenient, low cost and multi-useful.

8. References

J. Aracil and F. Gordillo (2004). Describing function method for stability analysis of PD and PI fuzzy controllers. *Fuzzy Sets and Systems* 143, 233-249.

K. Astrom and B. Wittenmark, (1995) *Adaptive control*, Reading. Addison-Wesley, MA.

L. Beji and Y. Bestaoui. (2005). Motion generation and adaptive control method of automated guided vehicles in road following. *IEEE TRANSACTIONS ON INTELLIGENT TRANSPORTATION SYSTEMS 6 (1) 113-123.*

K. Y. Cheng and Y. Y. Tzou, (2004). Fuzzy optimization techniques applies to the design of a digital PMSM servo drive. *IEEE Trans. On Power Electronics* 19, 1085-1099.

C. W. Chuang, M. S. Huang, K. Y. Chen, R. F. Fung. (2008), Adaptive vision-based control of a motor-toggle mechanism: Simulations and experiments. *Journal of Sound and Vibration, 312, 848–861*

R. F. Fung and R. T. Yang (2001). Motion control of an electrohydraulic actuated toggle mechanism. *Mechatronics,* 11, 939-946.

R. F. Fung, J. W. Wu and D. S. Chen. (2001). A variable structure control toggle mechanism driven by linear synchronous motor with joint coulomb friction, *Journal of Sound and Vibration,* 247, 741-753.

W. Gao and J. C. Hung, (1993). Variable structure control of nonlinear systems: a new approach.*IEEE Trans. On Industrial Electronics* 40, 45-55.

M. Hashimoto, F. Oba and T. Tomiie (1999). Mobile robot localization using color signboard. *Mechatronics* 9, 633-656.

M. S. Huang, K. Y. Chen, R. F. Fung (2008), Numerical and experimental identifications of a motor-toggle mechanism, *Applied Mathematical Modelling*

J. Y. Hung, W. Gao and J. C. Hung, (1993), Variable structure control. *IEEE Trans. On Industrial Electronics* 40, 2-22.

Y. Ju, S. K. Chang, B. Jong, S. Hong, M. H. Lee and F. Harashima. (2001). Vision based lateral control by yaw rate feedback. *The 27th Annual Conference of the IEEE Industrial Electronics Society* C, 2135-2138.

T. S. Lee, C. H. Lin and F. J. Lin, (2005). An adaptive H_∞ controller design for permanent magnet synchronous motor drives, Control Engineering Practice 13 425-439.

F. J. Lin, R. F. Fung and Y. S. Lin, (1997). Adaptive control of slider-crank mechanism motion: simulations and experiments, *International Journal of Systems Science* 28 1227-1238.

F. J. Lin, R. F. Fung and Y. C. Wang. (1997). Sliding mode and fuzzy control of toggle mechanism using pm synchronous servomotor drive. *IEE Proceedings Control Theory and Applications,* 144, 393-402.

F. J. Lin and R. J. Wai. (2002). Hybrid computed torque controlled motor-toggle servomechanism using fuzzy neural network uncertainty observer. *Neurocomputing,* 48, 403-422.

Z. B. Li, Z. L. Wang and J. F. Li. (2004). A hybrid control scheme of adaptive and variable structure for flexible spacecraft. *Aerospace Science and Technology,* 8, 423-430.

H. T. Moon, H. S. Kim and M. J. Youg (2003). A discrete-time predictive current control for PMSM. *IEEE Trans. On Power Electronics* 18, 464-472.

K. S. Narendra and A. M. Annaswamy, (1988). *Stable Adaptive System,* Prentice-Hall, Englewood Cliffs, NJ,

O. Nasisi and R. Carelli. (2003). Adaptive servo visual robot control, *Robotics and Autonomous Systems,* 43, 51-78.

J. H. Park and Y. J. Lee, (2002). Robust visual servoing for motion control of the ball on a plate. *Mechatronics* 13, 723-738.

I. Petrovic and M. Brezak (2002). Machine vision based control of the ball and beam. *Advanced Motion Control IEEE 7th International workshop* 3-5, 573-577.

R. Rahbari and C. W. D. Silva (2000). Fuzzy logic control of hydraulic system. *Industrial Technology Proceeding of IEEE International Conference* 2, 313-318.

J. J. E. Slotine and W. Li,(1991). *Applied nonlinear control,* Prentice-Hall, Englewood Cliffs, NJ.

J. J. E. Slotine and W. Li, (1988). Adaptive manipulator control a case study, *IEEE Transactions and Automatic Control* 33, 995-1003.

J. J. E. Slotine and W. Li, (1989) Composite Adaptive control of robot manipulators, *Automatica* 25, 509-519.

R. J. Wai, C. H. Lin and F. J. Lin. (2001). Adaptive fuzzy neural network control for motor-toggle servomechanism. *Mechatronics,* 11, 95-117.

R. J Wai. (2003). Robust fuzzy neural network control for nonlinear motor-toggle servomechanism. *Fuzzy sets and systems,* 139, 185-208.

J. Yong, C. S. Kim, J. Bae, S. Hong, M. H. Lee and F. Harashima, (2001). Vision based lateral control by yaw rate feedback. The 27th Annual Conference of the IEEE Industrial Electronics Society C, 2135-2138.

6

Online 3-D Trajectory Estimation of a Flying Object from a Monocular Image Sequence for Catching

Rafael Herrejon Mendoza, Shingo Kagami and Koichi Hashimoto
Tohoku University
Japan

1. Introduction

Catching a fast moving object can be used to describe work across many subfields of robotics, sensing, processing, actuation, and systems design. The reaction time allowed to the entire robot system: sensors, processor and actuators is very short. The sensor system must provide estimates of the object trajectory as early as possible, so that the robot may begin moving to approximately the correct place as early as possible. High accuracy must be obtained, so that the best possible catching position can be computed and maximum reaction time is available. 3D visual tracking and catching of a flying object has been achieved successfully by several researchers in recent years (Andersson; 1989)-(Mori et al.; 2004). There are two basic approaches to visual servo control: Position-Based Visual Servoing (PBVS), where computer techniques are used to reconstruct a representation of the 3D workspace of the robot, and actuator commands are computed with respect to the 3D workspace; and, Image-Based Visual Servoing (IBVS), where an error signal measured directly in the image is mapped to actuator commands.

In most of the research done in robotic catching using PBVS, the trajectory of the object is predicted with data obtained with a stereo vision system (Andersson; 1989)-(Namiki & Ishikawa; 2003), and the catching is achieved using a combination of light weight robots (Hove & Slotine; 1991) with fast grasping actuators (Hong & Slotine; 1995; Namiki & Ishikawa; 2003). A major difference exists between motion and structure estimation from binocular image sequences and that from monocular image sequences. With binocular image sequences, once the baseline is calibrated, the 3-D position of the object with reference with the cameras can be obtained.

Using IBVS, catching a ball has been achieved successfully in a hand-eye configuration with a 6 DOF robot manipulator and one CCD camera based on GAG strategy (Mori et al.; 2004).

Estimation of 3D trajectories from a monocular image sequence has been researched by (Avidan & Shashua; 2000; Cui et al.; 1994; Chan et al.; 2002; Ribnick et al.; 2009), among others, but to the best of our knowledge, no published work has addressed the 3-D catching of a fast moving object using monocular images with a PBVS system.

Our system (see Fig. 1) consists of one high speed stationary camera, a personal computer to calculate and predict the trajectory online of the object, and a 6 d.o.f. arm to approach the manipulator to the predicted position.

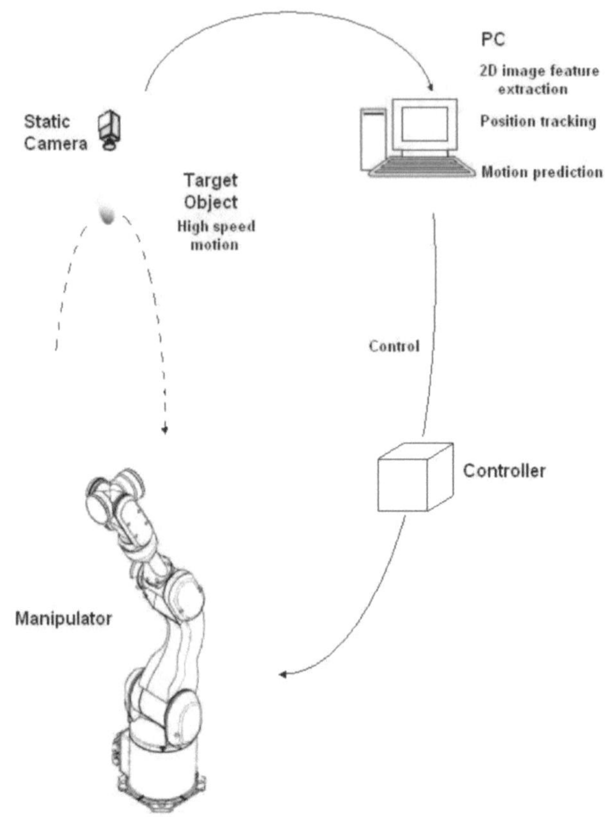

Fig. 1. System configuration used for catching.

The low level robot controller must be able to operate the actuator as close as possible to its capabilities, unlike conventional controllers. The robot must be able to be made to arrive not only at a specific place, but accurately at a specific time. The robot system must act before accurate data is available. The initial data is incorrect due to the inherent noise in the image, but if the robot waits until accurate data is obtained, there is very few time for motion.

2. Target trajectory estimation

Taking 3D points to a 2D plane is the objective of projective geometry. Due to its importance in artificial vision, work on this area has been used and developed thoroughly. The approach to determine motion consists of two steps: 1) Extract, match and determine the location of corresponding features, 2) Determine motion parameters from the feature correspondences. In this paper, only the second step is discussed.

2.1 Camera model

The standard pinhole model is used throughout this article. The camera coordinate system is assigned so as the x and y axis form the basis for the image plane, the z-axis is perpendicular to the image plane and goes through its optical center (c_u, c_v). Its origin is located at a distance f from the image plane. Using a perspective projection model, every 3-D point $\mathbf{P} = [X,Y,Z]^T$ on the surface of an object is deflated to a 2D point $\mathbf{p} = [u,v]^T$ in the image plane via a linear transformation known as the projection or intrinsic matrix **A**.

$$\mathbf{A} = \begin{bmatrix} -f_u & 0 & c_u \\ 0 & -f_v & c_v \\ 0 & 0 & 1 \end{bmatrix} \quad (1)$$

where f_u and f_v are conversion factors transforming distance units in the retinal plane into horizontal and vertical image pixels.
The projection of a 3D point on the retinal plane is given by

$$s\tilde{\mathbf{p}} = \mathbf{A}\tilde{\mathbf{P}} \quad (2)$$

where $\tilde{\mathbf{p}} = [u, v, 1]^T$ and $\tilde{\mathbf{P}} = [{}^c x, {}^c y, {}^c z, 1]^T$ are augmented vectors and s is an arbitrary scale factor. From this model, it is clear that any point in the line defined by the projected and original point produces the same projection on the retinal plane.

2.2 3D rigid-body motion

In this coordinate system, the camera is stationary and the scene is moving. For simplicity, assume that the camera takes images at regular intervals. As the rigid object move with respect to the camera, a sequence of images is obtained.
The motion of a rigid body in a 3D space has six degree of freedom. These are the three translation components of an arbitrary point within the object and the three rotation variables about that point. The translation component of the motion of a point at time t_i can be calculated with

$$x_i = C_1 + C_2 t_i + C_3 t_i^2 \quad (3)$$

$$y_i = C_4 + C_5 t_i + C_6 t_i^2 \quad (4)$$

$$z_i = C_7 + C_8 t_i + C_9 t_i^2 \quad (5)$$

where C_1,C_4,C_7 are initial positions, C_2,C_5,C_8 are velocities and C_3, C_6, C_9 are accelerations in the camera x, y, z axis respectively.

2.3 Observation vector

From (2), let the perspective of $\mathbf{P}_i$ be $\mathbf{p}_i = (u_i', v_i', 1)^T$. Its first two components u_i', v_i' represent the position of the point in image coordinates, and are given by

$$u_i' = -f_u \frac{x_i}{z_i} + c_u \quad (6)$$

$$v_i' = -f_v \frac{y_i}{z_i} + c_v. \quad (7)$$

If $u_i = u'_i - c_u$ and $v_i = v'_i - c_v$, (6, 7) can be expressed as

$$u_i = -f_u \frac{x_i}{z_i} \tag{8}$$

$$v_i = -f_v \frac{y_i}{z_i}. \tag{9}$$

Substituting (3, 4, 5) into (8) and (9) to obtain

$$u_i = -f_u \frac{C_1 + C_2 t_i + C_3 t_i^2}{C_7 + C_8 t_i + C_9 t_i^2} \tag{10}$$

$$v_i = -f_v \frac{C_4 + C_5 t_i + C_6 t_i^2}{C_7 + C_8 t_i + C_9 t_i^2}. \tag{11}$$

Reordering and multiplying (10) and (11) by a constant d such as $dC_9 = 1$, yields

$$\begin{aligned} d(C_7 u_i + C_8 u_i t_i + f_u C_1 + f_u C_2 t_i + f_u C_3 t_i^2) = \\ -dC_9 t_i^2 u_i, \end{aligned} \tag{12}$$

and

$$\begin{aligned} d(C_7 v_i + C_8 v_i t_i + f_v C_4 + f_v C_5 t_i + f_v C_6 t_i^2) = \\ -dC_9 t_i^2 v_i. \end{aligned} \tag{13}$$

We have the equation describing the state observation as follows

$$\mathbf{H}_i \mathbf{a}_i + \boldsymbol{\mu}_i = \mathbf{q}_i, \tag{14}$$

where $\boldsymbol{\mu}_i$ is a vector representing the noise in observation, $\mathbf{H}_i$ is the state observation matrix given by

$$\mathbf{H}_i = \begin{bmatrix} f_u & f_u t_i & f_u t_i^2 & 0 & 0 & 0 & u_i & u_i t_i \\ 0 & 0 & 0 & f_v & f_v t_i & f_v t_i^2 & v_i & v_i t_i \end{bmatrix}, \tag{15}$$

$\mathbf{a}_i$ is the state vector

$$\mathbf{a}_i = \begin{bmatrix} dC_1 & dC_2 & dC_3 & dC_4 & dC_5 & dC_6 & dC_7 & dC_8 \end{bmatrix}^T \tag{16}$$

and

$$\mathbf{q}_i = [-u_i t_i^2, -v_i t_i^2]^T \tag{17}$$

is the observation vector.

Considering one point in the space as the only feature to be tracked (the center of mass of an object), the issue of acquiring feature correspondences is dramatically simplified, but it is impossible to determine uniquely the solution. If the rigid object was n times farther away

from the image plane but translated at n times the speed, the projected image would be exactly the same.
In order to be able to calculate the motion, one constraint in motion has to be added. We consider the case of a not-self propelled projectile, in this case, the vector of acceleration is gravity.

$$C_3^2 + C_6^2 + C_9^2 = \frac{g^2}{4} \tag{18}$$

Equation 19 is a constraint given by the addition of the decomposition of the vector of gravity in its different components on each axis for a free falling object.
Substituting C_3,C_6 and C_9 from (14) and (17) into (19) yields

$$\frac{1}{d^2}a_3^2 + \frac{1}{d^2}a_6^2 + \frac{1}{d^2} = \frac{g^2}{4}. \tag{19}$$

From (20) the constant d can be calculated as

$$d = 2\sqrt{\frac{a_3^2 + a_6^2 + 1}{g^2}}. \tag{20}$$

2.4 Object trajectory estimation method

Recursive least squares is used to find the best estimate of the state from the previous state. The best estimate for time i is computed as

$$\hat{\mathbf{a}}_i = \hat{\mathbf{a}}_{i-1} + \mathbf{K}_i(\mathbf{q}_i - \mathbf{H}_i\hat{\mathbf{a}}_{i-1}). \tag{21}$$

where $\mathbf{K}_i$ is the gain matrix, $\mathbf{q}_i$ is the measurement vector for one point, and $\mathbf{H}_i$ is the projection matrix and given by the camera model and time.
The equation that describes the computation of the gain matrix is

$$\mathbf{K}_i = \mathbf{P}_i\mathbf{H}_i^T. \tag{22}$$

$\mathbf{P}_i$ is the covariance matrix for the estimation of the state i, and can be expressed mathematically as

$$\mathbf{P}_i = (\mathbf{P}_{i-1}^{-1} + \mathbf{H}_i^T\mathbf{H}_i)^{-1}. \tag{23}$$

The accuracy of the estimation depends of the number of points projected in the camera plane. Assuming we can observe enough points, the error from the calculated path and the projected path tends to zero.

2.5 Estimated trajectory accuracy

We evaluate the error by the sum of the squares of the 3-D euclidean distance between the simulated position (${}^c x(t)$, ${}^c y(t)$, ${}^c z(t)$) and the estimated position (${}^c\hat{x}(t)$, ${}^c\hat{y}(t)$, ${}^c\hat{z}(t)$), over the flying time interval i.e,

$$e_{xyz} = \sum_{t=0}^{T} ({}^{c}x(t) - {}^{c}\hat{x}(t))^2 + ({}^{c}y(t) - {}^{c}\hat{y}(t))^2 + ({}^{c}z(t) - {}^{c}\hat{z}(t))^2 \quad (24)$$

and in image coordinates, we evaluate the mean distance between the projected object trajectory and the reprojected estimated coordinate ($\hat{u}$, $\hat{v}$), which are functions of the estimated position, as follow

$$e_{uv}(N) = \frac{1}{N}\sum_{n=1}^{N}\sqrt{(u_n - \hat{u}_n)^2 + (v_n - \hat{v}_n)^2}. \quad (25)$$

3. Catching task

3.1 Constraints for the catching task

There are several constraints present for any robotic motion, but for catching there are some others that must be considered. Both of the types are included here for completion.

1. Initial Conditions. The initial arm position, velocity and acceleration are constrained to be their values at the time the target is first sighted
2. Catching Conditions. At the time of the catch, the end effector's position (x_r) has to match that of the ball. Thus at t_{catch}, $\mathbf{x}_r$ is constrained.
3. System Limits. The end effector's velocity and acceleration must stay below the limits physically acceptable to the system. The position, velocity and acceleration of the end effector must each be continuous. The end effector cannot leave the workspace
4. Freedom to change. When new vision information comes in, it should be possible to update the trajectory accordingly.

Two requirements are necessary for a particular trajectory matching solution have relevance to catching. One, the algorithm must require no prior knowledge of the trajectory such as starting position or velocity, and two, the algorithm can not be too computationally intensive.

3.2 Catching approaches

The processing of the computer images is time consuming, causing inherent delays in the information flow, when the position of a moving object is determined from the images, the computed value specifies the location of the object some periods ago. A time delay in the calculated position of the moving object is the main cause of difficulties in the visual-based implementation of the system. This problem can be avoided by predicting the position of the moving object.

There are two fundamental approaches to catching. One approach is to calculate an intercept point, move to it before the object arrives, wait and close at the appropriate time, the situation is analogous to a baseball catcher that positions the glove in the path of the ball, stopping it almost instantaneously. The other approach is to match the trajectory of the object in order to grasp the object with less impact and to allow for more time for grasping, like catching a raw egg, matching the movement of the hand with that of the egg. To be able to match the trajectory, it is expected that the robot end-effector can travel faster that the target within the robot's workspace.

3.3 Catch point determination

Depending on the initial angle and velocity, an object thrown at 1.7 meters from the base of the x-axis of the robot takes approximately 0.7-0.8 seconds to cover this distance. When the object enters the workspace of the robot, it typically travels at 3-6 m/s and the amount of time when the object is within the arm's workspace is about 0.20 seconds. Due to the maximum velocity of our arm (a six d.o.f industrial robot Mitsubishi PA10) being 1 m/s, it is physically impossible to match the trajectory of the object. Because of this constraint, we used move and wait approach for catching.

The catch point determination process begins by selection an initial prospective catch time. We assume that the closest point along the path of the object to the end-effector is when the z value of the object is equal to the robot's one.The time for the closest point is calculated solving 5 for t

$$t_{catch} = \frac{-C_8 + \sqrt{C_8^2 - (4C_9C_7 - Z_{r0})}}{2C_9}; \tag{26}$$

where Z_{r0} is the initial position of the robot's manipulator on camera coordinates, and C_7,C_8 and C_9 are the best estimated values obtained in 2.4. Note that the parabolic fit is updated with every new vision sample, therefore the position and time at which the robot would like to catch the changes during the course of the toss. Once t_{catch} has been obtained, it is just a matter of substituting its value in equations 3,4 to calculate the catching point.

3.4 Convergence criterion

When the mean square error e_{uv} in 26 is smaller than a chosen threshold (image noise + 1 pixel), and the error has been decreasing for the last 3 frames, we considered that the estimated path is close enough to the ground data and therefore the calculated rendez-vous point is valid. Prediction planning execution (PPE) strategy is started to move the robots end effector to the rendez-vous point.

3.5 Simulation results

Simulations for the task of tracking and catching a three dimensional flying target are described. At the initial time (t = 0), the initial position of the center of the manipulator end-effector is at (0.35, 0.33, 0.81) of the world coordinate frame. The speed of the manipulator is given by

$$\dot{x}_{robot} + \dot{y}_{robot} = 1m/s. \tag{27}$$

The object motion in world coordinates considered for this simulation (Fig. 2.a) is given by

$${}^{w}x(t) = 1.465 - 1.5t \tag{28}$$

$${}^{w}y(t) = 0.509 - 0.25t \tag{29}$$

$${}^{w}z(t) = 0.8 + 4.318t + \frac{1}{2}gt^2 \tag{30}$$

The coordinates of the object with respect to the camera can be calculated by

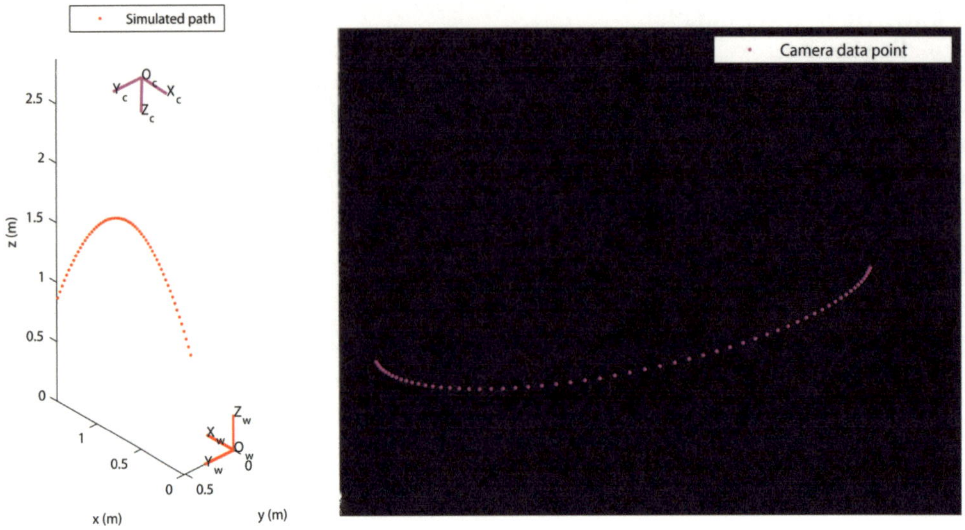

(a) (b)

Fig. 2. Path of the object in a) world and b) image coordinates

$$^{c}\mathbf{X}(t) = {}^{c}\mathbf{R}_{w}\,{}^{w}\mathbf{X}(t) + {}^{c}\mathbf{t}_{w} \tag{31}$$

where ${}^{c}\mathbf{R}_{w}$ is the rotation matrix from world to camera coordinates. First, a rotation about the x-axis, then about the y-axis, and finally the z-axis is considered. This sequence of rotations can be represented as the matrix product $\mathbf{R} = R_z(\phi)R_y(\theta)R_x(\psi)$.

The camera pose is given by rotating $\psi = 3.1806135$, $\theta = -0.0123876$ and $\phi = .0084783$ radians in the order previously stated. The translation vector is given by $\mathbf{t}$ = [0.889;–0.209;–2.853] meters. Substituting these parameters in (32), the object's motion in camera coordinates is given by

$$^{c}x_{sim}(t) = -0.599 + 1.374t + 0.137t^2 \tag{32}$$

$$^{c}y_{sim}(t) = 0.317 - 0.299t + 0.030t^2 \tag{33}$$

$$^{c}z_{sim}(t) = 2.091 - 4.356t + 4.898t^2. \tag{34}$$

The image of the simulated camera is a rectangle with a pixel array of 480 rows and 640 columns. The number of frames used is 60 at a sampling rate of 69 MHz, which accounts for a flying time of 0.87 seconds. The image coordinates (u,v) obtained using focal lengths f_u = 799, f_v = 799 and centers of image c_u = 267, c_v = 205 in (6,7) are shown in Fig. 2.b.

The object passes the catching point (0.42, 0.065, 0.81) at time t = 0.806. If the center of the manipulator end-effector can reach the catching point at the catching time, catching of the object is considered successful. Because the start of the actuation of the robot depends on the convergence criterion stated in 3.4, the success of the catching task is studied for image noise levels of 0, 0.5, 1 and 2 pixels.

Selection of an optimal convergence criteria to begin the robot motion is a difficult task. In Fig. 3, we can see that e_{uv} converges approximately 10 frames earlier than e_{xyz}, for all the

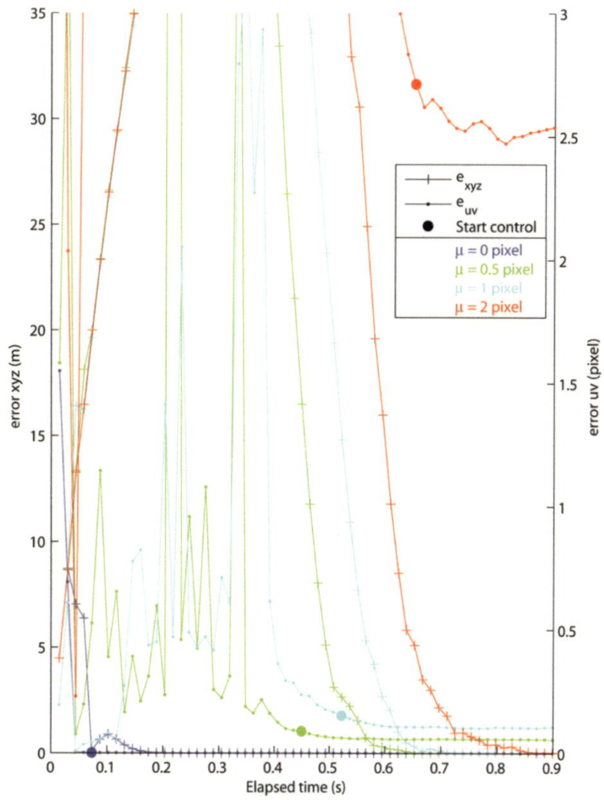

Fig. 3. Errors e_{xyz}, e_{uv} and trigger for servoing.

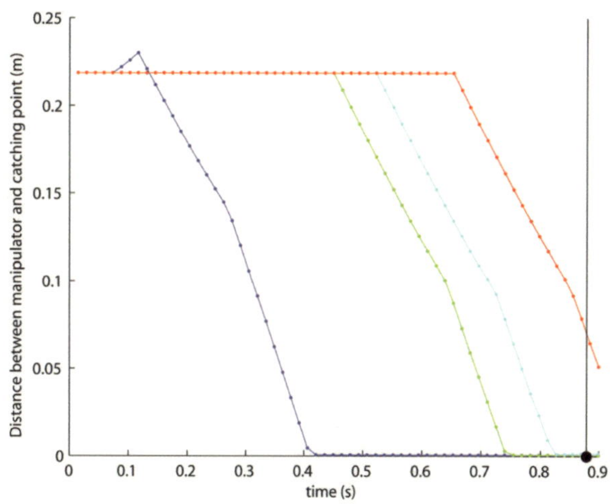

Fig. 4. Distance between manipulator and catching point

noise levels, but because e_{xyz} has not converged yet, triggering the start control flag at this moment would result in an incorrect catching position and the manipulator most probably lose valuable time back-tracking. It could be possible to wait until e_{xyz} is closer to convergence, but that would shorten the time for moving the manipulator. Our convergence

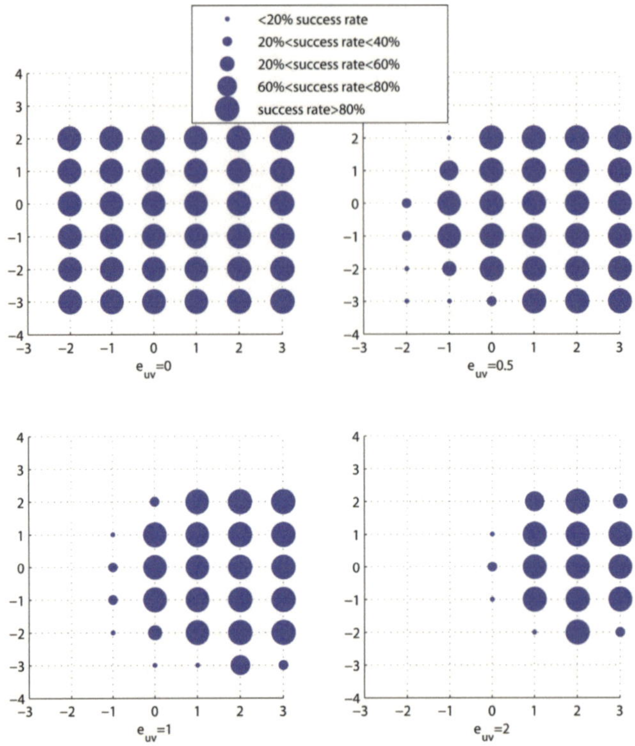

Fig. 5. Target catchable regions

criteria shows a reasonable balance within both stated situations. Fig. 4 shows the distance from the manipulator to the catching point. Judging from these results, we can see that the manipulator reaches the catching point at the catching time, when image noise is smaller than 2 pixels.

3.6 Target catchable region

In this section, we describe the target catchable region for the manipulator for each of the noise levels obtained by simulation. We consider several trajectories, landing grid points at time t = 0.806 from different initial positions. In these figures, the catching rate of the object is shown by size of the circle in the grid. As expected, the smaller the image noise is, the wider is the catching region. It was also found that trajectories that show a relative small change from their initial to final y-coordinates tend to converge faster than those with higher change rates.

4. Experimental results

Implementation of our visual servo trajectory control method was implemented to verify our simulation results. For this experiment, 58 images were taken with a Dragonfly Express Camera at 70 fps, the center of gravity of the object (a flipping coin) in the image plane (u,v) is used to calculate the trajectory. Camera calibration to obtain the intrinsic parameters was realized. Because the coin is turning, the calculated center of gravity varies accordingly to the image obtained, missing data is due to the observed coin projection in the image does

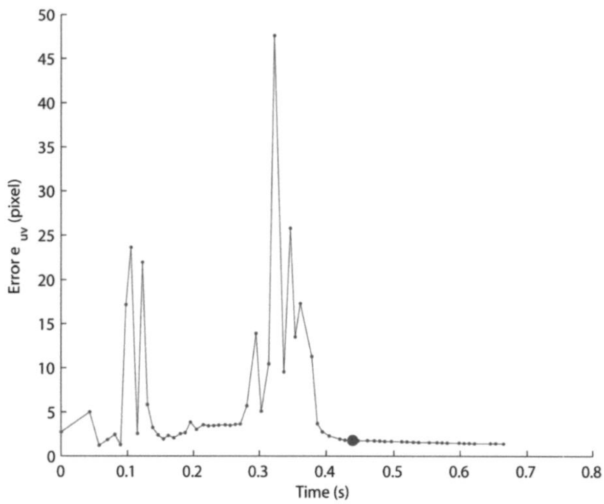

Fig. 6. Error e_{uv} and trigger for servoing.

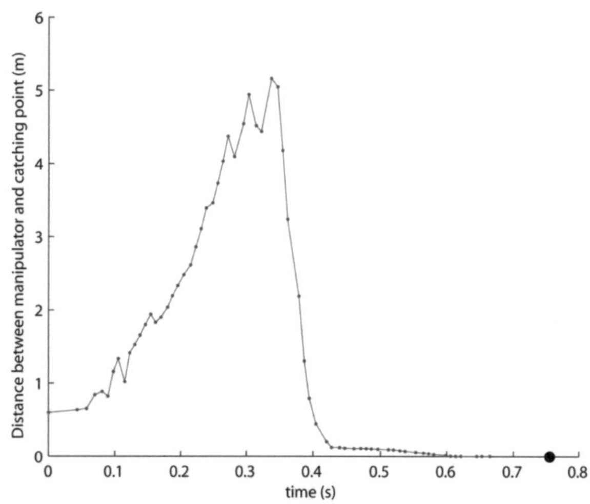

Fig. 7. Distance between manipulator and predicted catching point

not fulfill a minimum specified area, also, blur in the image causes errors in the calculation of the center of the coin. Error e_{uv} was found to be 1.4 pixels, we know from simulations the approximate catching range for this noise level. Experimental results are shown in Fig. 9 and Fig. 10, where it can be seen the movement of the robot to the catching point. Judging from these results, the robot performed the object catching task successfully. From Fig. 6 and Fig. 7, it is visible that the predicted catching point has already converged when control starts.

5. Conclusions and future work

This paper presented an implementation of ball catching task using a robot manipulator. We demonstrated that the robot can catch an object flying in three-dimensional space using recursive least squares (RLS) algorithm to extract and predict the position of the object from one feature correspondence from only a monocular vision system. The object trajectory path

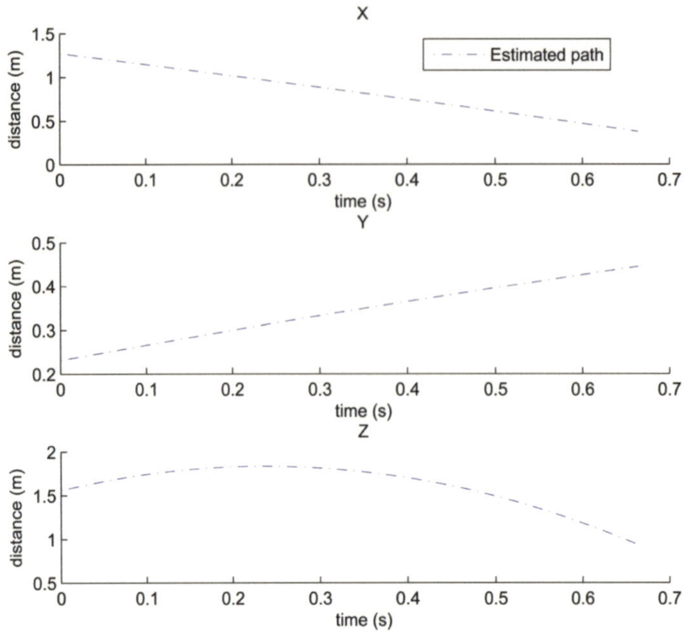

Fig. 8. Calculated path in each axis

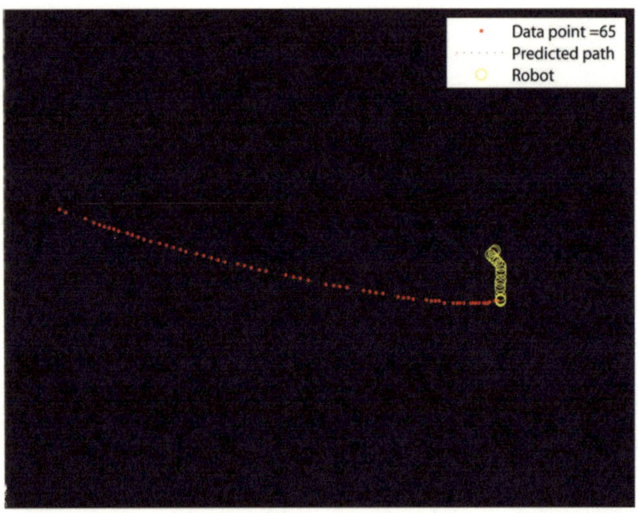

Fig. 9. The center of mass as observed by the camera

was obtained successfully even under high noise images. The recursive estimation technique presented in this paper has numerous advantages over other methods currently in use. First, using only one feature point, the issue of feature points correspondence is simplified. Another advantage is the recursive nature of the computations makes it suitable for real-time applications. Results on simulation and real imagery illustrate the performance of the estimator, and the feasibility of our estimation method for the catching task. Convergence of the path under image noise was studied and a satisfactory criteria was determined

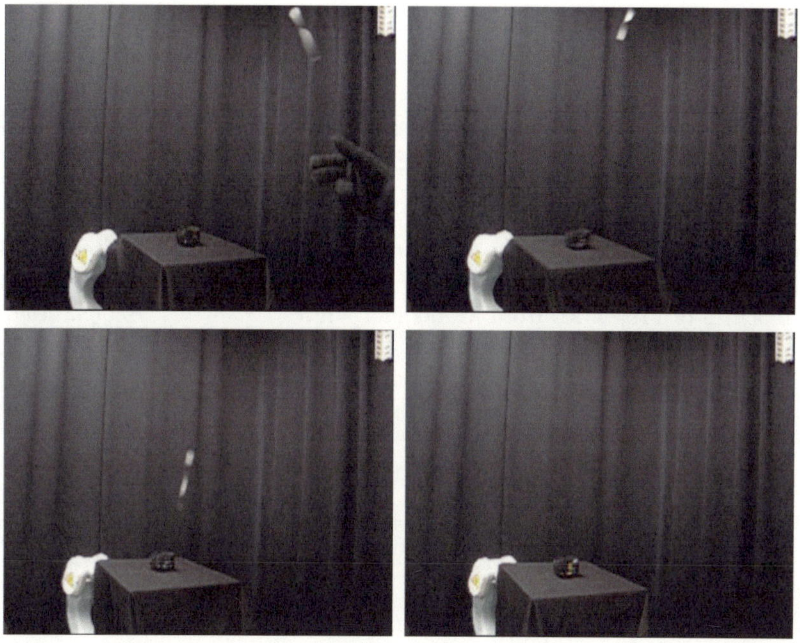

Fig. 10. Sequence of images of object catching

successfully for both simulations and experiments. Current research is directed towards the study of different control approaches to increase the catching range of the manipulator under noisy images.

6. References

Andersson, R. L. (1989). *A robot ping-pong player: Experimental in Real-Time Intelligent Control*, ATT Bell Laboratories, MIT Press.

Avidan, S. & A. Shashua, A. (2000). Trajectory Triangulation: 3D Reconstruction of Moving Points from a Monocular Image Sequence, *IEEE. Trans of Pat, An. and Mac. Int*, Vol. 22, pp. 348-357, 2000.

Chan, C.; Guesalaga, A.; &Obac, V. (2002). Robust Estimation of 3D Trajectories from a Monocular Image Sequence, *Int. journal of imaging sys. and tech.*, Vol. 12, pp. 128-137, 2002.

Cui, N.; Weng, J. J. & Cohen, P. (1994). Recursive-Batch Estimation of Motion and Structure from Monocular Image Sequences, *CVGIP: Image Understanding*, Vol. 59, pp. 154-170, 1994.

Frese, U.; Bauml, B.; Haidacher, S.; Schreiber, G.; Schaefer, I.; Hahnle, M. & Hirzinger, G. (2001). Off-the-Shelf Vision for a Robotic Ball Catcher, *Proc. IEEE/RSJ Intl. Conf. on Intelligent Robots and Systems*, Maui, 2001.

Hove, B. M. & Slotine, J.J.E. (1991). Experiments in Robotic Catching, *Proc. of American Control Conf*, Vol (1), pp. 380 - 385, Boston, MA, 1991.

Hong,W. & Slotine, J.J.E. (1995). Experiments in Hand-Eye Coordination Using Active Vision, *Proc. 4th Int. Symposium on Experimental Robotics*, Stanford, CA, 1995.

Namiki, A. & Ishikawa, M. (2003). Vision-Based Online Trajectory Generation and Its Application to Catching, *Control Problems in Robotics*, Springer-Verlag, pp. 249-264, Berlin, 2003.

Namiki, A. & Ishikawa, M. (2003). Robotic Catching Using a Direct Mapping from Visual Information to Motor Command, *Proc. IEEE Int. Conf. Robotics and Automation*, pp. 2400-2405, Taipei, Taiwan, 2003.

Mori, R.; Hashimoto,K. & Miyazaki, F. (2004). Tracking and Catching of 3D Flying Target based on GAG Strategy, *Proc. Int. Conf. Robotics and Automation*, pp. 4236-4241, 2004.

Ribnick, E.; Atev, S. & Papanikolopoulos, N. P. (2009). Estimating 3D Positions and Velocities of Projectiles from Monocular Views, *Trans. Pat. An. and Mach. Int.* Vol. 31(5), pp. 938-944, 2009.

7

Multi-Camera Visual Servoing of a Micro Helicopter Under Occlusions

Yuta Yoshihata, Kei Watanabe, Yasushi Iwatani and Koichi Hashimoto
Tohoku University
Japan

1. Introduction

Autonomous control of unmanned helicopters has the advantage that there is no need to develop skilled workers and has potential for surveillance tasks in dangerous areas including forest-fire reconnaissance and monitoring of volcanic activity. For vehicle navigation, the use of computer vision as a sensor is effective in unmapped areas. Visual feedback control is also suitable for autonomous takeoffs and landings, since precise position control is required at a neighborhood of the launch pad or the landing pad. Such applications have generated considerable interest in the vision based control community (Altug et al., 2005; Amidi et al., 1999; Ettinger et al., 2002; Mahony & Hamel, 2005; Mejias et al., 2006; Proctor et al., 2006; Saripalli et al., 2003; Shakernia et al., 2002; Wu et al., 2005; Yu et al., 2006).

The authors have developed a visual control system for a micro helicopter (Watanabe et al., 2008). The helicopter does not have any sensors that measure its position or posture. Two cameras are placed on the ground. They track four black balls attached to rods connected to the bottom of the helicopter. The differences between the current ball positions and given reference positions in the camera frames are fed to a set of PID controllers. It is not required that sensors for autonomous control are installed on the helicopter body, and we need no mechanical or electrical improvements of existing unmanned helicopters that are controlled remotely and manually.

In visual control, tracked objects have to be visible in the camera views, but tracking may fail due to occlusions. An occlusion occurs when an object moves across in front of a camera or when the background color happens to be similar to the color of a tracked object. Multicamera systems are suitable for designing a robust controller under occlusions, since even when a tracked object is not visible in a camera view, the other cameras may track it. The visual control system with two cameras proposed in (Yoshihata et al., 2007) is robust against temporary occlusions. If an occlusion is detected in a camera view then the other camera is used to control the helicopter. The positions of the invisible tracked objects in the image plane of the occluded camera are estimated by using the positions in the other image plane. The control method proposed in (Yoshihata et al., 2007) is called the camera selection approach in this paper.

This paper proposes another switched visual feedback control method that is called the image feature selection approach. It is robust against temporary and partial occlusions even when a tracked object is not visible in any of the camera views. We also use two cameras and two tracked objects for each camera. This configuration is redundant for helicopter control, but it is suitable for making a control system robust against occlusions. This paper assumes that at most one tracked object is occluded at each time, as a first step towards a unified framework that combines the image feature selection approach presented in this paper and the camera selection approach proposed in (Yoshihata et al., 2007). The errors between the current positions of the tracked objects and pre-specified references are used to compute the control input signals, when all the tracked objects are visible. If one of the tracked objects is invisible, then the controller uses the errors given by the other three tracked objects. The position of the occluded object is also estimated by using the other three tracked objects.

2. Experimental setup

The experimental system considered in this paper consists of a small helicopter and two stationary cameras as illustrated in Fig. 1. The helicopter does not have any sensors that measure the position or posture. It has four small black balls, and they are attached to rods connected to the bottom of the helicopter. The black balls are indexed from 1 to 4. The two cameras are placed on the ground and they look upward. Snapshots of the helicopter from the two cameras can be seen in Fig. 2. The camera configuration and the use of the redundant tracked objects enable a robust controller design under temporary and partial occlusions as described in Section 6.

The system takes 8.5 milli-seconds to make the control input signals from capturing images of the balls. This follows from the use of fast IEEE 1394 cameras, Dragonfly Express[1].

The small helicopter used in experiments is X. R. B-V2-lama developed by HIROBO (see Fig. 3). It has a coaxial rotor configuration. The two rotors share the same axis, and they rotate in opposite directions. The tail is a dummy. A stabilizer is installed on the upper rotor head. It mechanically keeps the posture horizontal.

Table 1 summarizes specifications of the system.

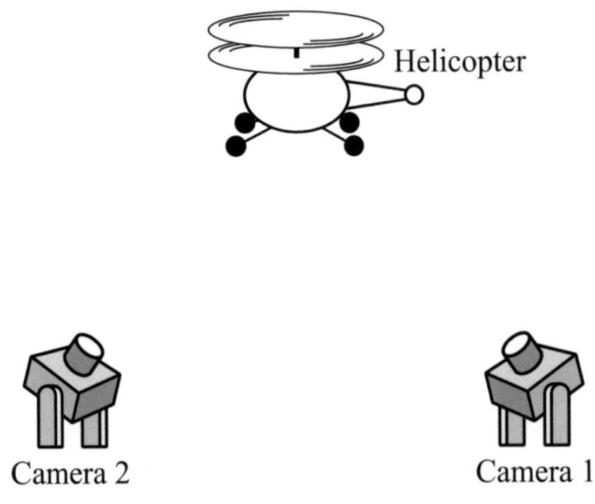

Fig. 1. System configuration.

[1] Dragonfly Express is a trademark of Point Grey Research Inc.

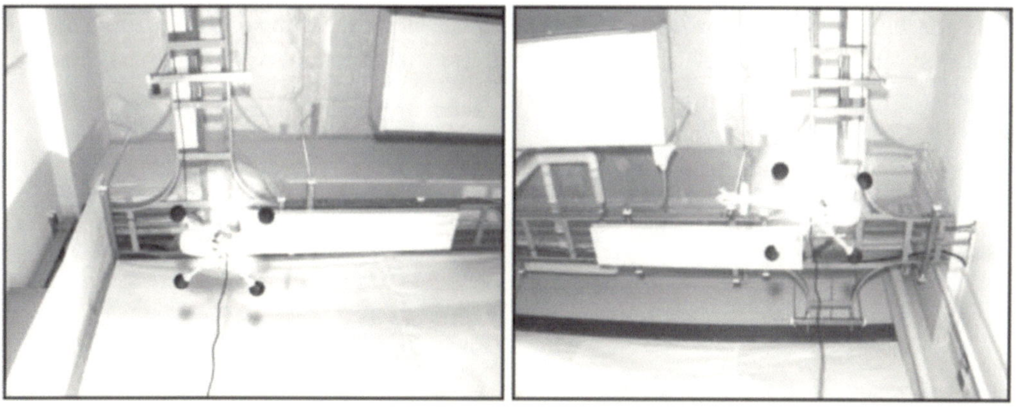

Fig. 2. Snapshots of the helicopter. The right one was captured from camera 1 and the left one from camera 2. The helicopter was controlled manually.

Fig. 3. X.R.B. with four black balls.

Length of the helicopter,	0.40 [m].
Height of the helicopter,	0.20 [m].
Rotor length of the helicopter,	0.35 [m].
Weight of the helicopter,	0.22 [kg].
Focal length of the lens,	0.0045 [m].
Camera resolution,	640 × 480 [pixels].
Pixel size,	7.4 [μm] × 7.4 [μm].

Table 1. Specifications of the system.

3. Mathematical preliminaries

3.1 Coordinate frames

Let Σ^w be the world reference frame and a coordinate frame Σ^b be attached to the helicopter body as illustrated in Fig. 4. The z^w axis is directed vertically downward. A coordinate frame

Σ^j is attached to camera j for j = 1, 2. The z^j axis lies along the optical axis of camera j. The axes x^w, x^1 and x^2 are parallel. The coordinate frame x^jy^j corresponds to the image frame of camera j, and it is denoted by Σ^{cj}.

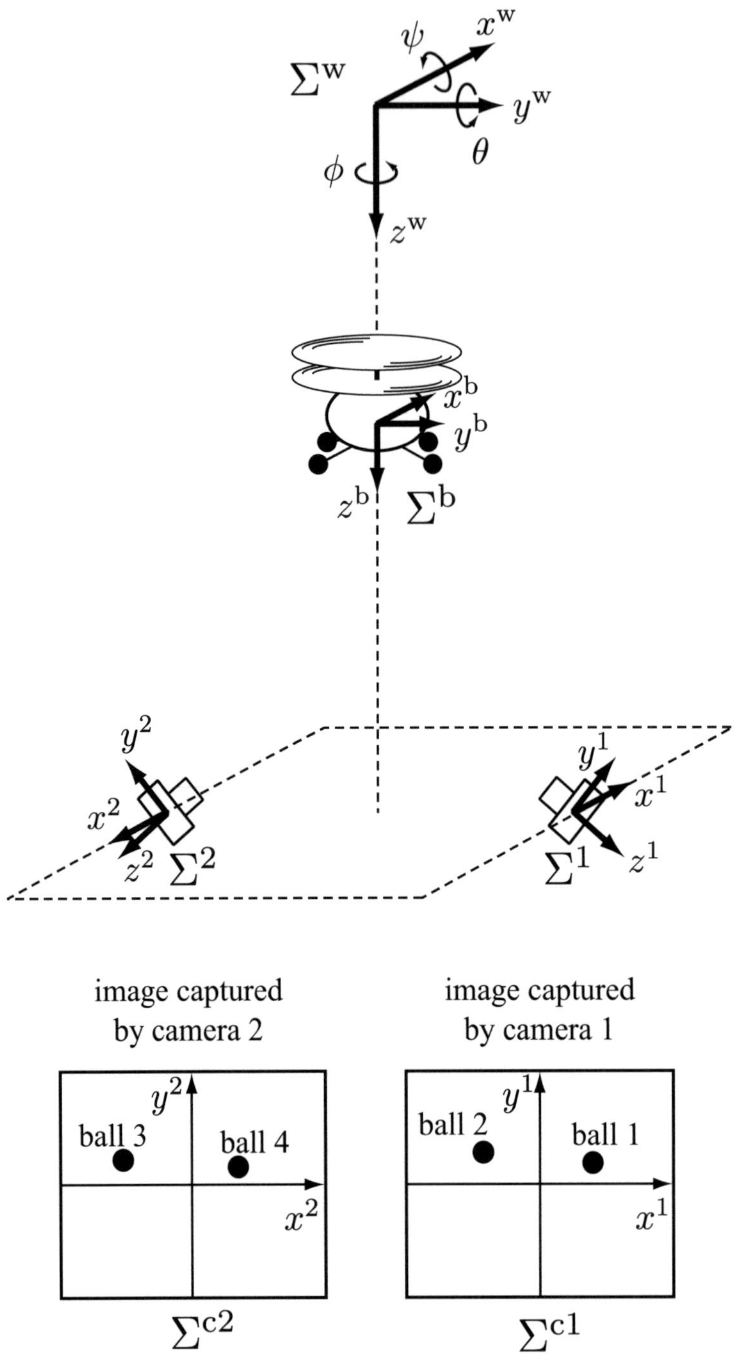

Fig. 4. Coordinate frames.

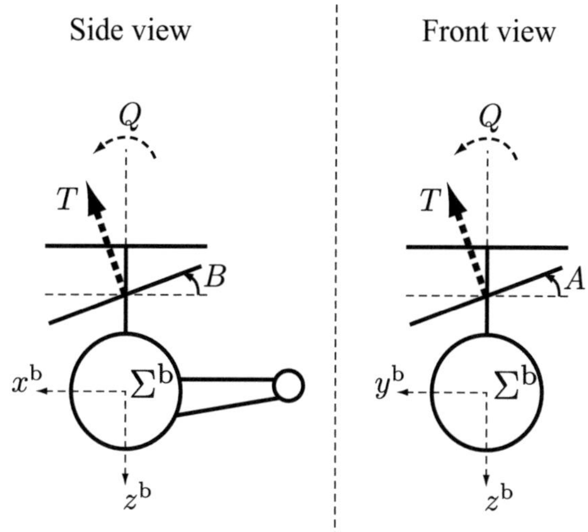

Fig. 5. The helicopter coordinate frame and input variables.

The helicopter position relative to the world reference frame Σ^w is denoted by (x, y, z). The roll, pitch and yaw angles are denoted by ψ, θ, φ, respectively. The following four variables are individually controlled by signals supplied to the transmitter (see Fig. 5):
B : Elevator, pitch angle of the lower rotor.
A : Aileron, roll angle of the lower rotor.
T : Throttle, resultant force of the two rotor thrusts.
Q : Rudder, difference of the two torques generated by the two rotors.
The corresponding input signals are denoted by V_B, V_A, V_T and V_Q. Note that x, y, z and ϕ are controlled by applying V_B, V_A, V_T and V_Q, respectively.

3.2 Mathematical preliminaries

In this paper, we make the following four assumptions:

1. It is supposed that

$$\theta(t) = 0, \psi(t) = 0, \forall\, t \geq 0, \tag{1}$$

   where recall that θ denotes the angle about y^w axis and ψ the angle about x^w axis.
2. The reference position relative to the world reference frame Σ^w is always set to $\mathbf{0}$. When the reference position is changed, the world reference frame is replaced and the reference position is set to the origin of the new world reference frame.
3. Camera 1 captures images of balls 1 and 2, and camera 2 takes images of balls 3 and 4.
4. At most one tracked object is occluded at each time.

Recall that the helicopter has the horizontal-keeping stabilizer. Both the angles θ and ψ converge to zero fast enough even when the body is inclined. Thus, the first assumption is not far from the truth in practice. We here define

$$\boldsymbol{r} = \begin{bmatrix} x & y & z & \phi \end{bmatrix}^{\mathrm{T}}. \tag{2}$$

Note that $\boldsymbol{r}$ means the vector of the generalized coordinates. Then, our goal is that $\boldsymbol{r}(t) \to \mathbf{0}$ as $t \to \infty$ from the first and second assumptions.

The third and fourth assumptions are made to consider a simple example in which visible image features should be selected from redundant features under temporary and partial occlusions. The assumptions are suitable for a first step towards a unified framework that combines the image feature selection approach presented in this paper and the camera selection approach proposed in (Yoshihata et al., 2007).

4. Image Jacobian

This section derives the image Jacobian that gives a relationship between the vector of the generalized coordinates $\boldsymbol{r}$ and the vector of the image features.

The position of the center of gravity of ball i in the image frame is denoted by $\xi_i = [\xi_{ix}, \xi_{iy}]^{\mathrm{T}} \in \mathbb{R}^2$ for $i = 1, \ldots, 4$. We define

$$\zeta_0 = \begin{bmatrix} \xi_1^{\mathrm{T}} & \xi_2^{\mathrm{T}} & \xi_3^{\mathrm{T}} & \xi_4^{\mathrm{T}} \end{bmatrix}^{\mathrm{T}}. \tag{3}$$

In addition, we set

$$\zeta_i = \begin{bmatrix} \xi_{\sigma_{i1}}^{\mathrm{T}} & \xi_{\sigma_{i2}}^{\mathrm{T}} & \xi_{\sigma_{i3}}^{\mathrm{T}} \end{bmatrix}^{\mathrm{T}}, \tag{4}$$

for $i = 1, \ldots, 4$, where

$$\sigma_{ik} \in \{1,2,3,4\} / \{i\}, \quad k = 1,2,3, \tag{5}$$

$$\sigma_{i1} < \sigma_{i2} < \sigma_{i3}. \tag{6}$$

The vector ζ_0 is used as the vector of image features, when all positions of the tracked balls can be measured correctly. On the other hand, ζ_i for $i = 1, 2, 3, 4$ implies the vector of visible image features when ball i is occluded. They enable us to give a switched controller that is robust against occlusions, if the fourth assumption holds or equivalently at most one tracked object is occluded. Details will be discussed in the next section.

Let ${}^{\mathrm{b}}\boldsymbol{p}_i \in \mathbb{R}^3$ denote the position of ball i in the frame Σ^{b}. The position of ball i in the frame Σ^j is denoted by

$$\boldsymbol{p}_i = \begin{bmatrix} x_i & y_i & z_i \end{bmatrix}^{\mathrm{T}} \in \mathbb{R}^3, \tag{7}$$

where $j = 1$ for $i = 1, 2$ and $j = 2$ for $i = 3, 4$. We have

$$\begin{bmatrix} \boldsymbol{p}_i \\ 1 \end{bmatrix} = {}^{j}\boldsymbol{H}_{\mathrm{w}}(\boldsymbol{r})\,{}^{\mathrm{w}}\boldsymbol{H}_{\mathrm{b}}(\boldsymbol{r}) \begin{bmatrix} {}^{\mathrm{b}}\boldsymbol{p}_i \\ 1 \end{bmatrix}, \tag{8}$$

where ${}^{j}\boldsymbol{H}_{\mathrm{w}}(\boldsymbol{r})$ and ${}^{\mathrm{w}}\boldsymbol{H}_{\mathrm{b}}(\boldsymbol{r})$ are the homogeneous transformation matrices from Σ^{w} to Σ^j and from Σ^{b} to Σ^{w}, respectively (see for example (Spong et al., 2005) for deriving the homogeneous transformation matrices). It then holds that

$$|z_i| \begin{bmatrix} \xi_i \\ 1 \end{bmatrix} = \boldsymbol{F} \begin{bmatrix} \boldsymbol{p}_i \\ 1 \end{bmatrix} \tag{9}$$

where

$$F = \begin{bmatrix} f & 0 & 0 & 0 \\ 0 & f & 0 & 0 \\ 0 & 0 & 1 & 0 \end{bmatrix}, \quad (10)$$

and f is the focal length of the lens (see for example (Ma et al., 2004) for the camera model). It is straightforward to verify that

$$\xi_i = \frac{f}{|z_i|}\begin{bmatrix} x_i \\ y_i \end{bmatrix}$$

$$=: \boldsymbol{\alpha}_i(\boldsymbol{r}). \quad (11)$$

We here define

$$\boldsymbol{\beta}_0(\boldsymbol{r}) = \begin{bmatrix} \boldsymbol{\alpha}_1^{\mathrm{T}}(\boldsymbol{r}) & \boldsymbol{\alpha}_2^{\mathrm{T}}(\boldsymbol{r}) & \boldsymbol{\alpha}_3^{\mathrm{T}}(\boldsymbol{r}) & \boldsymbol{\alpha}_4^{\mathrm{T}}(\boldsymbol{r}) \end{bmatrix}^{\mathrm{T}}, \quad (12)$$

$$\boldsymbol{\beta}_i(\boldsymbol{r}) = \begin{bmatrix} \boldsymbol{\alpha}_{\sigma_{i1}}^{\mathrm{T}}(\boldsymbol{r}) & \boldsymbol{\alpha}_{\sigma_{i2}}^{\mathrm{T}}(\boldsymbol{r}) & \boldsymbol{\alpha}_{\sigma_{i3}}^{\mathrm{T}}(\boldsymbol{r}) \end{bmatrix}^{\mathrm{T}}, \quad (13)$$

for $i = 1, \ldots, 4$, where σ_{i1}, σ_{i2} and σ_{i3} are defined by (5) and (6). The equations (12) and (13) provide transformations from the generalized coordinates $\boldsymbol{r}$ to the image features ζ_i.
We define

$$\boldsymbol{J}_i = \left.\frac{\partial \boldsymbol{\beta}_i}{\partial \boldsymbol{r}}\right|_{\boldsymbol{r}=0}. \quad (14)$$

Then it holds that

$$\dot{\zeta}_i = \boldsymbol{J}_i \dot{\boldsymbol{r}}, \quad (15)$$

at $\boldsymbol{r} = 0$. Each $\boldsymbol{J}_i$ ($i = 0, \ldots, 4$) is referred to as the image Jacobian.

5. Controller design

This paper proposes a switched visual feedback control system illustrated in Fig. 6, where ξ_i^{ref} denotes the image reference of ball i relative to the corresponding image frame Σ^{cj} and

$$\zeta_0^{\mathrm{ref}} = \begin{bmatrix} \xi_1^{\mathrm{ref}\,\mathrm{T}} & \xi_2^{\mathrm{ref}\,\mathrm{T}} & \xi_3^{\mathrm{ref}\,\mathrm{T}} & \xi_4^{\mathrm{ref}\,\mathrm{T}} \end{bmatrix}^{\mathrm{T}}, \quad (16)$$

$$\zeta_i^{\mathrm{ref}} = \begin{bmatrix} \xi_{\sigma_{i1}}^{\mathrm{ref}\,\mathrm{T}} & \xi_{\sigma_{i2}}^{\mathrm{ref}\,\mathrm{T}} & \xi_{\sigma_{i3}}^{\mathrm{ref}\,\mathrm{T}} \end{bmatrix}^{\mathrm{T}}, \quad (17)$$

for $i = 1, \ldots, 4$, where σ_{i1}, σ_{i2} and σ_{i3} are defined by (5) and (6). The system is an image-based visual servo system, since the proposed controller uses the image Jacobian derived in the previous section and the errors between the vector of the image features $\zeta_i(t)$ and the corresponding given reference ζ_i^{ref} to obtain the input signals. Image-based visual servo control is robust against model uncertainties (Hashimoto, 2003).

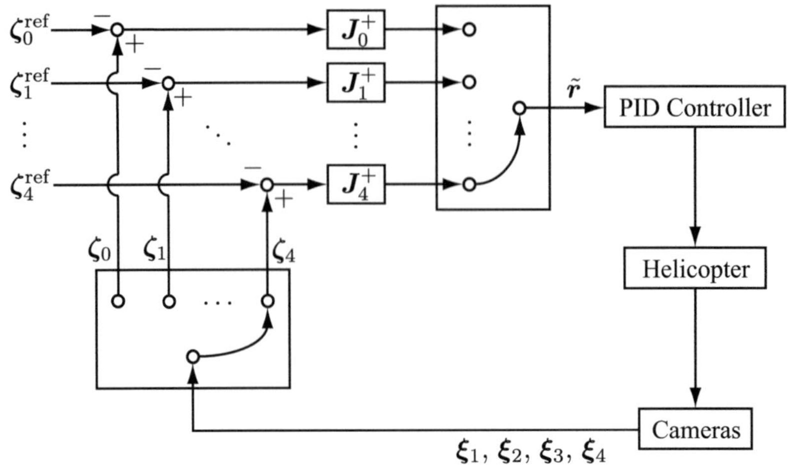

Fig. 6. Closed loop system.

The switch in the closed loop depends on which image feature ξ_i is invisible. The decision of switching will be described in Section 5.2. In this paper, the image feature ξ_i is labeled as 'normal', when the system decides that ξ_i is measured correctly. Similarly, ξ_i is labeled as 'occluded', when the system decides that ξ_i is not measured correctly.

5.1 Measurement of image features

An image feature $\xi_i(t)$ ($i = 1, \ldots, 4$) is given by the following manner. A binary data matrix at time t is first obtained from an image captured by camera j, and it is denoted by $\boldsymbol{I}_j(x^j, y^j)$ for $j = 1, 2$. The matrix $\boldsymbol{I}_j(x^j, y^j)$ has values of 1 for black and 0 for white. We then make a search window $\mathbb{S}_i$ whose center is defined as follows:

Normal case: It is set at $\xi_i(t-h)$, where h denotes the sampling time.
Occluded case: We estimate $\xi_i(t)$ by

$$J_0 J_i^+(\zeta_i(t) - \zeta_i^{\text{ref}}) + \zeta_0^{\text{ref}} =: \begin{bmatrix} \tilde{\xi}_1^{\text{T}} & \tilde{\xi}_2^{\text{T}} & \tilde{\xi}_3^{\text{T}} & \tilde{\xi}_4^{\text{T}} \end{bmatrix}^{\text{T}}, \tag{18}$$

where J_i^+ denotes the Moore-Penrose inverse of J_i^+. The center is set at $\tilde{\xi}_i$.
The size of the window $\mathbb{S}_i$ is given by a constant. We define an image data matrix by

$$\overline{\boldsymbol{I}}_{ji}(x^j, y^j) = \begin{cases} \boldsymbol{I}_j(x^j, y^j), & \text{for } (x^j, y^j) \in \mathbb{S}_i, \\ 0, & \text{otherwise}, \end{cases}$$

where $j = 1$ for $i = 1, 2$ and $j = 2$ for $i = 3, 4$. The image feature $\xi_i(t)$ is the center of mass of $\overline{\boldsymbol{I}}_{ji}(x^j, y^j)$.

5.2 Selection of image features

Let three constants δ, $m_{\min}$ and $m_{\max}$ be given. Let $m_i(t)$ denote the area, or equivalently the zero-th order moment, of the image data $\overline{\boldsymbol{I}}_{ji}(x^j, y^j)$. An occlusion is detected or cancelled for each image feature $\xi_i(t)$ in the following manner.

Normal case: If $m_{\min} \le m_i(t) \le m_{\max}$ holds, then $\xi_i(t)$ is labeled as 'normal' again. Otherwise, it is labeled as 'occluded'.
Occluded case: If it holds that $m_{\min} \le m_i(t) \le m_{\max}$ and

$$\| J_0^+(\zeta_0(t)-\zeta_0^{\mathrm{ref}}) - J_i^+(\zeta_i(t)-\zeta_i^{\mathrm{ref}}) \| < \delta, \tag{19}$$

then $\xi_i(t)$ is labeled as 'normal'. Otherwise, it is labeled as 'occluded' again.
If $\xi_i(t)$ is occluded for i, then $\zeta_i(t)$ is used at the next step. Otherwise, or equivalently if every image feature $\xi_i(t)$ is normal, then ζ_0 is used at the next step.

5.3 Control input voltages

We compute

$$\tilde{r}(t) = \begin{bmatrix} \tilde{x}(t) & \tilde{y}(t) & \tilde{z}(t) & \tilde{\phi}(t) \end{bmatrix}^{\mathrm{T}}$$

$$:= J_i^+(\zeta_i(t)-\zeta_i^{\mathrm{ref}}), \tag{20}$$

for ζ_i selected in the previous subsection. The input signals are given by a set of PID controllers of the form

$$V_B(t) = b_1 - P_1\tilde{x} - I_1\int_0^t \tilde{x}\mathrm{d}t - D_1\dot{\tilde{x}}, \tag{21}$$

$$V_A(t) = b_2 - P_2\tilde{y} - I_2\int_0^t \tilde{y}\mathrm{d}t - D_2\dot{\tilde{y}}, \tag{22}$$

$$V_T(t) = b_3 - P_3\tilde{z} - I_3\int_0^t \tilde{z}\mathrm{d}t - D_3\dot{\tilde{z}}, \tag{23}$$

$$V_Q(t) = b_4 - P_4\tilde{\phi} - I_4\int_0^t \tilde{\phi}\mathrm{d}t - D_4\dot{\tilde{\phi}}, \tag{24}$$

where b_i, P_i, I_i and D_i are constants for $i = 1, \ldots, 4$.

6. Experiment and result

The world reference frame Σ^{w} and the camera frames Σ^1 and Σ^2 are located as shown in Fig. 7. The controller gains are tuned to the values in Table 2. The positions of the four black balls in the frame Σ^{b} are given by

$${}^{\mathrm{b}}\boldsymbol{p}_1 = [0.1 \quad 0.1 \quad 0.04]^{\mathrm{T}}, \tag{25}$$

$${}^{\mathrm{b}}\boldsymbol{p}_2 = [-0.1 \quad 0.1 \quad 0.04]^{\mathrm{T}}, \tag{26}$$

$${}^{\mathrm{b}}\boldsymbol{p}_3 = [0.1 \quad -0.1 \quad 0.04]^{\mathrm{T}}, \tag{27}$$

$${}^{\mathrm{b}}\boldsymbol{p}_4 = [-0.1 \quad -0.1 \quad 0.04]^{\mathrm{T}}. \tag{28}$$

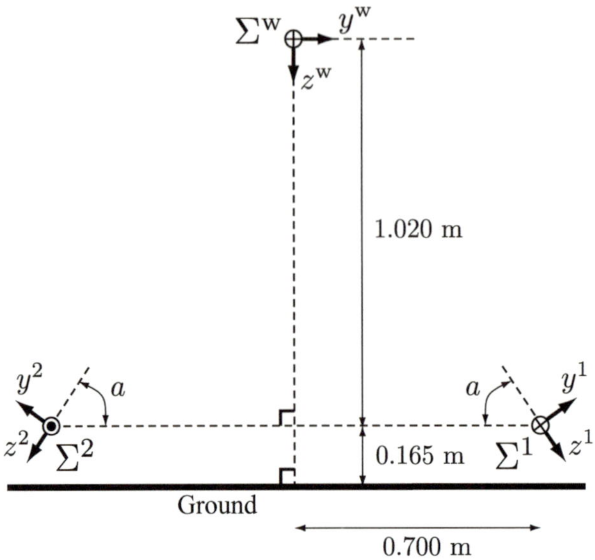

Fig. 7. Locations of the world reference frame Σ^w and the camera frames Σ^1 and Σ^2. The angle a is set to $a = 11\pi/36$.

	b_i	P_i	I_i	D_i
V_B	3.47	3.30	0.05	2.60
V_A	3.38	3.30	0.05	2.60
V_T	2.70	1.90	0.05	0.80
V_Q	1.92	3.00	0.05	0.08

Table 2. PID gains.

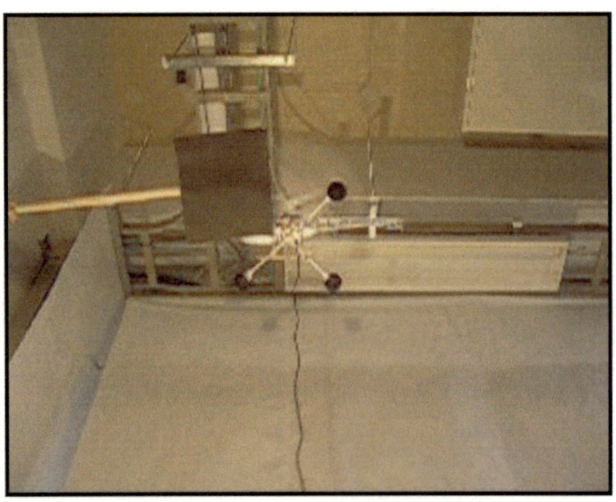

Fig. 8. A snapshot of helicopter flight under an occlusion. This was captured by a camera placed next to camera 2. Ball 3 was not captured correctly at this moment.

The image reference ζ_0^{ref} is set to

$$\zeta_0^{ref} = \begin{bmatrix} 84.6 & 10.5 & -21.1 & 16.1 & -65.6 & 41.9 & 43.4 & 40.9 \end{bmatrix}^T, \text{ (pixels)}. \quad (29)$$

This was obtained by an actual measurement.
Ball 1 or 3 was occluded temporarily and intentionally. Long time occlusions for around 10 seconds were presented twice for each ball. Short time occlusions were done four times for each ball, and they were successively done from ball 3 to 1. A snapshot of helicopter flight under an occlusion can be seen at Fig. 8.
Fig. 9 shows the x positions of balls 1 and 3 in the corresponding image planes. When an occlusion is detected, the value is set at -150 in the figure to make the plot easy to read. For example, ξ_3 was labeled 'occluded' from 15 to 25 seconds. It is seen that the number of occlusion detection is equivalent to the number of intentional occlusions.

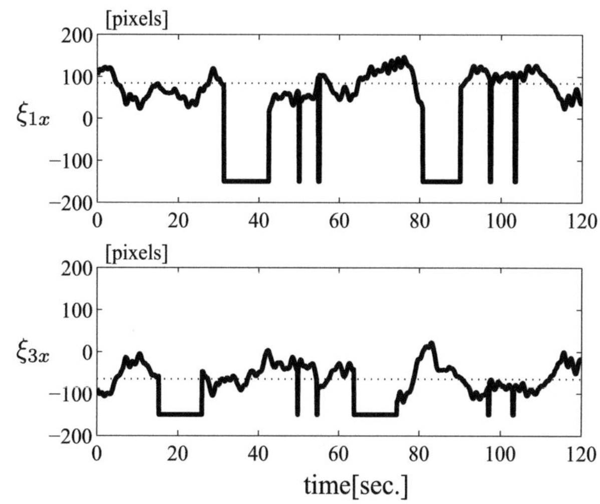

Fig. 9. Experimental result. Solid lines: Time profiles of the positions of image features. When an occlusion is detected, the value is set to -150. Dotted lines: Given references.

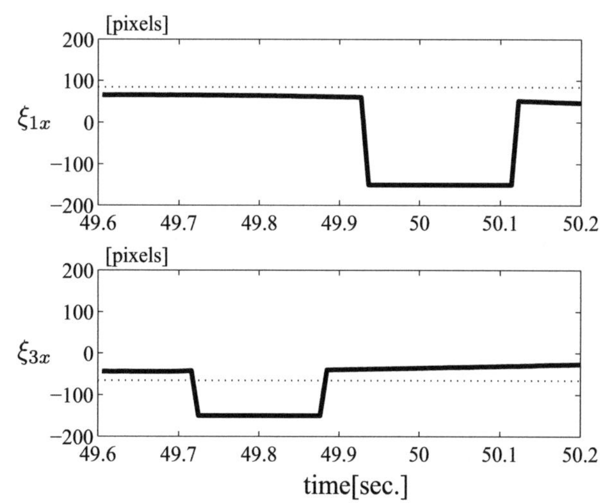

Fig. 10. Experimental result: Time profiles of the positions of image features. This is a closeup of Fig. 9 between 49.6 and 50.2 seconds.

Fig. 10 illustrates a closeup of Fig. 9 between 49.60 and 50.20 seconds. An occlusion is detected for ball 3 from 49.72 to 49.88 seconds. After 50 milli-seconds, an occlusion is detected for ball 1. Our system deals with such rapid change, since high-speed cameras are used.

Fig. 11 shows the generalized coordinates $\tilde{r}$ defined by (20). It is seen that the helicopter hovered in a neighborhood of the reference position. In particular, the z position is within 7 [cm] for all time.

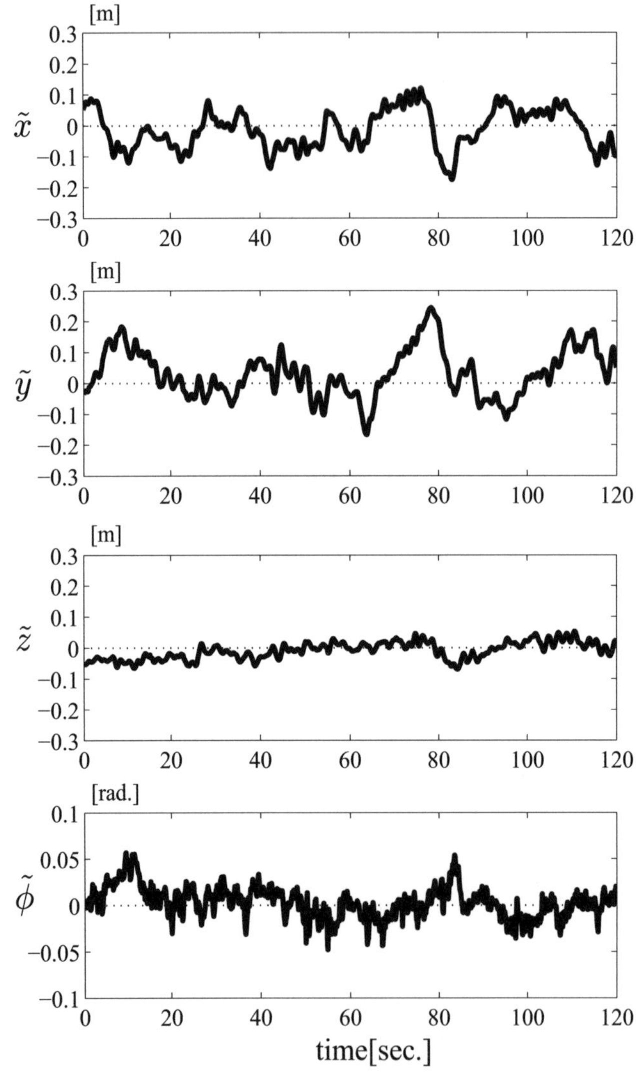

Fig. 11. Experimental result: Time profile of the generalized coordinates $\tilde{r}$.

Several movies can be seen at http://www.ic.is.tohoku.ac.jp/E/research/ helicopter/. They show stability, convergence and robustness of the system in an easy-to-understand way, while the properties may not be seen easily from the figures shown here.

7. Conclusion

This paper has presented a visual control system that enables a small helicopter to hover under temporary and partial occlusions. Two stationary and upward-looking cameras track four black balls attached to rods connected to the bottom of the helicopter. The differences between the current tracked object positions and pre-specified reference positions are fed to a set of PID controllers, when all the tracked objects are visible. If an occlusion is detected for a tracked object, the controller uses the errors given by the other three tracked objects. The system can keep the helicopter in a stable hover, and the proposed method is robust to temporary and partial occlusions even when a tracked object is not visible in any of the camera views.

8. References

Altug, E., Ostrowski, J. P. & Taylor, C. J. (2005). Control of a quadrotor helicopter using dual camera visual feedback, *International Journal of Robotics Research* 24(5): 329–341.

Amidi, O., Kanade, T. & Fujita, K. (1999). A visual odometer for autonomous helicopter flight, *Robotics and Autonomous Systems* 28: 185–193.

Ettinger, S. M., Nechyba, M. C., Ifju, P. G. & Waszak, M. (2002). Vision-guided flight stability and control for micro air vehicles, *Proc. IEEE/RSJ International Conference on Intelligent Robots and Systems*, Lausanne, Switzerland, pp. 2134–2140.

Hashimoto, K. (2003). A review on vision-based control of robot manipulators, *Advanced Robotics* 17(10): 969–991.

Ma, Y., Soatto, S., Koˇseck´a, J. & Sastry, S. S. (2004). *An Invitation to 3-D Vision: From Images to Geometric Models*, Springer-Verlag.

Mahony, R. & Hamel, T. (2005). Image-based visual servo control of aerial robotic systems using linear image features, *IEEE Trans. on Robotics* 21(2): 227–239.

Mejias, L. O., Saripalli, S., Cervera, P. & Sukhatme, G. S. (2006). Visual servoing of an autonomous helicopter in urban areas using feature tracking, *Journal of Field Robotics* 23(3): 185–199.

Proctor, A. A., Johnson, E. N. & Apker, T. B. (2006). Vision-only control and guidance for aircraft, *Journal of Field Robotics* 23(10): 863–890.

Saripalli, S., Montgomery, J. F. & Sukhatme, G. S. (2003). Visually-guided landing of an unmanned aerial vehicle, *IEEE Trans. on Robotics and Automation* 19(3): 371–381.

Shakernia, O., Sharp, C. S., Vidal, R., Shim, D. H., Ma, Y. & Sastry, S. (2002). Multiple view motion estimation and control for landing an unmanned aerial vehicle, *IEEE International Conference on Robotics and Automation*, Washington, DC, pp. 2793–2798.

Spong, M. W., Hutchinson, S. & Vidyasagar, M. (2005). *Robot Modeling and Control*,Wiley.

Watanabe, K., Yoshihata, Y., Iwatani, Y. & Hashimoto, K. (2008). Image-based visual PID control of a micro helicopter using a stationary camera, *Advanced Robotics* 22(2-3): 381- 393.

Wu, A. D., Johnson, E. N. & Proctor, A. A. (2005). Vision-aided inertial navigation for flight control, *AIAA Guidance, Navigation and Control Conference and Exhibit*, San Francisco, California, pp. 1669–1681.

Yoshihata, Y., Watanabe, K., Iwatani, Y. & Hashimoto, K. (2007). Visual control of a micro helicopter under dynamic occlusions, *The 13th International Conference on Advanced Robotics*, Jeju, Korea, pp. 785–790. Also, in Lee, S., Suh, I. H., & Kim, M. S., editors,

Recent Progress in Robotics: Viable Robotic Service to Human, pp. 185–197. LNCIS, Springer-Verlag (2008).

Yu, Z., Celestino, D. & Nonami, K. (2006). Development of 3D vision enabled small-scale autonomous helicopter, *2006 IEEE/RSJ International Conference on Intelligent Robots and Systems*, Beijing, Chaina, pp. 2912–2917.

8

Model Based Software Production Utilized by Visual Templates

Mika Karaila
Metso Automation Inc
Finland

1. Introduction

In the automation domain programs are written by engineers. Available programming languages are normally standard IEC 61131-3 or vendor specific visual language. Programming requires domain knowledge and programming skills. Reusing programs is often simple copy / clone a working solution. There are different kinds of solutions done to effective produce programs. In Metso Automation application programs are first modeled and second systematically reused. The principles are applicable to be used in other context.

2. Function block language

2.1 Introduction

The visual notation of FBL consists of symbols and lines connecting them. In FBL, symbols represent advanced functions. The core elements of FBL, function blocks, are sub-routines running specific functions to control a process. As an example, measuring the water level in a water tank could be implemented as a function block.

In addition to function blocks, FBL programs may contain port symbols (also called Publishers) for other programs to access function blocks and their values. The function block values are stored in parameters. As an analogy, the role of a function block in FBL is comparable to the role of an object in an object-oriented language. The parameters, which can be internal (private) or public, can, in turn, be compared to member variables. An internal parameter has its own local name that is not visible outside the program module. A public parameter can be an interface port with a local name or a direct access port with a globally unique name.

In addition to function blocks and ports, FBL programs may contain external data point symbols for subscribing data published by ports, external module symbols to represent external program modules, and I/O module symbols to represent physical input and output connections. An external data point is a reference to data that is located somewhere else. In distributed control systems, calculations are distributed to multiple processors. Therefore, if a parameter value is needed from another module, the engineer has to add an external data point symbol to the program. By using this symbol, data is actually transferred (if needed) from another processor to local memory.

From the FBL elements, the engineer can, for instance, build visual programs that control some equipment in a factory that is running the process. These processes are continuous and controlled in real-time.
Visual languages have been extensively studied in the literature (Mohamed, 2000, Burnett 1995, Shu 1988, Pressman 1997). As mentioned earlier, computer programs are usually written using textual languages, but in more sophisticated or domain-specific environments, programming can be done in a visual way, as in LabVIEW (Rahman, 1995). LabVIEW is originated in 1986, while the roots of FBL go back to 1988 (Karaila, 1989). FBL is not a standardized language as IEC 61131-6 language.

2.2 Background

In the late 1980´s the first implementation was done for FBL. The first target was to replace a textual programming language because graphical documentation was already at that time one of the customer's requirements. FBL was successfully taken into use and there were only a few programs that were written in textual format.
One of the most important design goals was to design both the programming environment and FBL for extensibility. This means that developers could easily extend the visual language by adding new graphical symbols to it. Such new symbols, for example, may represent new types in this strongly typed language. In fact, in FBL, users can add new symbols to the language even without adding any new code in the programming environment. The reuse of visual code in an integrated programming environment is powerful and efficient. The same kinds of notifications are done (Debbie, 1995). Developers have implemented an engineering environment that allows extensions and integration of third party tools. Further, new symbol classes or categories can also be added to FBL. This, however, requires modifications to the programming environment. Usability is important to engineering efficiency. For cost effectiveness, using a commercial solution was a good way to share code maintenance costs. As a drawing editor Metso has used commercial CAD program, which can be AutoCAD® Copyright 2009 Autodesk or BricsCAD Copyright 2001-2009 Menhirs NV. Both can be used for that purpose. In this way, developers were able to focus our own work on the application domain instead of graphical editor issues.

2.3 Main design goals and principles

Developers had the following goals in the development of FBL and the programming environment:

- Basic product configuration and a tool for customer projects.
- Both FBL and the programming environment must be flexible and possible to extend because it was known from the beginning that new features are coming/needed every year.
- Maintaining the language should be feasible, and adding new types and functions should be easy.
- Easy to use, because typical users have minimal programming skills.
- Easy to reuse written applications, because customer projects are very similar.
- Third party tools and products should be easy to be integrated with the programming environment.

FBL can be used to program basic automation and advanced quality controls. Metso's engineers can implement different kind of applications with FBL. As the amount of different

sub domains are integrated into FBL, the use of FBL is growing. Our customers will maintain and modify those FBL programs. Customer's people are typically automation engineers. They will come to the FBL training. They are responsible for maintenance and process design. They usually do not have any programming experience. Most of the time goes into environment training and main principles of the automation system. The FBL language itself is not so much used, only a few programs are made during the training that is typically one week long. This is one way to evaluate the learning curve of FBL. There are other studies about advances in data flow programming languages (Johnston, 2004). These indicate the same findings as developers have experienced such as, 'The data flow semantics of visual programming languages are intuitive for non-programmers to understand and thus improve communication between the customer and the developer'.
Design principles of the language are briefly summarized next.

- In the visual drawing, symbols used should represent both data and functionality. There will be an artifact in the system that can be mapped into a symbol. So each symbol will have some meaningful concrete function or element in the system. There will be very direct mapping from the eq. IO card symbol to a program physical IO card that will run a real electrical connection.
- Symbols are for creating communication to transfer signal data. One symbol that contains an output and can be connected by line to another symbol input to represent data-flow. Data-flow will be in this way explicit.
- Layout should be organized so that inputs will be on the left and outputs on the right. There will be immediate visual feedback during testing program values can be visualized.
- All of the above will create a combination that merges algorithm and user interface to one functional entity.

These four strategies: concreteness, directness, explicitness and immediate visual feedback are listed in (Burnett, 1999).

2.4 Basic symbols

Function block language contains thousands of symbols. The following is a categorized list of basic symbols:

- Administration part symbol for defining purpose of the diagram,
- Function part symbol for defining CPU and execution parameters,
- External reference symbol for transferring data outside module,
- Local data symbol for allocating memory for temporary signal data,
- Port symbol for defining access name for external reference, and
- Function block symbol for making signal operation / handling / calculation.

Basic symbols are just for data (memory location) and function block symbols with numbers are functions that are executing algorithms. Language is not making a memory location or register references, instead that is actually done in the program loading phase into execution. Binding is done as late as possible.
Administration symbols contain metadata about the program like process area, short description of the program and customer logo. The program itself is drawn inside the frame of the administration symbol defines. There are different sizes available and the program can be extended to multiple pages. Signal connections between the pages can be created by reference symbols.
Functional administration symbol defines execution interval and logical location in the system. This symbol is used to define a new module.

A port can be either an interface or a direct access port. Interface port name is a suffix for the name of the module. Direct access port name is a global name that must be unique in one system (factory level). Port is an access point to a memory location with the name.
External reference is in our terminology an external data point. It contains a name and communication parameters. In the principle name is a reference to the port, which is a named memory location. According to the communication parameters, data is transferred from the port and updated to an external data point. In this way communication takes care of values.
Local data point is inside the module and is needed only to store values between function blocks. It can be needed for storing a value between calculation function blocks.
Function blocks in Figure 1 encapsulate actual subprograms. Encapsulation protects memory allocation and safe execution. Function block always uses the same amount of memory. Execution is controlled by execution order (number between 1--9999) that is given for each function block symbol. All function blocks are sorted and executed in given order. Function block contains inputs, outputs and parameters. Inputs are read before the execution and parameters are used for the calculation and after execution outputs are set. In this way, users can only use these building blocks to define their own program.

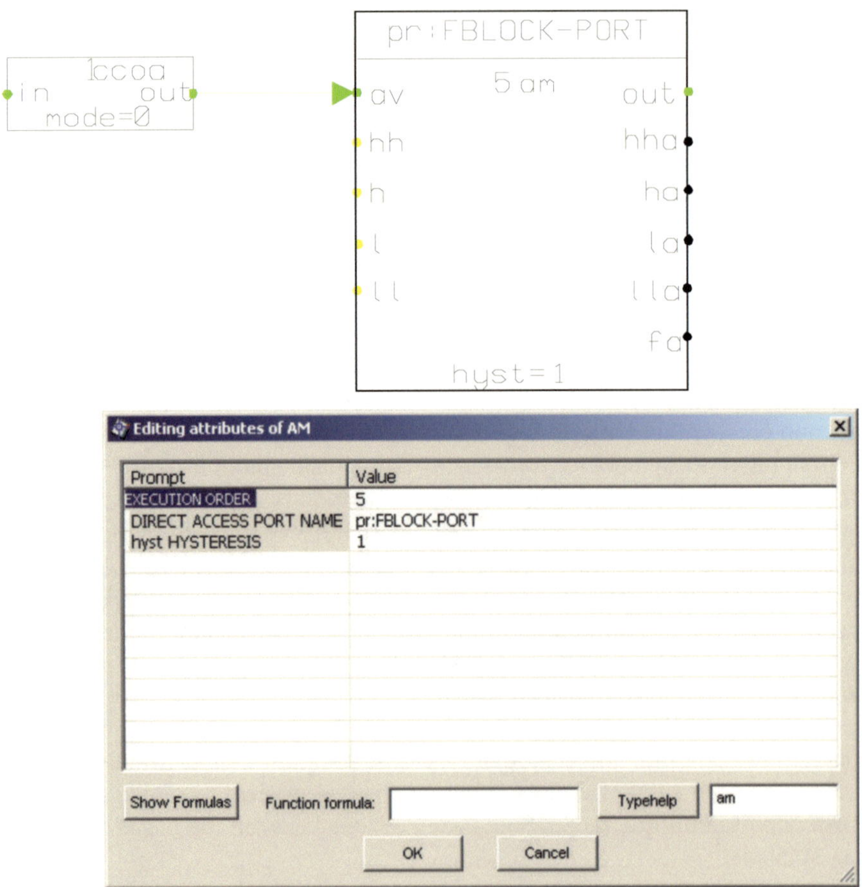

Fig. 1. Two function block symbols with the am symbol's parameter dialog.

Common function blocks are pid for controlling, logical and/or functions for boolean algorithm and calculations. Basic system function blocks are copy (ccox), select (disx), analog measurement (am, am2), binary measurement (bm) and device specific blocks like motor (mtr, mtre, mtr2) and valve (mgv, mgve, mgv2). More application specialized function blocks are for enthalpy calculation went and steam flow calculation (stfl).

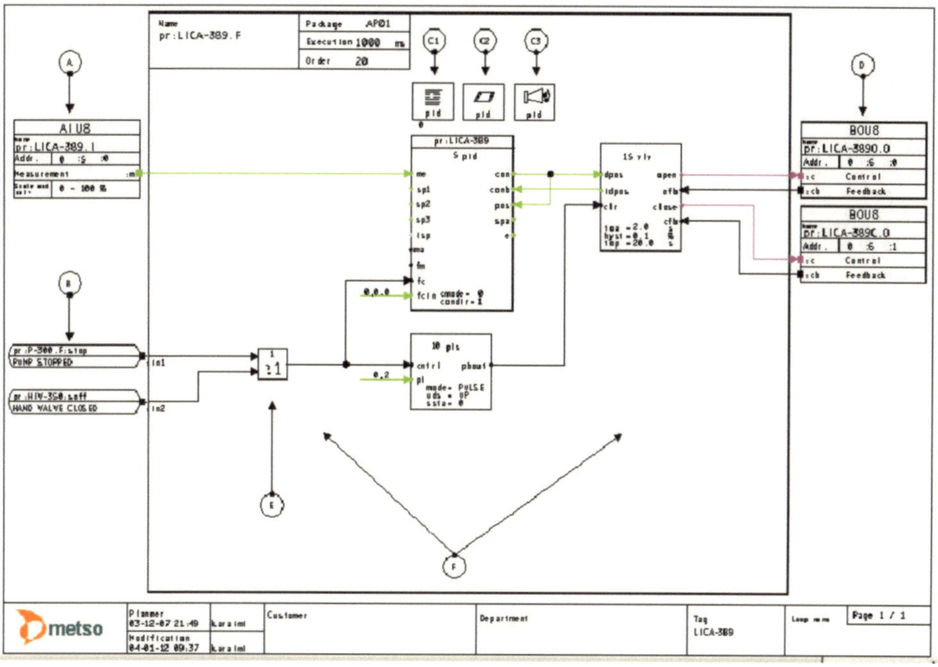

Fig. 2. FBL control loop program.

Figure 2 shows an example FBL program. Symbols A and D are standard input/output (I/O) symbols. Symbols C 1-3 contain texts and other operability and alarming parameter definitions (as priority and alarm group) for the control room functions. Operators in the control room look after the process status from the monitors. The process is constantly measured and run by the programs but people are still making decisions and performing actions (pushing buttons) to control the process. Symbol C3 is for alarm functions. Finally, F is the area for the actual control program. All other symbols representing function blocks and connections are in the same program as the other symbols are building their own individual programs. A function block is a basic subroutine running a specific function to control the process.

The graphical layout is to be read from left to right: inputs are on the left and outputs are on the right. Figure 2 represents a typical automation program in size and functionality. It gives a good overview for the user of one functional entity. The symbols inside one diagram are connected by lines, while connections outside one diagram are constructed using symbols that contain reference names, as shown in Figure 1 symbol B.

Figure 2 shows one Function Block Diagram that can be used to generate multiple textual files. Those files are from a one-page program to several pages long; each file is an individual program. In addition, variables that are connected by lines in a FBL program are stored in each file. Program modules are distributed in different places in the system. The Process Control Server (PCS) runs I/Os and control programs. Operator Server (OPS) and Alarm Processor (ALP), in turn, run other configuration functions. For example, in the control room OPS is for Human-Machine-Interface (HMI); the operator can change displays and look at different parts of the process and manipulate control parameters from the monitor windows).

2.5 Module symbols

FBL module symbols are application programs that can be distributed in the system. As an example, the I/O- symbol generates a small application program that can be loaded to the field bus controller. It will load needed parameters into the I/O- card and transfer data from the I/O- card to the field bus controller that will communicate with the actual controlling CPU unit that runs function blocks. In the same way the gateway symbol that connects an external device to the system using communication protocol is loaded into the CPU unit that has a serial or an Ethernet connection.
Symbols for creating a connection can be divided into two major groups:

- I/O-symbol to connect a physical field device. I/O card makes analog/digital transformation to an electrical signal.
- Gateway-symbol to connect a software component to another system using communication protocol.

Different kind of I/O-symbols are available, they represent I/O-card. It contains parameters like I/O-address, filtering and other signal processing parameters. Gateway-symbol contains address for accessing data through software protocol. The physical connection can be Ethernet, RS-485 or RS-232. The address depends on used protocol. In MODBUS (MODBUS) protocol addressing is register-based (address format examples 'reg 1001' or 'dw 10'). Signal data-flow is coming in principle the same way as with I/O-connection. The interface module is executed by the driver and the actual data is connected with the external data point to transfer the data from the driver to the application program.
The wiring from I/O-card connections to the field device connects signal flow electrically. From the I/O-card the signal is processed digitally and field bus transfers data between the I/O-card and CPU unit. This is physical distribution and the signal route is illustrated in Figure 3.
Module symbols are usually for defining parameters for user interface and alarm handling, like texts, alarm priority and alarm area. These are loaded to all operator stations and alarm servers.
These module symbols are used for defining

- Text data for user interface,
- User interface panels,
- Alarm handling parameters,
- Long time history data collection parameters, and
- Feedback simulation (action response in virtual environment).

These application programs listed above are not connected by lines as function blocks are connected. The connections are fixed and the user can give one connection name that creates all other needed connections as external data points. This reduces the amount of lines in the diagram. They are usually located near the corresponding function block symbol they are referring. Reference is done by using the same names in the symbols.

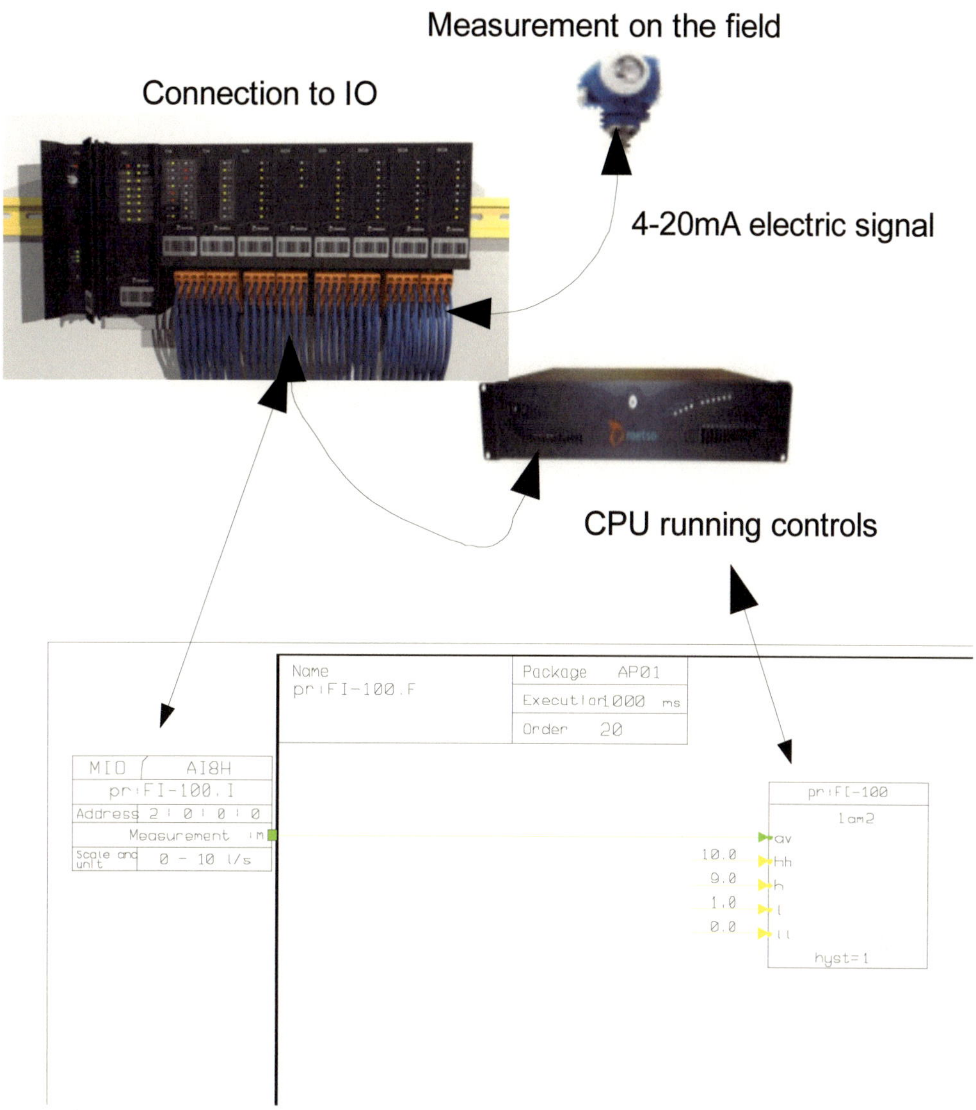

Fig. 3. I/O-signal data flow from the measuring device to the controlling CPU.

2.6 Connections and networks

The connection networks can be very simple point-to-point connections or very complex networks. The network structure solver will take all network connections together and find out the target connection. The target connection is the connection target for the rest of the network participants. In other words, the connection target is the named memory location that others will use.

Some examples of connection networks are shown in Figure 4:

- Point-to-point connection, where output is connected to input.
- Multiple connections, where lines can be connected together with a connection dot that will join underlying lines and creates a connection junction point.

* Connection references, where lines can be connected with symbols that contains reference name from other pages to the same logical connection network.

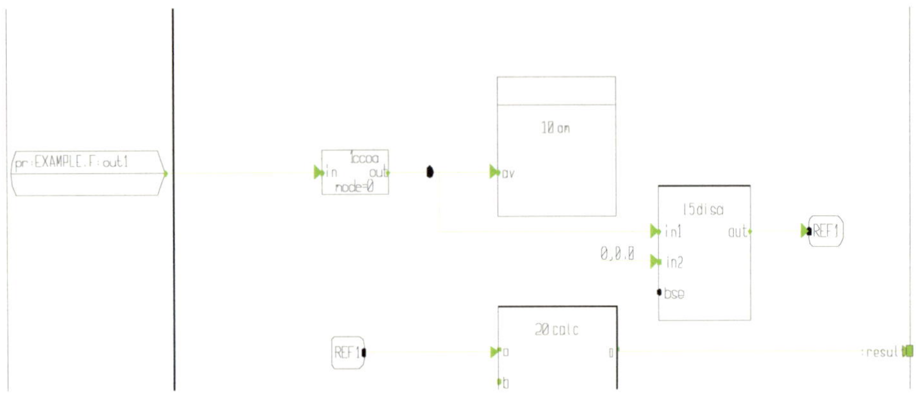

Fig. 4. Connection network examples.

Connection resolving must first always create the whole network from the sub-networks. After that it can run through the connection algorithm that finds the connection target. This is a very simplified explanation for the whole underlying system that contains a lot of specific rules for connection solving.

2.7 Strong typing

The system is strongly typed and simple basic types are represented by fixed colors. Only the basic and most common types are with color. Having too many colors would make it difficult for the user or programmer to distinguish the different types based on color (Whitley, 2001). Further, the benefits of using colors are diminished when printing the programs using a black and white printer; only some grey scales are available in that case or in some cases different line styles are used to indicate signal types (like dashed, dotted etc.).
Colors are used in connection points and connection lines. Color defines the type of signal data. Basic types are with color in the following way:

* Green (ana): indicates two values, value (float) and fault bits (uns16)
* Black (bin): indicates a true/false bit (bit 0) and fault bits (bits 1-15)
* Brown (binev): indicates bin and time stamp
* Blue (intl): indicates long integer and fault bits
* Cyan (ints): indicates short integer and fault bits
* Magenta (bo): indicates bin and pulse time (time)
* Red (fails): indicates fault bits (uns16, bit 1-15)
* Yellow (float): indicates plain real number (float)
* Gray (any): all other types (less used misc. types)

Note that the above are scalar types / array \& other multi--dimensional types are drawn by a thicker line but with the same color as the element type of the vector / table.

The user can draw the connection line freely by routing the line and then the program creates the arrow-head automatically at the end of the line to represent data flow direction. Connection lines can cross and if they are connected there is a connection dot in crossing that will connect signals together. In addition, there are special data types for the communication. The function blocks are also based on types that are composed as structures.
At Metso we have developed our own meta-language for defining all the needed structures. Types and also more complex structures such as function blocks are defined with this metalanguage. This metainformation is available from the type database. This can be used to build function block symbols with default layout. Default layout is to place inputs on the left side of symbol, parameters in the middle and outputs on the right.

3. Template mechanism of Function Block Language

3.1 Introduction

Domain specific modeling is used in different levels in FBL. All the function blocks are small models that reflect real physical devices or some needed functionality. A motor, for instance, is modeled as a function block named mtr. The same model can be used for all basic motors and pumps. Similar way valve model is a function block named mgv (magnetic valve). In this way, function blocks are created to solve basic problems in the domain; the name of the block is the name of the focused object. Function Block can be parameterized and connected to other FBL elements. It will read inputs, run itself according to the parameters and write output values. FBL also contains elements that are for user interface and alarm handling. Modeling hides many complex operations.

3.2 Meta template mechanism

Our solution is to use visual templates for efficient programming (Karaila & Systä, 2007). A visual template can e.g. be used to implement motor control. The motor template will contain a set of parameters that are used to create an application program instance.
The engineering tools and database separate data and presentation, Application has a presentation role and actual parameter data is in the database. Transformation attaches template and the result is the implementation. This mechanism works in the same way as in the web applications. The Excel integration gives an effective way to modify existing data in the database. For version upgrades it is possible to export data into one's own XML file. These facts are behind the optimal combination of FBL and framework to maximize effective programming.
Templates are used for example in C++ programming language and in web applications. C++ templates are considered 'type-safe'. The FBL template engine differs from traditional template engines because the FBL template is evaluated immediately in design time. C++ templates are expanded at compile-time. FBL templates can be parameterized using database interface and this kind of principle is also used in web applications. Many languages that are used in web programming like Java or Python have own template engine. These kinds of web servers use primary data from the database and produce interface as shown in Figure 5. This makes effective separation between the business data and presentation. Data can be easily maintained and presentation can be modified. In this way they are loosely coupled.

In the same way FBL templates have parameters in the database and the FBL template contains transformation information. In traditional C++ programming, people use a Standard Template Library (STL). Web based templating testing needs to a run generator to check the end result. In the same way in using STL, compiling is needed to validate the template. In the FBL, template functions are evaluated immediately and transformation is made.

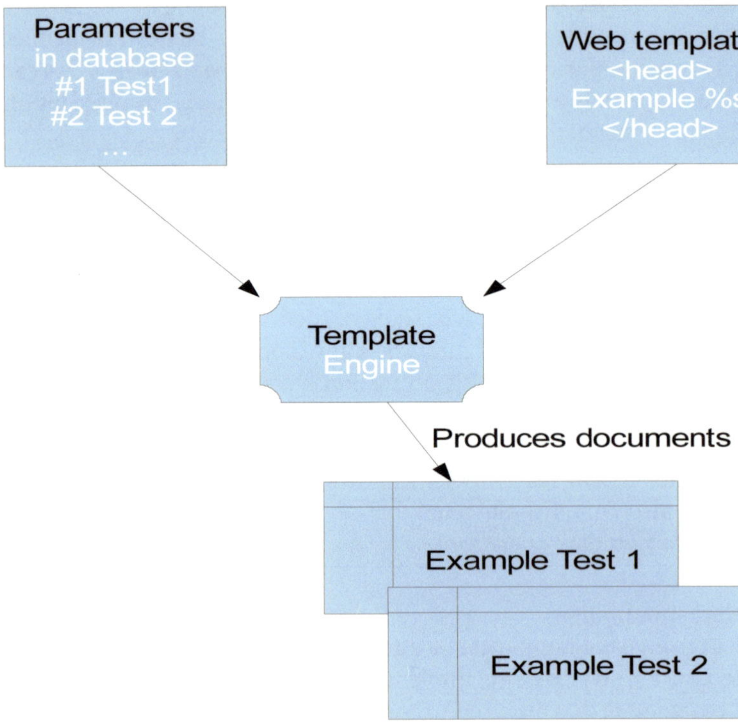

Fig. 5. Principle of web template engine.

Static metaprogramming (template metaprogramming) techniques in general are used to enable the customization of programs at compilation time. For instance, compilation of a program for different platforms can be made easier with such techniques like using generative programming (Czarnecki & Eisenecker, 2000). Static metaprogramming may, however, also be rather challenging. E.g. debugging is typically difficult due to the lack of proper tools. This, in turn, challenges the testing of static meta-programs. Processing and evaluation of template codes at compile-time causes an overhead, which, however, could and desirably does make the executable code more efficient. This overhead might have some significance in larger projects but is typically insignificant in smaller ones. In addition to efficiency, template meta-programming techniques support genericity and facilitate code minimization and maintenance. This is because the programmers can focus on designing and implementing general, perhaps architecture-level structures. FBL templates are used to define a common program structure for a family of application program instances. The templates are further used to create these instances which are called control loops in the terminology of the domain. One template can be used to create several program instances, up to 100 in practice. Each instance has its own identifier and parameter set. The program structure which is derived from the template is the same in each program instance. In essence, FBL templates are programs that contain data structures and encapsulated functions. Templates are built by first defining parameters that can later be used as an interface to create an instance from the template. Templates further contain formulas, in which the parameters are used. Evaluation of the formulas is automatic. In some cases, the evaluation may modify the program structure, as in conditional compiling, as a result. Formulas are used in FBL templates for evaluating mathematical expressions and for concluding logical truth-values. Each formula is a mini-language statement. The mini-language used is a simple language without real programming capabilities. For practical reasons, e.g. for easy editing and understanding, the mini-language formulas and expressions are compact and fit in one line. FBL language is generative and each template is actually meta-programmed using the mini-language.

Larger models are for modeling more complex functions that need more connections and generic parameters. These connections are to other modules and ports in the system. Parameters are model specific and can be used in multiple elements.

Our engineering tools and FBL editor are main elements in a DSM environment. FBL editor is used for model building and testing. Engineering tools are for managing templates and instances.

3.3 Working with templates

A template is a key component for effective software production. As an example, a basic measurement is needed in every project. But the measurement can be a temperature, a pressure or a level measurement. There is some variation between the measurements like the measurement range is different as the unit depends on physical measurement. The program has input with an address and a range with a unit. The alarm limits of the measurement can be set in programming phase to some initial values. The basic analog measurement template is the model that solves this problem. A template contains the model that can be parameterized and the instance is varied by these parameters. One measurement template can be used in all these different measurements if there are no other requirements.

In practice, a visual template is built with an FBL editor. It contains commands for creating a template. The next step is to make first a program that will contain all other needed parts.

After that, templating can start by the following steps:

- Create design members, these are parameters for a visual template,
- Define needed formulas, these use parameters defined above,
- Save a template, and
- Create an instance and test it (modify parameter values).

First, the user defines all the parameters needed. This can be done using a specific dialog shown in Figure 6.

Parameters work like a placeholder and follow the same syntax rules as Python variables except that they are preceded by $ enclosed in {}. Parameter example: ${var}. Parameter identifiers are case sensitive.

After this, the user can define the formula like in Excel to a separated field that will store the formula as shown in Figure 7. In the evaluation phase the formula is evaluated and the result is placed in the actual value field. The engineer can already see the current value that is calculated from the design parameter value. Formula evaluation is automatic and it helps the engineer to always see evaluated values.

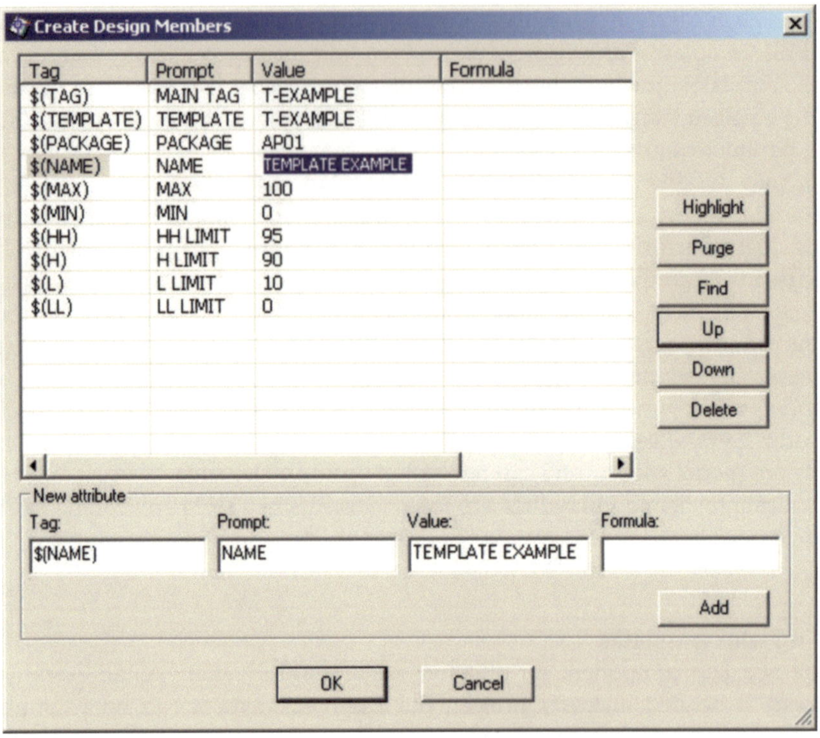

Fig. 6. Step 1: User defines first design parameters.

A complete parameter use example:

- Parameter identifier: \$(MYPARAMETER)
- Parameter value: Example text
- Usage: External datapoint, comment attribute
- Formula field: Test \$(MYPARAMETER)
- Comment field: Test Example text

After step three, template saving, the engineer can create a new FBL program instance from the template shown in Figure 8. Usually new instances are created by using Excel as a parameter entry interface. Template testing always needs multiple instances because otherwise there can be some non-formulated value or wrong formula that will create a non-unique identifier or overlapping address definition.

The FBL visual templating is implemented by mini-language that needs minimal programming. It can be extended when needed but the current functionality has been enough. Using these functions enables the user to meta-program FBL.

Template directives / functions are listed below. Some of them are domain specific.

- eval formula
- mathematical formulas
- strings and parameter value
- function-formula (conditional part, works like snippet)
- value reference (syntax for parameter, reference to outside)
- select formula
- prefix formula (special string handling with prefix)

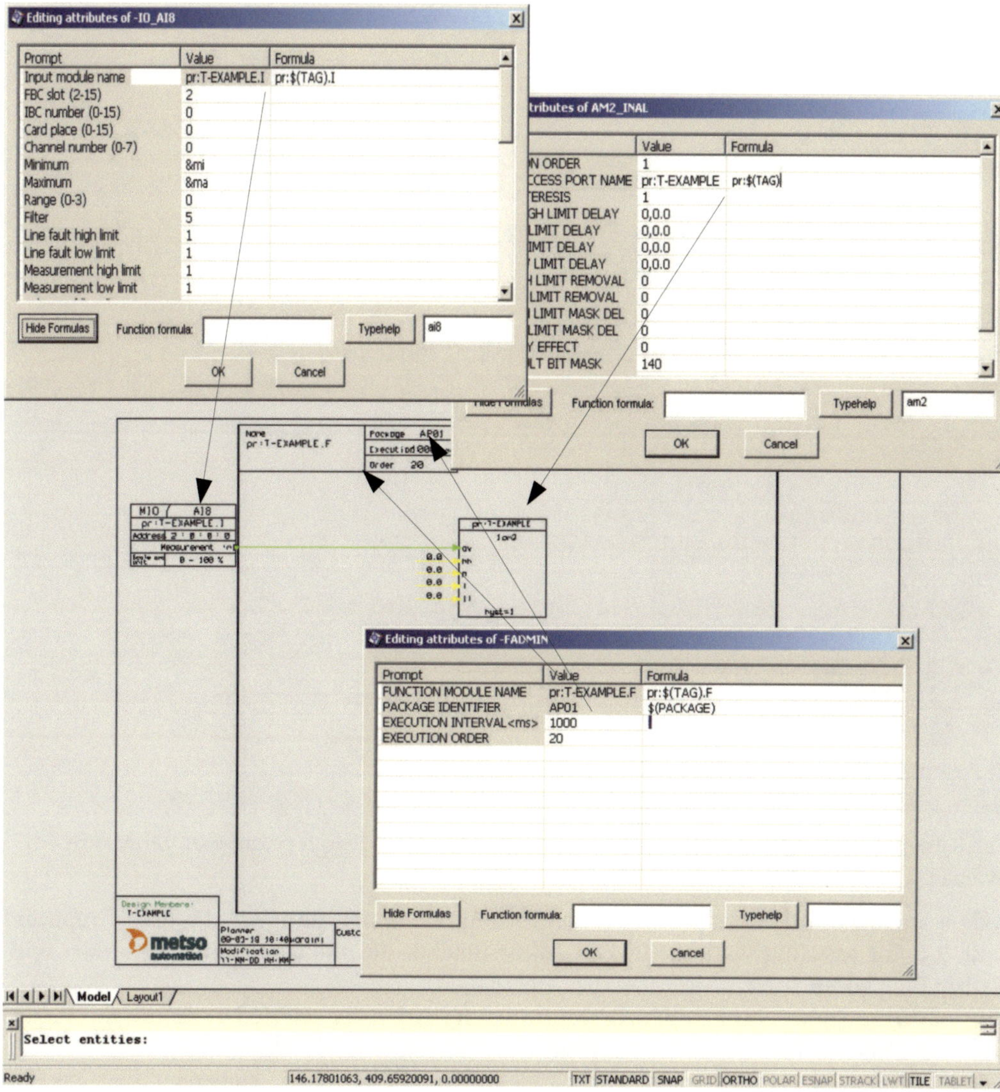

Fig. 7. Step 2: Formulas are defined in each needed location.

Eval is used in formulas to mark parts that will need mathematical evaluation. Otherwise all variables are evaluated as strings.

Mathematical formulas are evaluated according to standard evaluation order. Most of the basic calculations are implemented into the library.

Strings in the evaluation phase are replaced and formula evaluation result is in the value field. Value field is usually a symbol's attribute value but it can also be a comment text.

Function formula works like a snippet. Ordinarily, these are formally-defined operative units to incorporate into larger programming modules. In a visual template, function formula is always included into the template. The "code" amount is fixed but the connections and all parameters are evaluated inside elements belonging to the function formula. It can be turned on or off by a conditional statement. If the result is true, part of the

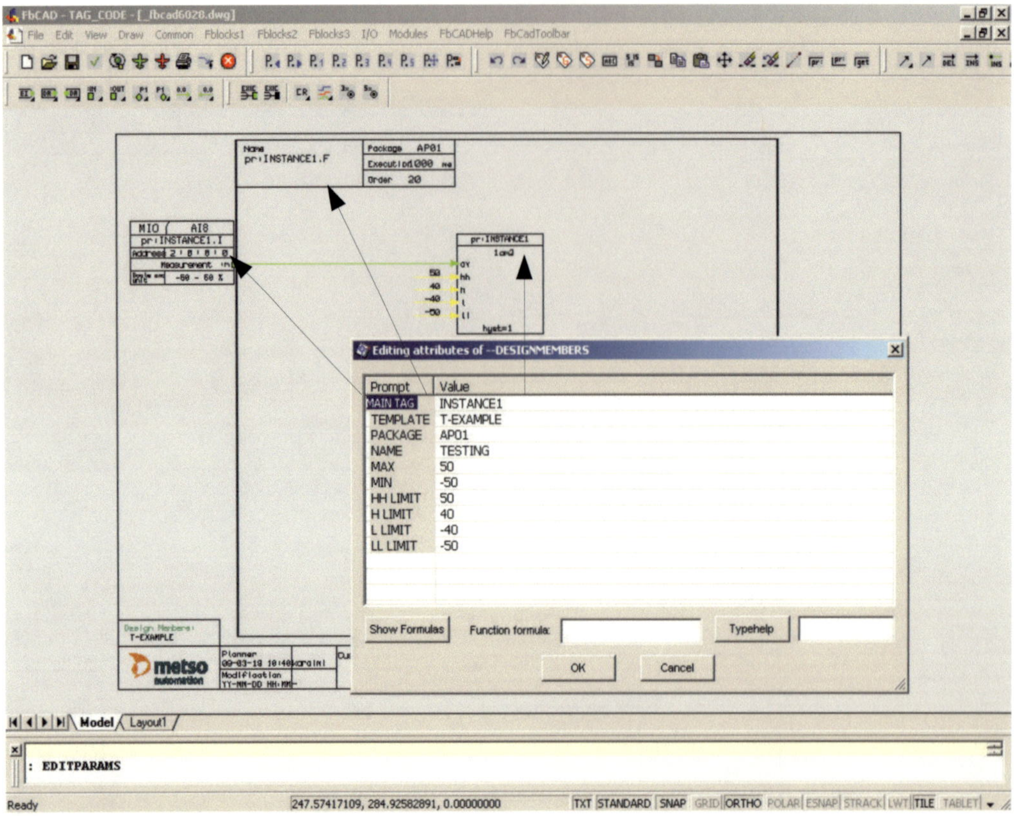

Fig. 8. Step 3: Testing template with new values. Modified design parameter values are evaluated and new values are visible in the diagram.

code is included, otherwise not. Function formula does not minimize the use of repeated code it is for selecting features. In FBL editor function formulas are usually marked with dashed blue boxes.

The following Figure 9 shows function-formula definition for selected elements and Figure 10 demonstrates action that hides a snippet.

Select formula can be used as 'switch...case' or 'if...then... else...' statement for selecting another value by given value. This is a kind of enumeration based transformation.

Prefix formula is used to minimize entering the full reference name. In automation domain, devices are named and in the programming phase it is easy to use a pure name without any prefix or suffix. This abstraction removes / hides programming details from the user.

In step one, shown in Figure 6, the user must first define design parameters that can be used as variables in formulas. Mandatory parameters are:

- TAG (instance identifier),
- PACKAGE (logical name for download target) and
- TEMPLATE (template identifier).

Usual parameters are MIN, MAX, UNIT, HH (high high alarm limit), H (high alarm limit), L (low limit), LL (lower low limit) and so on.

In step two, the user can look at properties of the symbol and add their own formula to calculate a new value.

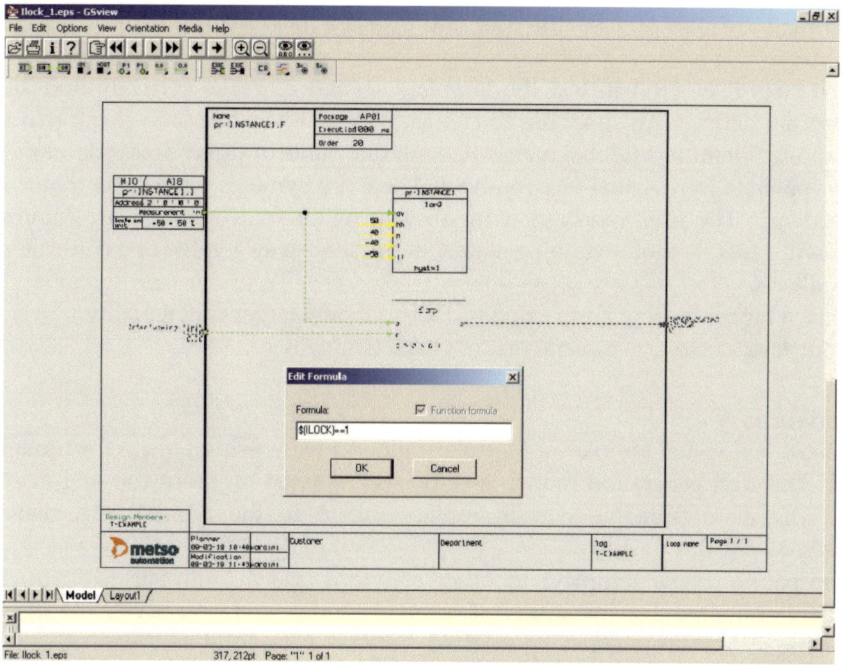

Fig. 9. Symbols are selected & active. Function formula defined for selected elements (lower function block and connections into it.

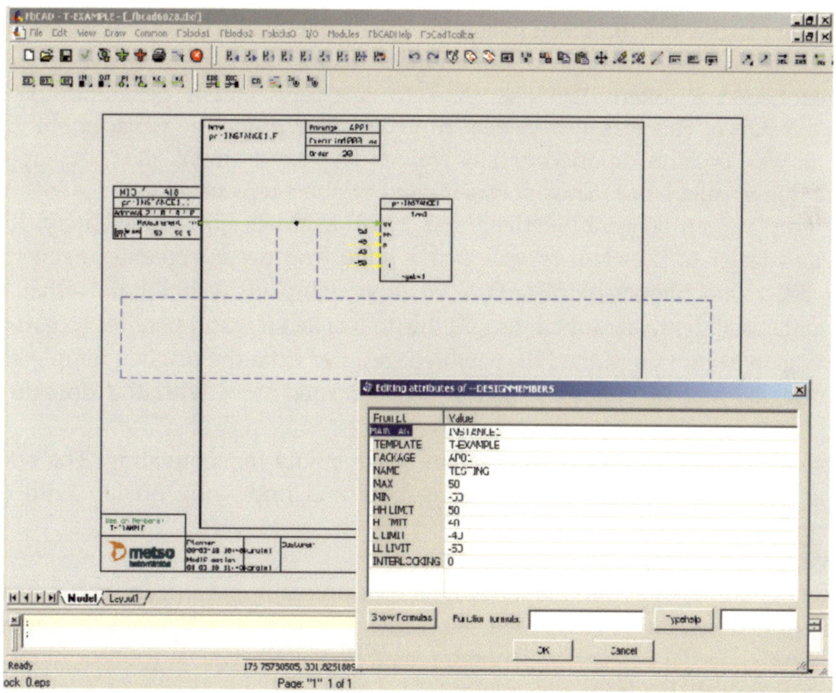

Fig. 10. Function formula 'hides' interlocking elements with the value 0. Elements can be activated with value 1 back to the diagram.

In the template creation process, the user has to save a diagram as a template into template storage.

In the last step, it is good to test the template so that it works correctly and all needed parameters are defined. The user has to create at least two instances to check that there are no overlapping identifiers (global names like module name or direct access name).

Testing is possible in a virtual environment. There are symbols for each actuator to create action feedback. The user can have a motor that will start from the start command and feedback will generate motor running status. In the same way a valve or a controller will get action feedback.

In this way, a higher level of abstraction is done to model larger functionality.

For this purpose Metso has implemented a visual template.

3.4 Experiences

Before Metso had visual templates, Metso's engineers were using typical for modeling FBL solutions. This first generation model is static and is based on more copying existing FBL diagram. The main principle was to replace tokens in the typical with real instance parameters.

When comparing visual template to other solutions, visual template is interactive and immediately evaluated. For instance, it is faster to modify and test. Before the final testing, the following actions are needed: specialized instance, compiling and loading into runtime environment.

Like in other 'Little Languages' (Deursen, 1998) visual templates contain small language, but gives an effective way to use metaprogramming.

The earlier way to create specialized instances was taking more time. An older template was named typical. A typical contained replacement tokens. Each parameterized value field actually contained a token. The user had to run replacement generation to get the specialized instance. This was always needed to test the typical. The replacement token was lost and it was possible to modify any value. The replacement did not support any transformation or calculation. Thus, it was limited to direct replacements.

A visual template can be parameterized and it will evaluate FBL immediately. It is more dynamic and faster to use than typical that is static and needs separate regeneration for updating FBL. One important difference to other template techniques is that the FBL instance contains all template functions and due to this fact it is still possible to parameterize again and again even though the FBL is edited to differ from the original template. Typical did not offer all the functionality that is implemented now with the domain specific formulas.

Mass production of FBL programs is the key productivity for templating. The new visual templating improves productivity by saving time and improving quality with standard project templates.

Productivity is measured in many places:

- Project department measurements (annual measurements existing, over 10 years).
- Value Added Reseller (VAR) partners, specific process area: 100 templates enough.
- In general, over 90 percent of programs made from a template (project library makes automatic calculation from each project).
- Excel or sheet as main parameter input method (data and implementation can be separated; engineering tools can separate data from implementation).

Applicability to domain and product family principles is very good. Existing loop can be turned into a template by a few steps. Template programming adds variables and additional function into existing FBL diagram. Template programming is interactive and the user can immediately test functionality.

In other template based languages, a template is separated and needs rendering / generation that will create an instance from a template. This requires extra maintenance. In our domain, instances contain all template formulas. This is a benefit for us even it can be in some other domains a disadvantage. The framework allows template changes / updates so that it keeps all matching parameter values untouched. This flexibility gives the freedom to change an original template and update it afterwards for all needed instances.

The instance of the template can inherit values from another instance by a reference formula. This reduces the amount of parameters that the user must enter. Referenced template parameters are read--only values. A value change in parent instance is propagated into all children. The purpose of the feature is to reduce parameter amount and automate parameter value propagation. As an example, one design parameter contains text that is used in the primary loop, but the same text is also used in its own history collection definition loop. In this case it is easy to make reference from a history loop to a primary loop. An engineer can change text in the primary loop and it is automatically propagated into the history loop. And in the history loop, an engineer does not have to enter text anymore. An additional positive effect comes to maintenance. It is better to split functionality into its own features and bind needed parameters together by referencing. For us, our FBL and its metaprogramming support makes visual templating a practical reuse technology.

End customers are becoming more demanding.

- Easy and fast to create from specification to template and implementation. Specifications are coming later and later. Or in some cases the customer or process expert defines automation functionality at the factory in the start--up phase.
- Easy to make modifications and take those into use just by changing or updating the template.

Even through the template functionality has been in existence now for some years there is still work to do with usability and metaprogramming. There is the need to teach this technique. The conversion tool will need some tuning even it can transform an old typical to a template.

Time will show the life cycle of the templates. There have already been cases that the project is first done with templates and delivered without the formulas. This kind of downgrading is sometimes needed to support old installed systems.

4. Reuse mechanisms

4.1 Introduction

Support for software reuse can be hard to utilize. Systematic reuse will require process, analysis, feedback for continuous improvement and knowledge management.

Traditional software reuse can be implemented by components and libraries. In the similar way FBL contains build-in functions that are Function Blocks. These are documented in system manuals and are used to implement application programs.

For effective application programming, the solution is to reuse application programs. It is harder because they do not usually contain extra documentation or they are not categorized into any hierarchical structure like build-in Function Blocks are in the libraries. The system

level reuse also actually exists in product level because the automation system is based on a software product line (Ommering, 2005).
Another need to reuse already made projects is to estimate the effort needed to implement the same kind of project. A project can be a part of an earlier project like just one or two process areas are similar in a new project. 'Similar' means that the process area like "Fresh water treatment" is implemented with the same process equipments and can be used in the new project as a starting point. This kind of search and pre-study is needed and used in our sales. If there is existing the same kind of implementation, project engineers can start redesign using the existing implementation (Karaila & Leppäniemi, 2004).
In FBL, three types of reuse occur, in three abstraction levels:

- Level 1 Function Block (system level),
- level 2 Template (model reuse), parameter reuse between the template instances, and
- Level 3 Function Group (model group reuse, higher abstraction level).

The modeling is more demanding than the system level reuse. The user has to first select the template which is not always as clear as selecting a function block. The basic level function blocks are documented and always available. Templates are currently documented only in intranet level and loaded separately as their own library.
In a search for finding a possible template, there are parameters that can be used to narrow search results. This needs domain knowledge. The reuse library offers all parameters and allows the user to use those in search criteria.
Another reuse level is to reuse just parameter values. This can be done in the template level. The parent - child parameter referencing helps to maintain consistency between the same problem entities that is implemented with multiple instances. The main instance, core loop contains all common parameters like name and alarm area. Each child is referenced into those common parameters. In this way, a change in common parameter is propagated into each child instance.
Function Group level utilizes the next level in abstraction hierarchy. Function Group can handle a set of instances that are template based in one Function Group diagram. Function Group diagram visualizes connections between the application programs.

4.2 Reuse in practice

Project library application search dialog in Figure 11 is the starting point for reuse. The search interface allows users to search application solutions according to saved metadata and performed analysis. The search can be focused on certain process areas and projects. More detailed search criteria can include e.g. the main function of the program (function block like pid-controller or motor controller), the IO card type used and the application creator.
Application data is shown in Figure 12. The general part contains metainformation about the project and program itself. The entity count, primary function block, template generated information and user question count are created in the analysis phase. The IO data is also extracted in analysis. In the file information fields, data is needed to access file and template. The template match is in this case 100\%. When no structural changes between the template and instance exist the match value equals 100. That is, only different parameter values may exist. Each structural chance diminishes match value by a certain amount. For example by deleting and adding one symbol the match value is decreased by two to 98.

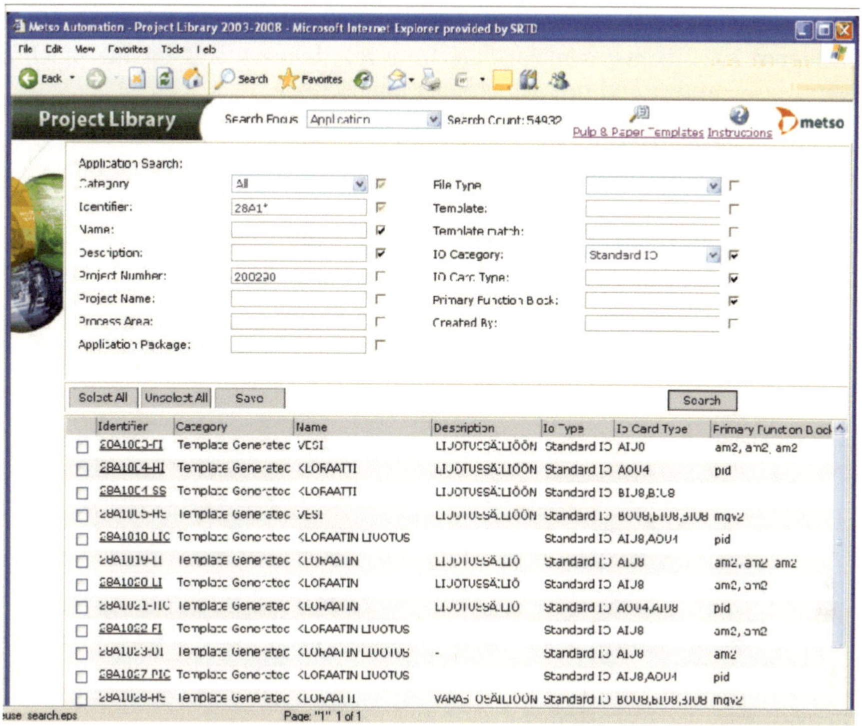

Fig. 11. Reuse library search dialog.

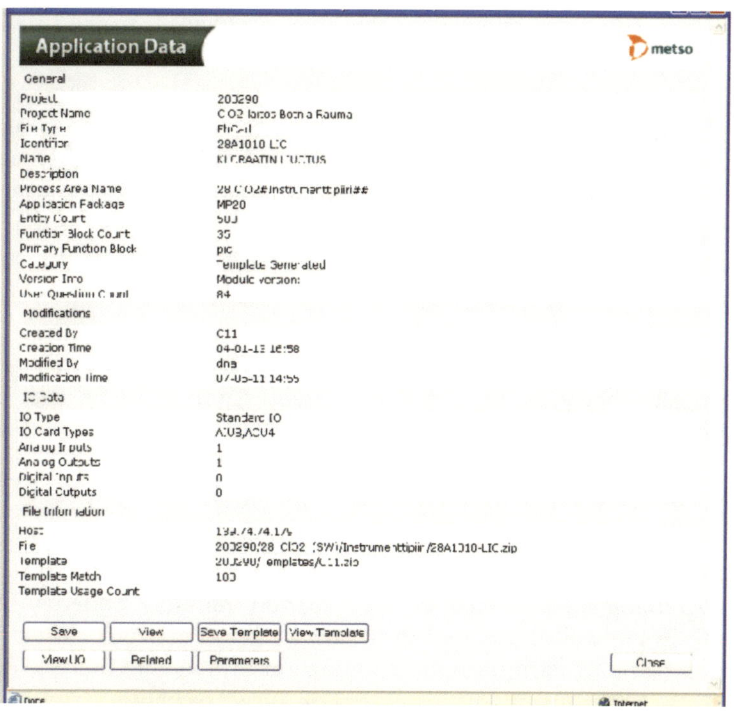

Fig. 12. Reuse library shows application data.

The search can be also focused on project, process area or template. The project data is shown in Figure 13. It contains major data from the delivery and for the practical reuse project team, main process and process supplier are needed.

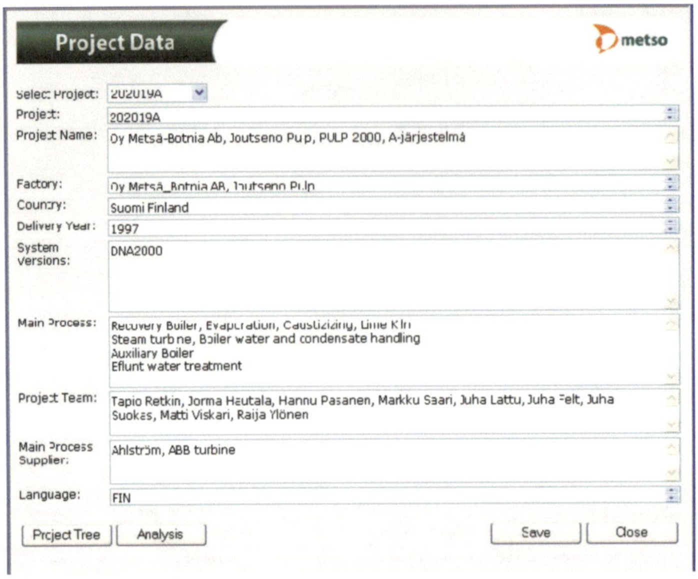

Fig. 13. Reuse library shows customer project specific data.

The user can search and navigate from the application to the template or to the related loops. User interface supports downloading multiple files together in one zip file.

4.3 Analysis

In reuse library, saving application will run FBL analyze that first creates a fingerprint from each application program. Fingerprint is a calculated value from the diagram entities. It is used to find similar diagrams faster. If the instance is template based, analysis will create a link to the template. In this way user can get the template easily. The project analyze will calculate summary information from the project. This information is used in estimating the project efficiency. Later the same information can be used to sell a new project. This makes better accuracy for estimating the cost of the new project.

The project library is for archiving projects, but it is actually a huge reuse library. It also contains the template library and its own special Quality Control library. This special library contains mainly handmade solutions that are needed for integrating some older actuator device into our system. The project library is integrated to the project delivery process. Each delivered project is archived into the project library for reuse.

4.4 Discussion

Traditional programming reuse analysis tries to find reusable patterns. Strategies for component analysis are well introduced in (Rothenberger et al., 2003). These practices are categorized to project similarity, reuse planning, measurement, process improvement, formalized process, management support, education, object technology and commonality of architecture.

Our project library and reuse model covers project similarity very well because it is one starting point in finding reusable FBL programs. The reuse process is planned. The analysis measures template usage and the feedback system with template library targets for improved templates. Because every project is archived into the project library in the same way, the process is formal and repeatable. The analysis also gives good numbers for the management. Knowledge management is not so visible in our process but the reuse and template based design are part of the project delivery process. The knowledge needed to successfully use the templates takes some time. The automation domain is based on product family and the basic architecture has remained solid. The technology is based on different solutions and the object technology is used in various places.

Evolution during last four years has not affected reuse. There are new IO cards and new function blocks. Domain specific language reuse in dynamic domain is discussed in (Korhonen, 2002). This focuses more on code generation and language principles than reusing actual applications. The project library internally uses XML in many places and it has worked as a good transformation base. This was originated partly from the first agent-based implementation. This solution offered easier maintenance for the whole reuse library because it allowed transformations and extensions.

The publication implemented agent-based software is currently a simpler java application. It no longer uses agents anymore. The search engine user interface was enhanced in 2008 and new features were added by user requests. One important feature is to search special applications, only 1-2 applications per project. These applications contain rare I/O-cards and can be found using the card type in the search criteria. In the same way, some special Function Blocks can be searched.

The project library for reuse is in active use. The current search request amount is still almost one thousand searches monthly. The main page contains the amount of searches. It shows the current value 54932 (end of 2008). This makes the last four years of use an average of 1000 searches per month. In the initial phase in 2004, the amount of metadata was less than 2 Gb. The current (measured in the end of 2008) amount of metadata in the library is over 3.5 Gb and there are millions of application programs stored in the file system.

The actual metadata in the reuse database is growing and there has now been added more data about process such as machinery supplier and project people. If the salesman compares similar kinds of processes they have to check the supplier to validate reuse possibility. For tacit information and other not formalized information about the project, people are listed in the database. This makes it possible that people can be contacted and a short discussion can solve other unclear things.

The metadata makes searches more exact and implements actually feature based reuse library as is discussed in (Park & Palmer, 1995). The key factor is to select features as adding primary function block and IO card type among other metainformation. But instead of reusing components as stated in the article, Metso reuses application programs and templates. This kind of reuse affects to both productivity and quality much better.

5. Maintenance and round-trip engineering

5.1 Introduction

The biggest parts of software life-cycle costs are shown to be due to maintenance activities (Sneed, 1996), (Jones, 1998) (Erlikh, 2000). The systems that have long life cycles and require high maintainability, a key for lower maintenance costs is quality. Maintenance can be supported by various reverse engineering techniques like comprehension and visualization. Software visualization techniques applied to software written in traditional, textual programming languages can be problematic to be linked with reengineering activities afterwards, especially if standard notations, such as UML (UML, 2009), are not used: if the reverse engineering tool uses a different notation than the one used in software design, mappings between the different notations are needed. Since the models and views constructed from the existing program are presented with the same language used for development, the reverse engineering activities can be conveniently mapped with re-engineering activities, therefore enabling full round-trip support.

FBL application programs are located at the customer's own factories. Those programs are modified when there are some changes needed. These are frequent changes that must be done quickly. Even though FBL evolves and a version is upgraded, old programs can be used without any major work. This is part of the maintenance work that requires compatibility.

The following goals have been set for FBL maintenance:

- application level implementation remains the same even when symbols are updated,
- better performance: faster open and save, switch to testing faster,
- better usability and
- modern outlook: style is according to operating system and CAD platform.

5.2 Reverse and forward engineering

Reverse engineering activities aim at constructing representations and models of the subject software systems in another form or at a higher level of abstraction (Chikofsky & Cross, 1990). New representations are constructed after identifying the system's components and their interrelations.

Clustering in traditional reverse engineering methods can be constructed, for instance, by taking advantage of the syntax of the programming language used, by using software product metrics to identify highly cohesive clusters, or by using existing software architecture models and mapping them with the lower level details. In Java, for instance, package hierarchies can be used to structure classes and interfaces of the system. These hierarchies can be extracted by automated means. However, there are no guarantees that the packages contain sets of classes that conceptually form subsystems or components. Software product metrics used for identifying subsystems typically measure inter couplings and intra cohesion of the sets of software elements. These methods can only give educated guesses for clustering. Architectural models used in top-down reverse engineering approaches provide a good way to form a clustering. However, such high-level models do rarely exist and the construction of mappings with lower level software elements is typically difficult. In Metso's case, program uses the syntax of the language to construct high-level models for the FBL programs (Karaila & Systä, 2005).

In FBL, abstraction can be done by creating a new symbol from the existing application program. In Figure 14, a low-level FBL program is shown. For generating an abstract view to this program, the details of the program are filtered out and only the input and output symbols are preserved. An abstracted view is shown in the lower part of the same Figure 14 as one symbol. The abstracted program is called Function Group, indicating that one symbol contains several functions (function blocks and IOs). The symbol has two input points on the left: HLIM1 and LLIM1. These inputs limit values to form interlock interfaces H, H1 and

L. On the right there are five outputs HH, LL, H, L and H1. The outputs, in turn, are for interlocking and for different limit thresholds. If the measurement is over H value then the function group generates a high interlocking. If the value is even bigger and goes over HH value, then the function block generates a higher high limit. Correspondingly, the function group will generate low and lower low limits as signal value goes below a given limit. Parameters are captured inside the symbol. Program visualization creates new symbols on the fly for each abstracted component.

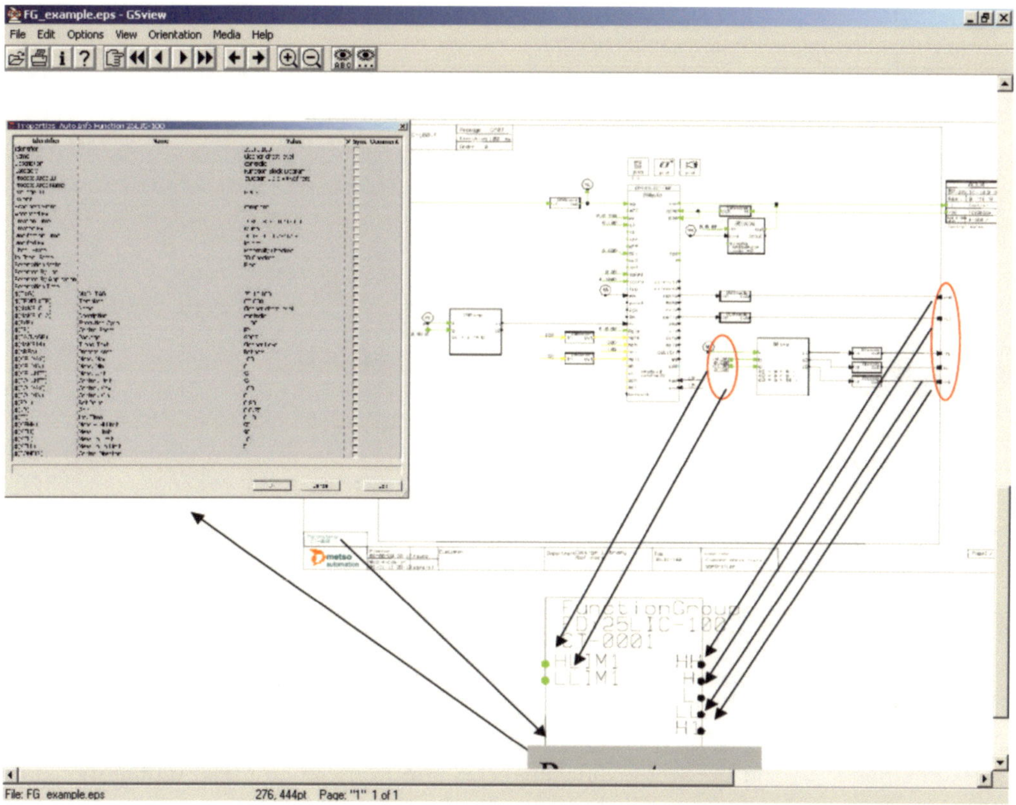

Fig. 14. Function Group example: parameters, implementation and symbol.

When compared to traditional reverse engineering techniques, a function group can be considered to correspond to a subsystem. Unlike in traditional approaches where various heuristics or metrics are used to help clustering program elements to subsystems, FBL syntax and information stored in the database are used to extract high-level views. This difference is significant: when reverse engineering FBL programs, the abstractions are always "correct", not educated guesses: the abstractions can be used for forward engineering activities as such. The differences between high-level views can only be due to different information filtering actions, not caused by different clustering. This makes reverse engineering of FBL programs significantly easier than reverse engineering programs written in traditional programming languages. On the other hand, this also means that the reverse engineering activities can be conveniently integrated with forward engineering activities, providing full round-trip support.

After constructing the higher-level function groups, they can be connected to each other. In FBL, internal communication connections are drawn inside modules by lines, while for external connections the engineer has to give a name. These external connections are stored in the database. To visualize external connections, database information is used to connect symbols as shown in Figure 15.

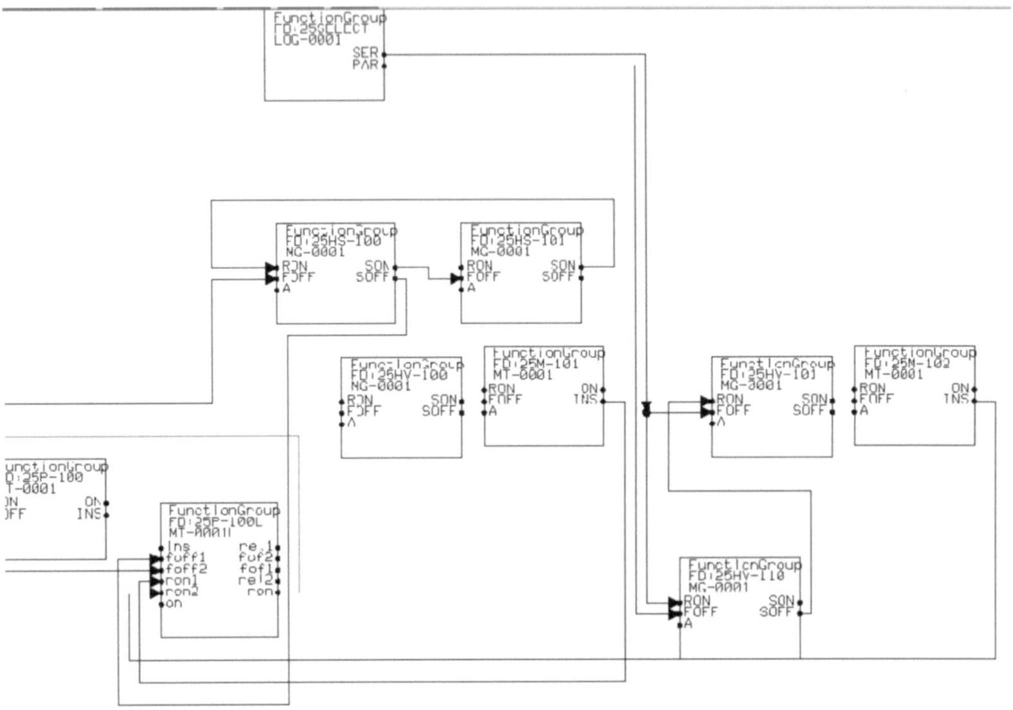

Fig. 15. Function group abstraction from FBL refiner programs.

To limit the size of the group of function group symbols, the engineer can select only a part of information stored in the whole database. This selection can be based on the metadata stored as well. In the domain FBL has been used, reasonable many of a large group of modules are from the same process area. In Figure 15, for instance, 10 symbols depicted are from the Refiner process area. Each function group symbol has a function that will need a user interface. Each device motor or valve has its own instance in both. Controller and selection logic are represented but the only one that is pure software is the interlocking logic. It is instantiated in the function group, but not in the normal user interface. The interlocking is in own display that the operator can open on demand.

In the Refiners process wood is mechanically cut / bladed to fibers. This mixture of paper fibers and water is pulp. Paper machines make paper from the pulp. The Refiner process is controlled by human operators from the display like the one shown in Figure 16.

Reverse engineering and data analysis techniques are used to get an overview of FBL programs. The environment can be used to generate high-level visual programs automatically.

A typical problem in this step is the layout. As indicated in studies, e.g. by (Storey et al.1997), the quality of layouts may have a significant impact on program understanding.

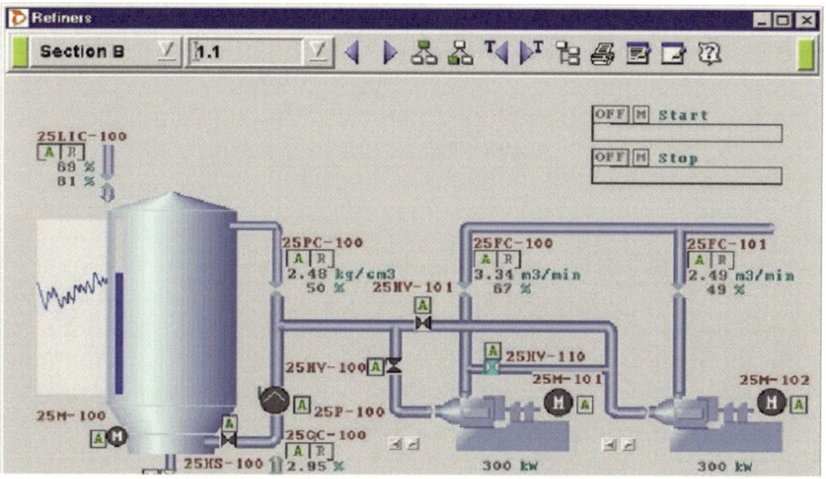

Fig. 16 Refiner user interface, operator display for controlling process.

According to our experiences, this also applies to visual programs. A commonly used solution for placing symbols is to use some automatic spatial spacing and auto-routing methods. The layouts of FBL programs have some fixed properties. The FBL programs are always read from left to right: inputs are on the left and outputs are on the right. The layout problem thus mainly concerns the rest of the FBL program. The solution selected for lay outing FBL programs is semiautomatic. The engineer needs to show a place for each symbol which is created automatically on the fly. Even though this approach requires manual intervention, it also has its advantages. The same tool environment is used for viewing and reverse engineering on the one hand and for programming on the other. Namely, the processes of forward and reverse engineering are not separated. In fact, the engineer is typically programming at the same time as analyzing a reusable (reverse engineered) solution. To be able to reuse the existing program, one has to learn the program structure first. After inserting all symbols needed, the engineer can activate a function that completes drawing with auto-routed connection lines. This feature is really powerful because in a normal case the engineer has to write each external data point / port connection manually in each FBL program. Now he can modify symbols and connections and in this way re-design the solution, e.g., to be more common and easier to understand.

5.3 Template maintenance

Trends in our template variation will focus on isolating IO from basic templates. This will reduce maintenance work that is needed. If a template contains some additional features like IO (standard IO, ACN IO, and LIS IO) and a new connection is implemented like FF IO, then all templates should be updated in case the IO is included inside the template. This is one fact that suggests separating IO from the core template. An example of separation is shown in Figure 17 that contains core templates in the middle and IO templates in the lower part. Other auxiliary features are placed in the upper part in own templates, like start and restart.

Figure 18 explains IO template in more detail. The tag application contains IO template and CORE. Communication is in its own part. This allows changes in application both in design time and in runtime easier. The flexibility is better because the new IO templates can be used without changes in the CORE templates. This will help in the future as new IO cards are designed and taken into use with IO templates.

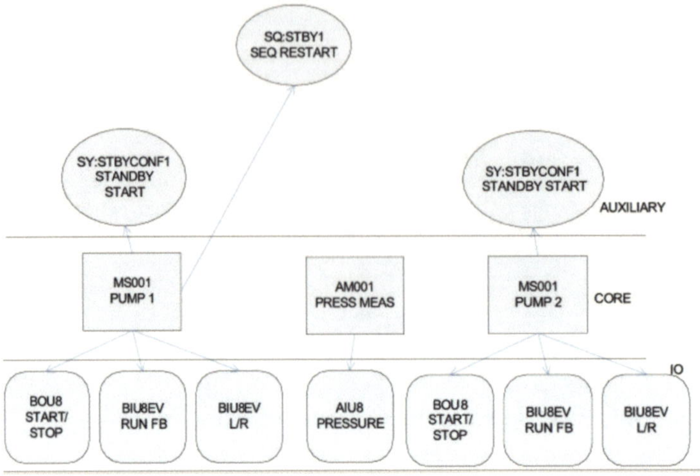

Fig. 17. Template separation levels: IO, core, auxiliary.

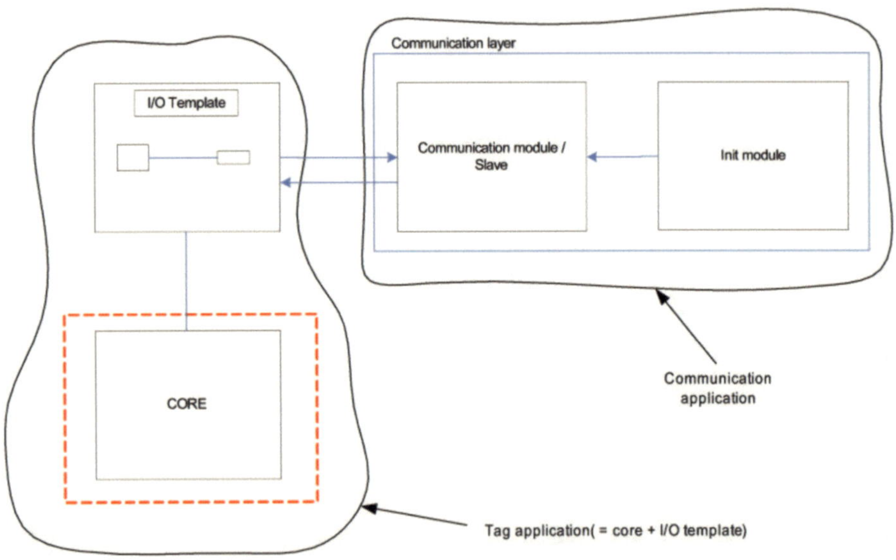

Fig. 18. Template modularization aims for managed variation and easier maintenance.

5.4 Discussion

According to the experiences on FBL and its programming environment at Metso Automation, in a combined reverse and forward engineering environment for visual programming, the role of layouts becomes quite important. Since the program analysis activities are often followed by forward engineering activities, the layouts constructed when analyzing programs should be "correct" and usable from the point of view of forward engineering activities. Also, since the engineer needs to understand the programs before being able to re-engineer or reuse them, semi-automated approaches for constructing layouts have shown to be quite feasible.

Re-engineering existing program instances means that they can be changed by extending or modifying them. For instance, new function blocks can be added, parameter values of existing programs can be changed, or connections between function groups can be changed. The engineer can thus create new programs that were first extracted from the database using reverse engineering techniques: he first creates a group of modules which are then visualized with the aid of reverse engineering techniques and finally re-engineered and/or reused.

For increasing the degree of reuse and thus decreasing the development times, reusing existing function groups instead of modifying individual programs is preferred. This assumes that the existing function groups are general enough to be usable in various programs. In many cases, the structure of the program itself is reusable but the differences occur in parameter values. For enabling reuse in such cases, a concept of a template has been introduced to FBL. The function group can use a template as a symbol to instantiate it. In this way, function groups are built from specialized templates.

The architecture layering and template mechanism gives us good tools for managing maintenance. At the template level, the model gives new maintenance needs as variation points but it needs more metainformation from the context (Cuccuru et al., 2007). There are sub-domain specific features in the templates such as power plant automation needs more accurate time stamps and chemical process automation requires more statistical data. The measurement template needs its own variation to fit from paper machine temperature measurement to oil refining temperature measurement. The oil refining measurement is more demanding and needs parallel measurements and statistical validation to insure reliability and robustness. This kind of knowledge management is needed in the future.

The long history can be used to reflect and analyze different maintenance activities. Normal maintenance activities focus on updating existing symbols and templates. From time to time people find bugs, which also call for maintenance. Sometimes cosmetic changes are also needed, like new better looking symbols or new layout that will make a program easier to read.

One practical issue is to support application maintenance. In the system level framework, tools can help a lot in this work. But designers have also had some bad experiences like making a modification in existing function block structure will make a big maintenance effort. After this designers have kept old function block structures untouched. It is better to create a new function block. A new function block can replace an old symbol if the connection points are matching. The framework can run a script that will automate the work. In exactly the same way, templates are versioned. A base template will be left untouched and a new template will be extended. An instantiated template can be easily upgraded to the new version. This is the normal method in customer projects. The project engineer can make a better template and changes / updates will keep all existing parameters. This is an efficient working method that improves quality.

6. Summary

FBL is a visual domain specific language that heavily relies on the usage of templates and meta-programming. FBL has been developed for writing automation control programs. Based on several years of practical use, it has proved to be easy to learn and adapted by its users.

Despite their undeniable benefits, template meta-programming techniques also have some drawbacks. Many compilers historically have quite poor support for templates. The use of templates can, in fact, make code somewhat less portable. Further, when errors are detected in template codes, most of the compilers produce confusing, unhelpful error messages. This can make templates difficult to develop. Debuggers also often have difficulties in working with templates.

A large group of methods and tool support for visual domain-specific programming is available. For example, (TRACE MODE, 2009), supports several IEC 6-1131/3 standard languages that can also be used to program control systems and business applications. One of the languages, namely Function Block Diagram (FBD), resembles FBL. Another toolkit is the Generic Modeling Environment (GME, 2009) that supports creating domain-specific modeling and program synthesis environments. In (Fröhlich et al., 2002), propose a meta-modeling based approach to provide and enforce modeling rules relevant for specific types of conceptual models used in automation domain, e.g. industrial plants or control systems. MetaEdit+ (Luoma et al., 2005 & MetaCase, 2006), in turn, supports meta-modeling for defining new domain-specific modeling languages and provides CASE-tool support for their use. While these approaches are partly related to ours, in this paper we have discussed yet new ideas, aspects, and working methods that are novel in using visual domain-specific languages.

In reference (Czarnecki, 2000), points out the following goals of generative programming: (i) decreasing the conceptual gap between program code and domain concepts, (ii) high reusability and adaptability, (iii) simplified managements of many variations of a component, and (iv) increased efficiency. In our case, where a visual domain-specific language FBL is used, all these generative programming goals can be achieved.

First, FBL as a visual language is intuitive. Moreover, custom symbols and icons can be used when programming certain types of applications. This provides a nice and customer-friendly way to map domain concepts with program elements. Second, templates have a significant role in FBL programs. A typical programming scenario includes selection of an appropriate template and its customization to a real program. A specific template library has been constructed and is constantly updated to better support programmers. In practice, the degree of reuse is very high. In new projects that are utilizing templates to a full extent, almost 100% of application programs are implemented by using templates. On the other hand, there are still projects that do not use any templates. New templates can, however, be easily constructed by comparing similarities of existing programs. i.e., new families of programs can be identified. This also supports the management of the programs belonging to this family. Finally, having ready-made templates can increase efficiency.

Reuse library developed has enabled an efficient way for users to archive and share implemented solutions and knowledge. The current java-based application solution filing process together with search tool has proven to be an efficient and practical solution.

The current content management database size exceeded 3.5 Giga bytes (2008). Database contains over hundreds of projects and links together over 62 Giga bytes of compressed files (1.2 million files). The usage of search tool has become a part of application engineers working manners. Approximately 1000 searches are performed monthly.

The analyses and template-matching processes implemented have allowed Metso to study more the real problem of finding a higher abstraction level for mass customization. Reuse helps sales and pre-design is started usually from the reuse library.

The software quality and usability has been improved based on internal measurements carried out at Metso and based on feedback from satisfied customers. In the programming environment, there has been a steady evolution and a desire to improve it. User group feedback has been collected to make further improvements, in a similar way to works presented in (Costagliola et al., 2002, Cox et al., 1997, Smedley & Cox, 1997).

The same environment that is used for development is also used for reverse engineering and maintaining FBL programs, thus providing a full round-trip support. The implemented environment together with the information on existing FBL programs gives engineers better understanding on the large existing group of FBL modules and their connections. The same kind of presentation of control diagrams and applications for interlocking are actually presented in German energy sector. This association of power and heat generating utilities is named VGB (German appreviation from Vereinigung der Großkraftwerksbetreiber; VGB, 2009). The documentation of the whole factory and its processes (water system or power generation) are normally written according to association guidance that is quite close to Metso's function group. Similar standard is System Control Diagram that is specified in Norway (SCD, 2009). The symbols and principles are almost the same as in programming with Function Groups.

To a great extent, the future design of controls could be carried out using function groups. Engineers who design advanced controls are seldom interested in details, but would rather like to program at higher level of abstraction, namely using function groups. The engineering environment indeed allows that. The actual experiences of the environment are still under study. Function groups are constructed for different processes to compare control structures and patterns that are used. From these existing solutions we will find out most common building blocks by statistical analysis using metadata stored in a reuse library. During last five years more entities and diagrams have been used in projects than before. The complexity of the programs has been almost at the same level. The conclusion of this five years trend is that the automation level is increasing steadily. Therefore, there is more implementation work in each project. The experiences gained so far indicate that similar physical processes with the same kind of machinery are easier to understand and reuse as high-level models, namely as packages with function groups in our case. Similar experiences have been presented (Wilkening et al., 1995). This also supports understanding on how to combine hardware and software as complete products (Holz, 2003). The experiences gained have shown that FBL and the engineering environment used is a flexible, practical, and well suited for the domain it is designed for, namely automation industry. We further believe that many of the features and advantages of the proposed FBL environment can be useful in traditional reverse engineering environments. In fact, features and benefits of an engineering framework corresponding to one discussed have been presented (Tilley, 1998). One of the most valuable parts of the proposed work is a possibility to reuse and re-engineer existing solutions. Unlike what is often used in traditional reverse engineering environments, semi-automated methods for constructing layouts have shown to be quite useful and feasible in the FBL environment. The semi-automated layout encourages the engineer to gradually learn the program, which is in any case required before he is able to re-engineer or reuse it. In addition, the usage of metadata has shown to be quite useful for querying the program database and to support program comprehension and analysis, especially concerning the evolution of the programs. Similar advantages could also be gained in traditional reverse engineering and program analysis tool support. We believe that

traditional reverse engineering environments could provide more advanced support for using metadata than what is currently available.
To summarize, the development of the template meta-programming support for FBL proceeded as follows. After the first release, fast feedback from the users had to be utilized in order to increase usability. Metso development team focused development on mini-language functionality in order to match our domain requirements. After that, the tools were modified to support different kinds of maintenance activities. The most important factor was always efficiency. Development team has learnt that getting feedback continuously from the users is crucial for successful maintenance and further development of FBL and its programming environment. These maintenance and development activities should and will continue as long as FBL is in use.
Future research and development will focus on further enhancing support for template meta-programming, e.g. by extending the template mini-language and by providing the additional means to raise the abstraction level of programming. Modern techniques and programming principles can be applied to the automation domain. Visual programming requires own specialized support that can be tuned to fit into the language and domain.

7. References

Burnett M., A. G. & Lewis, T. G. (1995) Visual Object-Oriented Programming Manning Publications Co. Greenwich, 280.

Burnett M. M., Webster, J. G. (ed.) (1999) Visual Programming *In Encyclopedia of Electrical and Electronics Engineering*, John Wiley & Sons Inc., New York.

Chikofsky E. and Cross J. (1990), Reverse Engineering and Design Recovery: A Taxonomy, *IEEE Software*, 7, 1, 1990, pp. 13-17.

Costagliola G., Francese R., Risi M., Scanniello G. (2002), A Component-Based Visual Environment Development Process, *In The Proc. of Software Engineering and Knowledge Engineering (SEKE'02)*, pp.327-334.

Cox P.T., Smedley T.J., Garden J., and McManus M. (1997), Experiences with Visual Programming in a Specific Domain - Visual Language Challenge, *In The Proc. of IEEE 1997 Symposium on Visual Languages (VL '97).*

Cuccuru, A.; Mraidha, C.; Terrier, F. & Gérard, S. (2007) Templatable Metamodels for Semantic Variation Points Model Driven Architecture- Foundations and Applications, *Model Driven Architecture - Foundations and Applications*, Springer, 68-82.

Czarnecki, K. & Eisenecker, U. (2000) Generative Programming: Methods, Tools, and Applications Addison-Wesley Professional.

Deursen, A. V. (1998) Little Languages: Little maintenance?

Debbie K. Carter, Albert D. Baker, W. B. A. (1995) I-I-Con: A Visual communications paradigm to integrate industrial control system engineering, *ISA Transactions, Elsevier Science Ltd.*, 34 (2), 153-163.

Erlikh L., (2000) Leveraging legacy system dollars for E-business, *IEEE IT Pro*, pp. 17-23.

Fröhlich P., Hu Z., and Schoelzke M. (2002), Imposing Modeling Rules on Industrial Applications through Meta-modeling, ER 2001 Workshops, HUMACS, DASWIS, ECOMO, and DAMA, LNCS 2465, pp. 166-182.

GME (Last visited September 200), Institute for Software Integrated Systems, The Generic Modeling Environment, http://www.isis.vanderbilt.edu/projects/gme/.

Hotz, L, Krebs, T. Günter, A.(2003) A Knowledge-based Product Derivation Process and some Ideas how to Integrate Product Development (position paper), *Workshop on Software Variability Management, Groningen, The Netherlands)*, February 13-14, 2003.

Johnston, W. M.; Hanna, J. R. P. & Millar, R. J. (2004). Advances in dataflow programming languages, *ACM Comput. Surv. 36, pp 1-34.*

Jones T.C., (1998) Estimating Software Costs, McGraw Hill.

Karaila, M. and Leppäniemi, A. (2004), Multi-Agent Based Framework for Large Scale Visual Program Reuse, *IFIP, Volume 159*/2005, 91-98.

Karaila M., Systä T. (2005), On the Role of Metadata in Visual Language Reuse and Reverse Engineering - An Industrial Case *Electronic Notes in Theoretical Computer Science, 2005, Volume 137, Issue 3*, 29-41.

Karaila, M. and Systä, T. (2007), Applying Template Meta-Programming Techniques for a Domain-Specific Visual Language - An Industrial Experience Report, *ICSE 2007.*

Korhonen, K. (2002), A case study on reusability of a DSL in a dynamic domain 2nd OOPSLA Workshop on Domain Specific Visual Languages.

Luoma J., Kelly S., Tolvanen J.P., (2005) Defining Domain-Specific Modeling Languages: Collected Experiences, *In Proc. of the 4th OOPSLA Workshop on Domain-Specific Modeling (DSM'04)*, LNCS 3714, Springer, pp. 198-209.

MetaCase, (2006) Domain-Specific Modeling with MetaEdit+, http://www.metacase.com/.

Mohamed E. Fayad, R. E. J. (ed.) (2000) Domain-Specific Application Frameworks. *Frameworks Experience by Industry* Wiley, 681

MODBUS, http://www.modbus.org/ Last visited September 2008.

Pressman, R. S. (1997) Software Engineering a Practioner's Approach McGraw-Hill.

Ommering, R. (2005) Software Reuse in Product Populations Software Engineering, *IEEE Transactions on*, 31, 537-550.

Park, S. & Palmer, J. D. (1995) A feature based reuse library Springer Berlin / Heidelberg, 1995, Volume 945/1995, 493-500.

Rahman Jamal, L. W. (1995). The Applicability of the Visual Programming Language LabVIEW to Large Real-World Applications, *In the Proc. of the 11th International IEEE Symposium on Visual Languages, IEEE Computer Society Washington, DC, US.*

Rothenberger, M. A.; Dooley, K. J.; Kulkarni, U. R. & Nada, N. (2003) Strategies for Software Reuse: A Principal Component Analysis of Reuse Practices *IEEE Transactions on Software Engineering, IEEE Computer Society*, 2003, 29, 825-837.

SCD, The Standardization Organizations in Norway, I-005 System Control Diagrams (Last visited September 2009), http://www.standard.no/.

Shu, N. C. Visual Programming Book Van Nostrand Reinhold Company. New York, 1988

Smedley T.J. and Cox P.T. (1997), Visual Languages for the Design and Development of Structured Objects, *Journal of Visual Languages and Computing, 8,* pp. 57-84.

Sneed maintenance costs, H. Sneed, (1996) Encapsulating Legacy Software for Use in Client/Server Systems, *In The Proc. of WCRE 1996*, pp. 104-119.

Storey M.-A.D. , K. Wong, F.D. Fracchia and H. A. Müller , (1997) On Integrating Visualization Techniques for Effective Software Exploration, *In Proc. of IEEE*

Symposium on Information Visualization (InfoVis'97), Phoenix, Arizona, U.S.A., 1997, pp. 38-45.

Tilley S (1998), A Reverse-Engineering Environment Framework. *Technical Report CMU/SEI-98-TR-005*, April 1998, 44 pages.

TRACE MODE, AdAstrA Research Group, (Last visited September 2009) TRACE MODE (IEC6-1131/3, http://www.tracemode.com/products/overview/IEC61131/,

UML, (Last visited September 2009), http://www.uml.org/.

VGB, Association of power and heat generating utilities. (Last visited September 2009), http://www.vgb.org/.

Whitley K.N., A. F. B. (2001) Visual Programming in the Wild: A Survey of LabVIEW Programmers *Journal of Visual Languages and Computing*, 2001, 12, 435-472.

Wilkening D.E., Loyall J. P., Pitarys M. J. and Littlejohn K. (1995), A Reuse Approach for Reengineering. *Journal of Systems Software 30*, pp. 117-125.

9

Visual Servoing for UAVs

Pascual Campoy, Iván F. Mondragón,
Miguel A. Olivares-Méndez and Carol Martínez
Universidad Politécnica de Madrid (Computer Vision Group)
Spain

1. Introduction

Vision is in fact the richest source of information for ourself and also for outdoors Robotics, and can be considered the most complex and challenging problem in signal processing for pattern recognition. The first results using Vision in the control loop have been obtained in indoors and structured environments, in which a line or known patterns are detected and followed by a robot (Feddema & Mitchell (1989), Masutani et al. (1994)). Successful works have demonstrated that visual information can be used in tasks such as servoing and guiding, in robot manipulators and mobile robots (Conticelli et al. (1999), Mariottini et al. (2007), Kragic & Christensen (2002).)

Visual Servoing is an open issue with a long way for researching and for obtaining increasingly better and more relevant results in Robotics. It combines image processing and control techniques, in such a way that the visual information is used within the control loop. The bottleneck of Visual Servoing can be considered the fact of obtaining robust and on-line visual interpretation of the environment, which can be usefully treated by control structures and algorithms. The solutions provided in Visual Servoing are typically divided into Image Based Control Techniques and Pose Based Control Techniques, depending on the kind of information provided by the vision system that determine the kind of references that have to be sent to the control structure (Hutchinson et al. (1996), Chaumette & Hutchinson (2006) and Siciliano & Khatib (2008)). Another classical division of the Visual Servoing algorithms considers the physical disposition of the visual system, yielding to eye-in-hand systems and eye-to-hand systems, that in the case of Unmanned Aerial Vehicles (UAV) can be translated as on-board visual systems (Mejias (2006)) and ground visual systems (Martínez et al. (2009)).

The challenge of Visual Servoing is to be useful in outdoors and non-structured environments. For this purpose the image processing algorithms have to provide visual information that has to be robust and works in real time. UAV can therefore be considered as a challenging testbed for visual servoing, that combines the difficulties of abrupt changes in the image sequence (i.e. vibrations), outdoors operation (non-structured environments) and 3D information changes (Mejias et al. (2006)). In this chapter we give special relevance to the fact of obtaining robust visual information for the visual servoing task. In section (2).we overview the main algorithms used for visual tracking and we discuss their robustness when they are applied to image sequences taken from the UAV. In sections (3). and (4). we analyze how vision systems can perform 3D pose estimation that can be used for controlling whether the camera platform or the UAV itself. In this context, section (3). analyzes visual pose estimation using multi-camera ground systems, while section (4). analyzes visual pose estimation obtained from onboard cameras. On the other hand, section (5)., shows two position based control applications for UAVs. Finally section (6). explodes the advantages of fuzzy control techniques for visual servoing in UAVs.

2. Image processing for visual servoing

Image processing is used to find characteristics in the image that can be used to recognize an object or points of interest. This relevant information extracted from the image (called features) ranges from simple structures, such as points or edges, to more complex structures, such as objects. Such features will be used as reference for any visual servoing task and control system.

On image regions, the spatial intensity also can be considered as a useful characteristic for patch tracking. In this context, the region intensities are considered as a unique feature that can be compared using correlation metrics on image intensity patterns.

Most of the features used as reference are interest points, which are points in an image that have a well-defined position, can be robustly detected, and are usually found in any kind of images. Some of these points are corners formed by the intersection of two edges, and others are points in the image that have rich information based on the intensity of the pixels. A detector used for this purpose is the Harris corner detector (Harris & Stephens (1988)). It extracts corners very quickly based on the magnitude of the eigenvalues of the autocorrelation matrix. Where the local autocorrelation function measures the local changes of a point with patches shifted by a small amount in different directions. However, taking into account that the features are going to be tracked along the image sequence, it is not enough to use only this measure to guarantee the robustness of the corner. This means that good features to track (Shi & Tomasi (1994)) have to be selected in order to ensure the stability of the tracking process. The robustness of a corner extracted with the Harris detector can be measured by changing the size of the detection window, which is increased to test the stability of the position of the extracted corners. A measure of this variation is then calculated based on a maximum difference criteria. Besides, the magnitude of the eigenvalues is used to only keep features with eigenvalues higher than a minimum value. Combination of such criteria leads to the selection of the good features to track. Figure 1(a) shows and example of good features to track on a image obtained on a UAV.

The use of other kind of features, such as edges, is another technique that can be applied on semi-structured environments. Since human constructions and objects are based on basic geometrical figures, the Hough transform (Duda & Hart (1972)) becomes a powerful technique to find them in the image. The simplest case of the algorithm is to find straight lines in an image that can be described with the equation $y = mx + b$. The main idea of the Hough transform is to consider the characteristics of the straight line not as image points x or y, but in terms of its parameters m and b, representing the same line as $y = \left(-\frac{\cos\theta}{\sin\theta}\right)x + \left(\frac{r}{\sin\theta}\right)$ in the parameter space, that is based on the angle of the vector from the origin to this closest point on the line (θ) and distance between the line and the origin (r). If a set of points form a straight line, they will produce sinusoids that cross at the parameters of that line. Thus, the problem of detecting collinear points can be converted to the problem of finding concurrent curves. To apply this concept just to points that might be on a line, some pre-processing algorithms are used to find edge features, such as the Canny

edge detector (Canny (1986)) or the ones based on derivatives of the images obtained by a convolution of image intensities and a mask (Sobel I. (1968)). These methods have been used in order to find power lines and isolators in an UAV inspection application (Mejías et al. (2007)).
The problem of tracking features can be solved with different approaches. The most popular algorithm to track features and image regions, is the Lucas-Kanade algorithm (Lucas & Kanade (1981)) which have demonstrated a good performance for real time with a good stability for small changes. Recently, feature descriptors have been successfully applied on visual tracking, showing a good robustness for image scaling, rotations, translations and illumination changes, eventhough they are time expensive to calculate. The generalized Lucas Kanade algorithm is overviewed on subsection 2.1, where it is applied for patch tracking and also for optical flow calculation, using the sparse L-K (subsection 2.1.1) and pyramidal L-K (subsection 2.1.2) variations. On subsection 2.2, features descriptors are introduced and used for robust matching, as explained on subsection 2.3

2.1 Appearance tracking

Appearance-based tracking techniques does not use features. They use the intensity values of a 'patch' of pixels that correspond to the object to be tracked. The method to track this patch of pixels is the generalized L-K algorithm, that works under three premises: first, the intensity constancy: the vicinity of each pixel considered as a feature does not change as it is tracked from frame to frame; second, the change in the position of the features between two consecutive frames must be minimum, so that the features are close enough to each other; and third, the neighboring points move in a solidarity form and have spatial coherence.
The patch is related to the next frame by a warping function that can be the optical flow or another model of motion. Taking into account the previously mentioned L-K premisses, the problem can be formulated in this way: lets define X as the set of points that form the patch window or template image T, where $\mathbf{x} = (x,y)^T$ is a column vector with the coordinates in the image plane of a given pixel and $T(\mathbf{x}) = T(x,y)$ is the grayscale value of the images a the locations $\mathbf{x}$. The goal of the algorithm is to align the template T with the input image I (where $I(\mathbf{x}) = I(x,y)$ is the grayscale value of the images a the locations $\mathbf{x}$). Because T transformed must match with a sub-image of I, the algorithm will find the set of parameters $\mu = (\mu_1,\mu_2, ...\mu_n)$ for a motion model function (e.g., Optical Flow, Affine, Homography) $W(\mathbf{x};\mu)$, also called the warping function. The objective function of the algorithm to be minimized in order to align the template and the actual image is equation 1:

$$e(\mathbf{W}) = \sum_{\forall \mathbf{x} \epsilon X} (I(W(\mathbf{x};\mu) - T(\mathbf{x}))^2 w(\mathbf{x}) \tag{1}$$

where $w(\mathbf{x})$ is a function to assign different weights to the comparison window. In general $w(\mathbf{x}) = 1$. Alternatively, w could be a Gaussian function to emphasize the central area of the window. This equation can also be reformulated to make it possible to solve for track sparse feature as is explained on section 2.1.1.
The Lucas Kanade problem is formulated to be solved in relation to all features in the form of a least squares' problem, having a closed form solution as follows.
Defining $w(\mathbf{x}) = 1$, the objective function (equation 1) is minimized with respect to μ and the sum is performed over all of the pixels $\mathbf{x}$ on the template image. Since the minimization process has to be made with respect to μ, and there is no lineal relation between the pixel

position and its intensity value, the Lucas-Kanade algorithm assumes a known initial value for the parameters μ and finds increments of the parameters $\delta\mu$. Hence, the expression to be minimized is:

$$\sum_{\forall \mathbf{x} \epsilon X} (I(W(\mathbf{x};\mu + \delta\mu) - T(\mathbf{x}))^2 \tag{2}$$

and the parameter actualization in every iteration is $\mu = \mu + \delta\mu$. In order to solve equation 2 efficiently, the objective function is linearized using a Taylor Series expansion employing only the first order terms. The parameter to be minimized is $\delta\mu$. Afterwards, the function to be minimized looks like equation 3 and can be solved like a "least squares problem" with equation 4.

$$\sum_{\forall \mathbf{x} \epsilon X} (I(W(\mathbf{x};\mu) + \nabla I \frac{\partial W}{\partial \mu} \delta\mu - T(\mathbf{x}))^2 \tag{3}$$

$$\delta\mu = H^{-1} \sum_{\forall \mathbf{x} \epsilon X} (\nabla I \frac{\partial W}{\partial \mu})^T (T(\mathbf{x}) - I(W(\mathbf{x};\mu)) \tag{4}$$

where H is the Hessian Matrix approximation,

$$H = \sum_{\forall \mathbf{x} \epsilon X} (\nabla I \frac{\partial W}{\partial \mu})^T (\nabla I \frac{\partial W}{\partial \mu}) \tag{5}$$

More details about this formulation can be found in (Buenaposada et al. (2003) and Baker and Matthews (2002)), where some modifications are introduced in order to make the minimization process more efficient, by inverting the roles of the template and changing the parameter update rule from an additive form to a compositional function. This is the so called ICIA (Inverse Compositional Image Alignment) algorithm, first proposed in (Baker and Matthews (2002)). These modifications where introduced to avoid the cost of computing the gradient of the images, the Jacobian of the Warping function in every step and the inversion of the Hessian Matrix that assumes the most computational cost of the algorithm.

2.1.1 Sparse Lucas Kanade

The Lucas Kanade algorithm can be applied on small windows around distinctive points as a sparse technique. In this case, the template is a small window (i.e., size of 3, 5, 7 or 9 pixels) and the warping function is defined by only a pure translational vector. In this context, the first assumption of the Lucas-Kanade method can be expressed as given a point $\mathbf{x}_i = (x, y)$ at time t which intensity is $I(x, y, t)$ will have moved by v_x, v_y and Δt between the two image frames, the following equation can be formulated:

$$I(x,y,t) = I(x + v_x, y + v_x, \Delta t) \tag{6}$$

If the general movement can be consider small and using the Taylor series, equation 6 can be developed as:

$$I(x + v_x, y + v_y, t + \Delta t) = I(x,y,t) + \frac{\partial I}{\partial x} v_x + \frac{\partial I}{\partial y} v_y + \frac{\partial I}{\partial t} \Delta t + H.O.T. \tag{7}$$

Because the higher order terms $H.O.T.$ can being ignored, from equation we found that:

$$\frac{\partial I}{\partial x}v_x + \frac{\partial I}{\partial y}v_y + \frac{\partial I}{\partial t}\Delta t = 0 \tag{8}$$

where v_x,v_y are the x and y components of the velocity or optical flow of $I(x,y,t)$ and $I_x = \frac{\partial I}{\partial x}$, $I_y = \frac{\partial I}{\partial y}$ and $I_t = \frac{\partial I}{\partial t}$ are the derivatives of the image at point $p = (x,y,t)$

$$I_x v_x + I_y v_y = -I_t \tag{9}$$

Equation 9 is known as the *Aperture Problem* of the optical flow. It arises when you have a small aperture or window in which to measure motion. If motion is detected in this small aperture, it is often that it will be seeing as a edge and not as a corner, causing that the movement direction can not be determined. To find the optical flow another set of equations is needed, given by some additional constraint.

The Lucas-Kanade algorithm forms the additional set of equation assuming that there is a local small window of size $m \times m$ centered at point $p = (x,y)$ in which all pixels moves coherently. If the windows pixel are numerates as $1...n$, with $n = m^2$, a set of equations can be found:

$$\begin{aligned} I_{x_1}v_x + I_{y_1}v_y &= -I_{t_1} \\ I_{x_2}v_x + I_{y_2}v_y &= -I_{t_2} \\ &\vdots \\ I_{x_n}v_x + I_{y_n}v_y &= -I_{t_n} \end{aligned} \tag{10}$$

Equation 10 have more than two equations for the two unknowns and thus the system is over-determined. A systems of the form $\mathbf{Ax} = \mathbf{b}$ can be former as equation 12 shows.

$$\begin{gathered} \mathbf{A} = \begin{bmatrix} I_{x_1} & I_{y_1} \\ I_{x_2} & I_{y_2} \\ \vdots & \vdots \\ I_{x_n} & I_{y_n} \end{bmatrix}, \quad \mathbf{x} = \begin{bmatrix} v_x \\ v_y \end{bmatrix}, \quad \mathbf{b} = \begin{bmatrix} -I_{t_1} \\ -I_{t_2} \\ \vdots \\ -I_{t_n} \end{bmatrix} \\ \text{or} \\ \begin{bmatrix} I_{x_1} & I_{y_1} \\ I_{x_2} & I_{y_2} \\ \vdots & \vdots \\ I_{x_n} & I_{y_n} \end{bmatrix} \begin{bmatrix} v_x \\ v_y \end{bmatrix} = \begin{bmatrix} -I_{t_1} \\ -I_{t_2} \\ \vdots \\ -I_{t_n} \end{bmatrix} \end{gathered} \tag{11}$$

The least squares method can be used to solve the over determined system of equation 12, finding that the optical flow can be defined as:

$$\begin{gathered} \mathbf{A}^T\mathbf{A}\mathbf{x} = \mathbf{A}^T\mathbf{b} \\ \text{or} \\ \begin{bmatrix} v_x \\ v_y \end{bmatrix} = (\mathbf{A}^T\mathbf{A})^{-1}\mathbf{A}^T\mathbf{b} \end{gathered} \tag{12}$$

2.1.2 Pyramidal L-K

On images with high motion, good matched features can be obtained using the Pyramidal Lucas-Kanade algorithm modification (Bouguet Jean Yves (1999)). It is used to solve the problem that arise when large and non-coherent motion are presented between consecutive frames, by firsts tracking features over large spatial scales on the pyramid image, obtaining an initial motion estimation, and then refine it by down sampling the levels of the images in the pyramid until it arrives to the original scale.

The overall pyramidal tracking algorithm proceeds as follows: first, a pyramidal representation of an image I of size $widthpixels \times heightpixels$ is generated. The zeroth level is composed by the original image and defined as I^0, then pyramids levels are recursively computed by dawnsampling the last available level (compute I^1 form I^0, then I^2 from I^1 and so on until I^{L_m} form I^{L-1})). Typical maximum pyramids Levels L_m are 2, 3 and 4. Then, the optical flow is computed at the deepest pyramid level L_m. Then, the result of that computation is propagated to the upper level $L_m - 1$ in a form of an initial guess for the pixel displacement (at level $L_m - 1$). Given that initial guess, the refined optical flow is computed at level $L_m - 1$, and the result is propagated to level $L_m - 2$ and so on up to the level 0 (the original image).

2.2 Feature descriptors and tracking

Feature description is a process to obtain interest points in the image which are defined by a series of characteristics that make it suitable for being matched on image sequences. This characteristics can include a clear mathematical definition, a well-defined position in image space and a local image structure around the interest point. This structure has to be rich in terms of local information contents that has to be robust under local and global perturbations in the image domain. These robustness includes those deformations arising from perspective transformations (i.e, scale changes, rotations and translations) as well as illumination/brightness variations, such that the interest points can be reliably computed with high degree of reproducibility.

There are many feature descriptors suitable for visual matching and tracking, from which Scale Invariant Feature Transform (SIFT) and Speeded Up Robust Feature algorithm (SURF) have been the more widely use on the literature and are overview in sections 2.2.1 and 2.2.2.

2.2.1 SIFT features

The SIFT (Scale Invariant Feature Transform) detector (Lowe (2004)) is one of the most widely used algorithms for interest point detection (called keypoints in the SIFT framework) and matching. This detector was developed with the intention to be used for object recognition. Because of this, it extracts keypoints invariant to scale and rotation using the gaussian difference of the images in different scales to ensure invariance to scale. To achieve invariance to rotation, one or more orientations based on local image gradient directions are assigned to each keypoint. The result of all this process is a descriptor associated to the keypoint, which provides an efficient tool to represent an interest point, allowing an easy matching against a database of keypoints. The calculation of these features has a considerable computational cost, which can be assumed because of the robustness of the keypoint and the accuracy obtained when matching these features. However, the use of these features depends on the nature of the task: whether it needs to be done fast or accurate. Figure 1(b) shows and example of SIFT keypoints on an aerial image taken with an UAV.

SIFT features can be used to track objects, using the rich information given by the keypoints descriptors. The object is matched along the image sequence comparing the model template (the image from which the database of features is created) and the SIFT descriptor of the current image, using the nearest neighbor method. Given the high dimensionality of the keypoint descriptor (128), its matching performance is improved using the Kd-tree search algorithm with the Best Bin First search modification proposed by Lowe (Beis and Lowe (1997)). The advantage of this method lies in the robustness of the matching using the descriptor, and in the fact that this match does not depend on the relative position of the template and the current image. Once the matching is performed, a perspective transformation is calculated using the matched Keypoints, comparing the original template with the current image.

2.2.2 SURF features

Speeded Up Robust Feature algorithm (Herbert Bay et al. (2006)) extracts features from an image which can be tracked over multiple views. The algorithm also generates a descriptor for each feature that can be used to identify it. SURF features descriptor are scale and rotation invariant. Scale invariance is attained using different amplitude gaussian filters, in such a way that its application results in an image pyramid. The level of the stack from which the feature is extracted assigns the feature to a scale. This relation provides scale invariance. The next step is to assign a repeatable orientation to the feature. The angle is calculated through the horizontal and vertical Haar wavelet responses in a circular domain around the feature. The angle calculated in this way provides a repeatable orientation to the feature. As with the scale invariance the angle invariance is attained using this relationship. Figure 1(c) shows and example of SURF features on an aerial image.

SURF descriptor is a 64 element vector. This vector is calculated in a domain oriented with the assigned angle and sized according to the scale of the feature. Descriptor is estimated using horizontal and vertical response histograms calculated in a 4 by 4 grid. There are two variants to this descriptor: the first provides a 32 element vector and the other one a 128 element vector. The algorithm uses integral images to implement the filters. This technique makes the algorithm very efficient.

The procedure to match SURF features is based on the descriptor associated to the extracted interest point. An interest point in the current image is compared to an interest point in the previous one by calculating the Euclidean distance between their descriptor vectors.

(a) (b) (c)

Fig. 1. Comparison between features point extractors. Figure 1(a) are features obtained using Good Features to Track, figure 1(b) are keypoints obtained using SIFT (the green arrows represents the keypoints orientation and scale) and figure 1(c) are descriptors obtained using SURF (red circles and line represents the descriptor scale and angle).

2.3 Robust matching

A set of corresponding or matched points between two images are frequently used to calculate geometrical transformation models like affine transformations, homographies or the fundamental matrix in stereo systems. The matched points can be obtained by a variety of methods and the set of matched points obtained often has two error sources. The first one is the measurement of the point position, which follows a Gaussian distribution. The second one is the *outliers* to the Gaussian error distribution, which are the mismatched points given by the selected algorithm. These outliers can severely disturb the estimated function, and consequently alter any measurement or application based on this geometric transformation. The goal then, is to determine a way to select a set of *inliers* from the total set of correspondences, so that the desired projection model can be estimated with some standard methods, but employing only the set of pairs considered as inliers. This kind of calculation is considered as *robust estimation,* because the estimation is tolerant (robust) to measurements following a different or unmodeled error distribution (outliers).

Thus, the objective is to filter the total set of matched points in order to detect and eliminated erroneous matched and estimate the projection model employing only the correspondences considered as inliers. There are many algorithms that have demonstrated good performance in model fitting, some of them are the Median of Squares (LMeds) (Rousseeuw & Leroy (1987)) and Random Sample Consensus (RANSAC) algorithm (Fischer & Bolles (1981)). Both are randomized algorithms and are able to cope with a large proportion of outliers.

In order to use a robust estimation method for a projective transformation, we will assume that a set of matched points between two projective planes (two images) obtained using some of the methods describe in section (2). are available. This set includes some unknown proportion of outliers or bad correspondences, giving a series of matched points $(x_i, y_i) \leftrightarrow (x'_i, y'_i)$ for $i = 1...n$, from which a perspective transformation must be calculated, once the outliers have been discarded.

For discard the outliers from the set of matched points, we use the RANSAC algorithm (Fischer & Bolles (1981)). It achieves its goal by iteratively selecting a random subset of the original data points by testing it to obtain the model and evaluating the model consensus, which is the total number of original data points that best fit the model. The model is obtained using a close form solution according to the desired projective transformation (an example is show on section 2.3.1). This procedure is then repeated a fixed number of times, each time producing either a model which is rejected because too few points are classified as inliers, or a refined model. When total trials are reached, the algorithm return the projection model with the largest number of inliers. The algorithm 1 shows a the general steps to obtain a robust transformation. Further description can be found on (Hartley & Zisserman (2004), Fischer & Bolles (1981)).

2.3.1 Robust homography

As an example of the generic robust method described above, we will show its application for a robust homography estimation. It can be viewed as the problem of estimating a 2D projective transformation that given a set of points $\overline{\mathbf{x}}_i$ in $\mathbb{P}^2$ and a corresponding set of points $\overline{\mathbf{x}}'_i$ in $\mathbb{P}^2$, compute the 3x3 matrix $\mathbf{H}$ that takes each $\overline{\mathbf{x}}_i$ to $\overline{\mathbf{x}}'_i$ or $\overline{\mathbf{x}}'_i = \mathbf{H}\,\overline{\mathbf{x}}_i$. In general the points $\overline{\mathbf{x}}_i$ and $\overline{\mathbf{x}}'_i$ are points in two images or in 2D plane surfaces.

Algorithm 1 Projective Transformation estimation using RANSAC

Require: Set of matched points $\mathbf{x}_i = (x_i, y_i) \leftrightarrow \mathbf{x'}_i = (x'_i, y'_i)$ for $i = 1 \ldots n$
  Define s = Minimum set of points to estimate the minimal solution.
  Define p = Probability that al least one of the random samples is free form outliers
  Define t = distance threshold to consider a point as an inlier for some model.
  Define ϵ = Initial probability that any selected point is an outlier.
  Define $Concesus$ = Desired number of minimum Inliers based on the total number of matched points
  Calculate the maximum number os samples $N = \log(1-p)/\log(1-(1-\epsilon)^s)$
  while $N > Trials$ **do**
    Randomly select s pairs of matched points
    Calculate the minimal solution for the model under test, using selected s points
    $inliers = 0$
    for $i = 0$ to n **do**
      Calculate the distance $d^2_{transfer} = d(\mathbf{x'}_i, \mathbf{H}\mathbf{x}_i)^2 + d(\mathbf{x}_i, \mathbf{H}^{-1}\mathbf{x'}_i)^2$
      if $d_{transfer} < t$ **then**
        $inliers = inliers + 1$
      end if
    end for
    if $inliers > Concensus$ **then**
      Calculate the final projective transformation using all inliers points
      $Concensus = inliers$
    end if
    recalculate $\epsilon = 1 - (inliers/n)$
    recalculate $N = \log(1-p)/\log(1-(1-\epsilon)^s)$
    $Trials = Trials + 1$
  end while

Taking into account that the number of degrees of freedom of the projective transformation is eight (defined up to scale) and because each point to point correspondences $(x_i,y_i) \leftrightarrow (x'_i, y'_i)$ gives rise to two independent equations in the entries of **H**, is enough with four correspondences to have a exact solution or minimal solution. If more than four points correspondences are given, the system is over determined and **H** is estimated using a minimization method. So, in order to use the algorithm 1, we define the minimum set of points to be $s = 4$.

If matrix **H** is written in the form of a vector $\mathbf{h} = [h_{11}, h_{12}, h_{13}, h_{21}, h_{22}, h_{23}, h_{31}, h_{32}, h_{33}]^t$ the homogeneous equations $\overline{\mathbf{x}}' = \mathbf{H}\overline{\mathbf{x}}$ for n points could be formed as $\mathbf{Ah} = 0$, with **A** a $2n \times 9$ matrix defined by equation 13:

$$A = \begin{bmatrix} x_1 & y_1 & 1 & 0 & 0 & 0 & -x_1x'_1 & -y_1x'_1 & -x'_1 \\ 0 & 0 & 0 & x_1 & y_1 & 1 & -x_1y'_1 & -y_1y'_1 & -y'_1 \\ \vdots & \vdots & \vdots & \vdots & \vdots & \vdots & \vdots & \vdots & \vdots \\ x_n & y_n & 1 & 0 & 0 & 0 & -x_nx'_n & -y_nx'_n & -x'_n \\ 0 & 0 & 0 & x_n & y_n & 1 & -x_ny'_n & -y_ny'_n & -y'_n \end{bmatrix} \quad (13)$$

In general, equation 13 can be solved using three different methods (the inhomogeneous solution, the homogeneous solution and non-linear geometric solution) as explained in

Criminisi et al. (1999). The most widely use of these methods is the inhomogeneous solution. In this method, one of the nine matrix elements is given a fixed unity value, forming an equation of the form **A′h′** = **b** as is shown in equation 14.

$$\begin{bmatrix} x_1 & y_1 & 1 & 0 & 0 & 0 & -x_1x_1' & -y_1x_1' \\ 0 & 0 & 0 & x_1 & y_1 & 1 & -x_1y_1' & -y_1y_1' \\ \vdots & \vdots & \vdots & \vdots & \vdots & \vdots & \vdots & \vdots \\ x_n & y_n & 1 & 0 & 0 & 0 & -x_nx_n' & -y_nx_n' \\ 0 & 0 & 0 & x_n & y_n & 1 & -x_ny_n' & -y_ny_n' \end{bmatrix} \begin{bmatrix} h_{11} \\ h_{12} \\ h_{13} \\ h_{21} \\ h_{22} \\ h_{23} \\ h_{31} \\ h_{32} \end{bmatrix} = \begin{bmatrix} x_1' \\ y_1' \\ \vdots \\ x_n' \\ y_n' \end{bmatrix} \tag{14}$$

The resulting simultaneous equations for the 8 unknown elements are then solved using a Gaussian elimination in the case of a minimal solution or using a pseudo-inverse method in case of an over-determined system Hartley and Zisserman (2004).
Figure 2 shows an example of a car tracking using a UAV, in which SURF algorithm, is used to obtain visual features, and the RANSAC algorithm is used for outliers rejection.

Fig. 2. Robust Homography Estimation using SURF features on a car tracking from a UAV. Up: Reference template. Down: Scene view, in which are present translation, rotation, and occlusions.

3. Ground visual system for pose estimation

Multi-camera systems are considered attractive because of the huge amount of information that can be recovered and the increase of the camera FOV (Field Of View) that can be obtained with these systems. These characteristics can help solving common vision problems such as occlusions, and can offer more tools for control, tracking, representation of objects, object analysis, panoramic photography, surveillance, navigation of mobile vehicles, among other tasks. However, in spite of the advantages offered by these systems, there are some applications where the hardware and the computational requirements make a multi-camera solution inadequate, taking into account that the larger the number of cameras used, the greater the complexity of the system is.

For example, in the case of pose estimation algorithms, when there is more than one camera involved, there are different subsystems that must be added to the algorithm:

- Camera calibration
- Feature Extraction and tracking in multiple images
- Feature Matching
- 3D reconstruction (triangulation)

Nonetheless, obtaining an adequate solution for each subsystem, it could be possible to obtain a multiple view-based 3D position estimation at real-time frame rates.

This section presents the use of a multi-camera system to detect, track, and estimate the position and orientation of a UAV by extracting some onboard landmarks, using the triangulation principle to recovered their 3D location, and then using this 3D information to estimate the position and orientation of the UAV with respect to a *World Coordinate System*. This information will be use later into a UAV's control loop to develop positioning and landing tasks.

3.0.2 Coordinate systems

Different coordinate systems are used to map the extracted visual information from $\Re^2$ to $\Re^3$, and then to convert this information into commands to the helicopter. This section provides a description of the coordinate systems and their corresponding transformations to achieve vision-based tasks.

There are different coordinate systems involved: the *Image Coordinate System* (X_i), that includes the *Lateral* (X_f) and *Central Coordinate Systems* (X_u) in the image plane, the *Camera Coordinate System* (X_c), the *Helicopter Coordinate System* (X_h), and an additional one: the *World Coordinate System* (X_w), used as the principal reference system to control the vehicle (see figure 3).

- ***Image* and *Camera Coordinate Systems***

The relation between the *Camera Coordinate System* and the *Image Coordinate System* is taken from the *"pinhole"* camera model. It states that any point referenced in the *Camera Coordinate System* $\mathbf{x_c}$ is projected onto the image plane in the point $\mathbf{x_f}$ by intersecting the ray that links the 3D point $\mathbf{x_c}$ with the center of projection and the image plane. This mapping is described in equation15, where $\mathbf{x_c}$ and $\mathbf{x_f}$ are represented in homogenous coordinates.

$$\begin{bmatrix} nx_f \\ ny_f \\ n \end{bmatrix} = \begin{bmatrix} f_x & 0 & c_x & 0 \\ 0 & f_y & c_y & 0 \\ 0 & 0 & 1 & 0 \end{bmatrix} \begin{bmatrix} x_c \\ y_c \\ z_c \\ 1 \end{bmatrix} \tag{15}$$

$$\mathbf{x_f} = \mathbf{K^k}[\mathbf{I}|\mathbf{0}]\mathbf{x_c}$$

The matrix $\mathbf{K^k}$ contains the intrinsic camera parameters of the kth camera, such as the coordinates of the center of projection (c_x, c_y) in pixel units, and the focal length (f_x, f_y), where

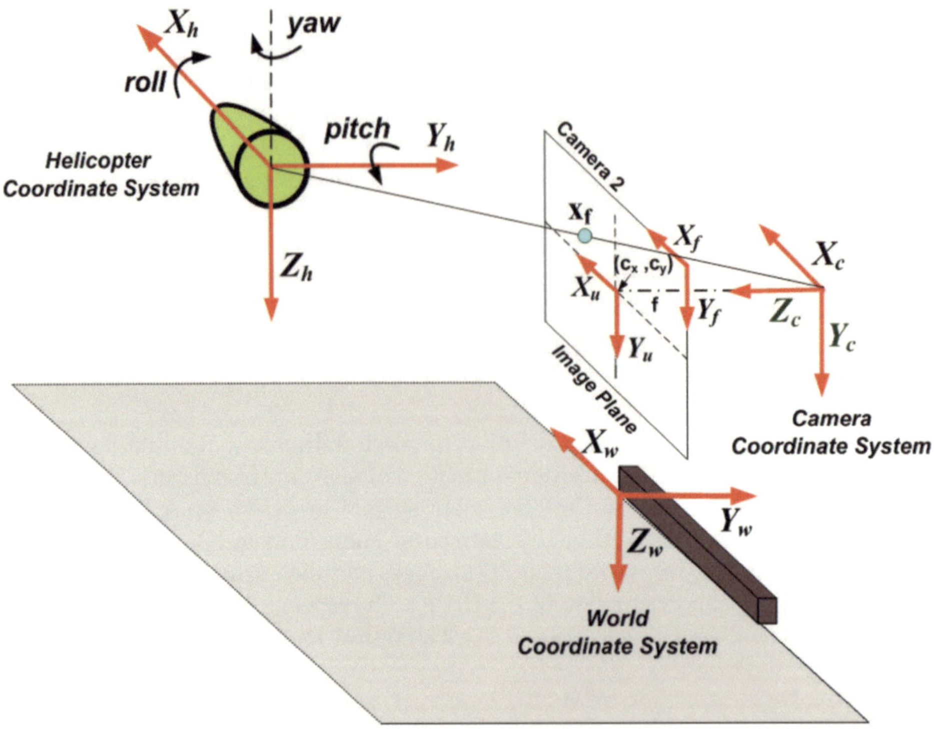

Fig. 3. Coordinate systems involved in the pose estimation algorithm.

$f_x = fm_x$ and $f_y = fm_y$ represent the focal length in terms of pixel dimensions, being m_x and m_y the number of pixels per unit distance.

The above-mentioned camera model assumes that the world point, the image point, and the optical center are collinear; however, in a real camera lens there are some effects (lens distortions) that have to be compensated in order to have a complete model. This compensation can be achieved by the calculation of the distortion coefficients through a calibration process (Zhang (2000)), in which the intrinsic camera parameters, as well as the radial and tangential distortion coefficients, are calculated.

- *Camera* and *World Coordinate Systems*

Considering that the cameras are fixed, these systems are related by a rigid transformation that allows to define the pose of the kth camera in a *World Coordinate Frame*. As presented in equation (16), this transformation is defined by a rotation matrix $\mathbf{R}^k$ and a translation vector $\mathbf{t}^k$ that link the two coordinate systems and represent the extrinsic camera parameters. Such parameters are calculated through a calibration process of the trinocular system.

$$\mathbf{x_c} = \begin{bmatrix} \mathbf{R}^k & \mathbf{t}^k \\ \mathbf{0}^T & 1 \end{bmatrix} \mathbf{x_w} \qquad (16)$$

- *World* and *Helicopter Coordinate Systems*

The *Helicopter Reference System*, as described in figure 3, has its origin at the center of mass of the vehicle and its correspondent axes: X_h, aligned with the helicopter's longitudinal axis; Y_h, transversal to the helicopter; and Z_h, pointing down. Considering that the estimation of the helicopter's pose with respect to the *World Coordinate System* is based on the distribution

of the landmarks around the *Helicopter Coordinate System*, and that the information extracted from the vision system will be used as reference to the flight controller, a relation between those coordinate systems has to be found.
In figure 3, it is possible to observe that this relation depends on a translation vector that defines the helicopter's position (**t**), and on a rotation matrix **R** that defines the orientation of the helicopter (*pitch, roll* and *yaw* angles). Considering that the helicopter is flying at low velocities ($< 4m/s$), *pitch* and *roll* angles are considered ≈ 0, and only the *yaw* angle (θ) is taken into account in order to send the adequate commands to the helicopter.
Therefore, the relation of the *World* and the *Helicopter Coordinate Systems* can be expressed as follows:

$$\begin{bmatrix} x_w \\ y_w \\ z_w \\ 1 \end{bmatrix} = \begin{bmatrix} cos(\theta) & -sin(\theta) & 0 & t_x \\ sin(\theta) & cos(\theta) & 0 & t_y \\ 0 & 0 & 1 & t_z \\ 0 & 0 & 0 & 1 \end{bmatrix} \begin{bmatrix} x_h \\ y_h \\ z_h \\ 1 \end{bmatrix} \tag{17}$$

Where (t_x, t_y, t_z) will represent the position of the helicopter $(x_{w_{uav}}, y_{w_{uav}}, z_{w_{uav}})$ with respect to the *World Coordinate System*, and θ the helicopter's orientation.

3.1 Feature extraction

The backprojection algorithm proposed by *Swain* and *Ballar* in (Swain & Ballard (1991)) is used to extract the different landmarks onboard the UAV. This algorithm finds a *Ratio* histogram Rh_i^k for each landmark i in the kth camera as defined in equation 18:

$$Rh_i^k(j) = min\left[\frac{Mh_i(j)}{Ih^k(j)}, 1\right] \tag{18}$$

This ratio Rh_i^k represents the relation between the bin j of a model histogram Mh_i and the bin j of the histogram of the image Ih^k which is the image of the k_{th} camera that is being analyzed. Once Rh_i^k is found, it is then backprojected onto the image. The resulting image is a gray-scaled image, whose pixel's values represent the probability that each pixel belongs to the color we are looking for.
The location of the landmarks in the different frames are found using the previous-mentioned algorithm and the *Continuously Adaptive Mean Shift* (*CamShift*) algorithm (Bradski (1998)). The *CamShift* takes the probability image for each landmark i in each camera k and moves a search window (previously initialized) iteratively in order to find the densest region (the peak) which will correspond to the object of interest (colored-landmark i). The centroid of each landmarks ($\bar{x}_i^k$, $\bar{y}_i^k$) is determined using the information contained inside the search window to calculate the zeroth ($m_{i_{00}}^k$), and first order moments ($m_{i_{10}}^k$, $m_{i_{01}}^k$), (equation 19). These centroids found in the different images (as presented in figure. 4) are then used as features for the 3D reconstruction stage.

$$\bar{x}_i^k = \frac{m_{i_{10}}^k}{m_{i_{00}}^k}; \quad \bar{y}_i^k = \frac{m_{i_{01}}^k}{m_{i_{00}}^k} \tag{19}$$

When working with overlapping FOVs in a 3D reconstruction process, it is necessary to find the relation of the information between the different cameras. This process is known as

Fig. 4. Feature Extraction. Different features must be extracted from images taken by different cameras. In this example color-based features have been considered.

feature matching. This is a critical process, which requires the differentiation of features in the same image and also the definition of a metric which tells us if the feature i in image $\mathbf{I}^1$ is the same feature i in image $\mathbf{I}^2$ (image -I- of camera k).
However, in this case, the feature matching problem has been solved taking into account the color information of the different landmarks; so that, for each image $\mathbf{I}^k$ there is a matrix $\mathbf{F}^{\mathbf{k}}_{\mathbf{4\times2}}$ that will contain the coordinates of the features i found in this image. Then, the features are matched by grouping only the characteristics found (the central moments of each landmark) with the same color, that will correspond to the information of the cameras that are seing the same landmarks.

3.1.1 3D reconstruction

Assuming that the intrinsic parameters ($\mathbf{K}^k$) and the extrinsic parameters ($\mathbf{R}^k$ and $\mathbf{t}^k$) of each camera are known (calculated through a calibration process), the 3D position of the matched landmarks can be recovered by intersecting in the 3D space the backprojection of the rays from the different cameras that represent the same landmark.
The relation of the found position of each landmark, expressed in the *Lateral Coordinate System* (image plane), with the position expressed in the *Camera Coordinate System*, is defined as:

$$x^k_{f_i} - c^k_x = f^k_x \frac{x^k_{c_i}}{z^k_{c_i}}, \quad y^k_{f_i} - c^k_y = f^k_y \frac{y^k_{c_i}}{z^k_{c_i}} \tag{20}$$

where ($x^k_{f_i}$, $y^k_{f_i}$) is the found position of each landmark expressed in the image plane, ($x^k_{c_i}$, $y^k_{c_i}$, $z^k_{c_i}$) represent the coordinates of the landmark expressed in the *Camera Coordinate System*, (c^k_x, c^k_y) the coordinates of the center of projection in pixel units, and (f^k_x, f^k_y) the focal length in terms of pixel dimensions.
If the relation of the 3D position of landmark i with its projection in each *Camera Coordinate System* is defined as:

$$\mathbf{x}^{\mathbf{k}}_{\mathbf{c_i}} = \begin{bmatrix} \mathbf{R}^{\mathbf{k}} & \mathbf{t}^{\mathbf{k}} \\ \mathbf{0}^{\mathrm{T}} & 1 \end{bmatrix} \mathbf{x}_{\mathbf{w_i}} \tag{21}$$

Then, integrating equation 21 and equation 20, and reorganizing them, it is possible to obtain the following equations:

$$x_{u_i}^k = f^k \frac{r_{11}^k x_{w_i} + r_{12}^k y_{w_i} + r_{13}^k z_{w_i} + t_x^k}{r_{31}^k x_{w_i} + r_{32}^k y_{w_i} + r_{33}^k z_{w_i} + t_z^k} \tag{22}$$

$$y_{u_i}^k = f^k \frac{r_{21}^k x_{w_i} + r_{22}^k y_{w_i} + r_{23}^k z_{w_i} + t_y^k}{r_{31}^k x_{w_i} + r_{32}^k y_{w_i} + r_{33}^k z_{w_i} + t_z^k} \tag{23}$$

Where $x_{u_i}^k$ and $y_{u_i}^k$ represent the coordinates of landmark i expressed in the *Central Camera Coordinate System* of the kth camera, r^k and t^k are the components of the rotation matrix $\mathbf{R^k}$ and the translation vector $\mathbf{t^k}$ that represent the extrinsic parameters, and x_{w_i}, y_{w_i}, z_{w_i} are the 3D coordinates of landmark i.
From equations 22 and 23 we have a linear system of two equations and three unknowns with the following form:

$$\begin{aligned}
(x_{c_i}^1 r_{31}^1 - r_{11}^1)x_{w_i} + (x_{c_i}^1 r_{32}^1 - r_{12}^1)y_{w_i} + (x_{c_i}^1 r_{33}^1 - r_{13}^1)z_{w_i} &= \\
&(t_x^1 - x_{c_i}^k t_z^1) \\
(y_{c_i}^1 r_{31}^1 - r_{21}^1)x_{w_i} + (y_{c_i}^1 r_{32}^1 - r_{22}^1)y_{w_i} + (y_{c_i}^1 r_{33}^1 - r_{23}^1)z_{w_i} &= \\
&(t_y^1 - y_{c_i}^k t_z^1) \\
\mathbf{Ac} = \mathbf{b}
\end{aligned} \tag{24}$$

If there are at least two cameras seeing the same landmark, it is possible to solve the overdetermined system using the least squares method whose solution will be equation 25, where the obtained vector $\mathbf{c}$ represents the 3D position (x_{w_i}, y_{w_i}, z_{w_i}) of the ith landmark:

$$\mathbf{c} \approx \mathbf{A}^{+}\mathbf{b} = (\mathbf{A}^{\mathrm{T}}\mathbf{A})^{-1}\mathbf{A}^{\mathrm{T}}\mathbf{b} \tag{25}$$

Once the 3D coordinates of the landmarks onboard the UAV have been calculated, the UAV's position ($\mathbf{x_{w_{uav}}}$) and its orientation with respect to *World Coordinate System* can be estimated using the 3D position found and the landmark's distribution around the *Helicopter Coordinate System* (see figure 5). The helicopter's orientation is defined only with respect to the Z_h axis (Yaw angle θ) and it is assumed that the angles, with respect to the other axes, are considered to be ≈ 0 (helicopter on hover state or flying at low velocities < 4 m/s). Therefore, equation 17 can be formulated for each landmark.
Reorganizing equation 17, considering that cθ= cos(θ), sθ= sin(θ), $x_{w_{uav}} = t_x$, $y_{w_{uav}} = t_y$, $z_{w_{uav}} = t_z$, and formulating equation 17 for all the landmarks detected, it is possible to create a system of equations of the form $\mathbf{Ac} = \mathbf{b}$ as in equation 26, with five unknowns: cθ, sθ, $x_{w_{uav}}$, $y_{w_{uav}}$, $z_{w_{uav}}$. If at least the 3D position of two landmarks is known, this system of equations can be solved as in equation 25, and the solution $\mathbf{c}$ is a 4×1 vector whose components define the orientation (*yaw* angle) and the position of the helicopter expressed with respect to a *World Coordinate System*.

$$\begin{bmatrix} x_{h_1} & -y_{h_1} & 1 & 0 & 0 \\ y_{h_1} & x_{h_1} & 0 & 1 & 0 \\ 0 & 0 & 0 & 0 & 1 \\ \vdots & \vdots & \vdots & \vdots & \vdots \\ x_{h_i} & -y_{h_i} & 1 & 0 & 0 \\ y_{h_i} & x_{h_i} & 0 & 1 & 0 \\ 0 & 0 & 0 & 0 & 1 \end{bmatrix} \begin{bmatrix} c\theta \\ s\theta \\ x_{w_{uav}} \\ y_{w_{uav}} \\ z_{w_{uav}} \end{bmatrix} = \begin{bmatrix} x_{w_1} \\ y_{w_1} \\ z_{w_1} - z_{h_1} \\ \vdots \\ x_{w_i} \\ y_{w_i} \\ z_{w_i} - z_{h_i} \end{bmatrix} \tag{26}$$

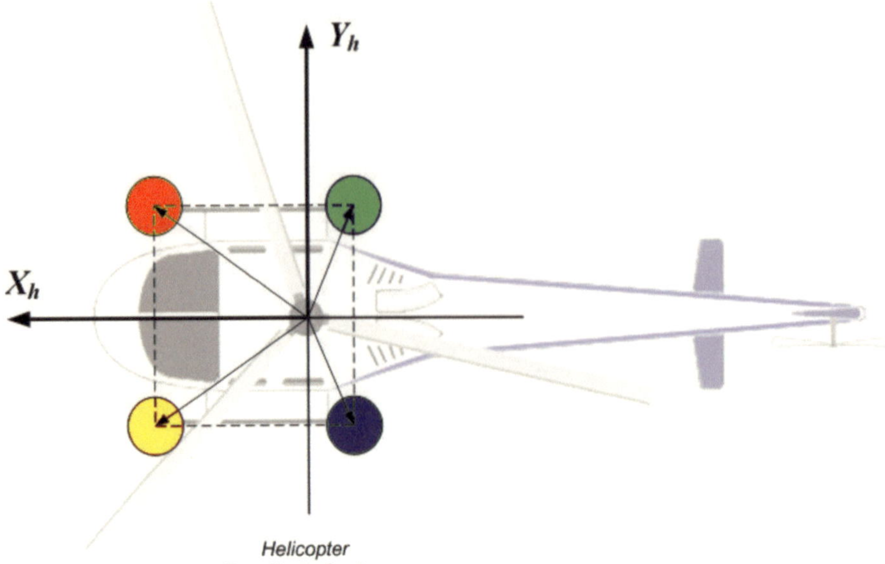

Fig. 5. Distribution of landmarks. The distribution of the landmarks in the *Helicopter coordinate system* is a known parameter used to extract the helicopter position and orientation with respect to the *World coordinate system*.

In figures: 6(a), 6(b), 6(c) and 6(d), it is possible to see an example of the UAV's position estimation using a ground-based multi camera system (see Martínez et al. (2009) for more details). In these figures, the vision-based position and orientation estimation (red lines) is also compared with the estimation obtained by the onboard sensors of the UAV (green lines).

4. Onboard visual system for pose estimation

In this section, a 3D pose estimation method based on projection matrix and homographies is explained. The method estimates the position of a world plane relative to the camera projection center for every image sequence using previous frame-to-frame homographies and the projective transformation at first frame, obtaining for each new image, the camera rotation matrix $\mathbf{R}$ and a translational vector $\mathbf{t}$. This method is based on the propose by Simon *et. al.* (Simon et al. (2000), Simon & Berger (2002)).

4.1 World plane projection onto the Image plane

In order to align the planar object on the world space and the camera axis system, we consider the general pinhole camera model and the homogeneous camera projection matrix, that maps a world point $\mathbf{x}_w$ in $\mathbb{P}^3$ (projective space) to a point $\mathbf{x}^i$ on i^{th} image in $\mathbb{P}^2$, defined by equation 27:

$$s\mathbf{x}^i = \mathbf{P}^i\mathbf{x}_w = \mathbf{K}[\mathbf{R}^i|\mathbf{t}^i]\mathbf{x}_w = \mathbf{K}\begin{bmatrix}\mathbf{r}_1^i & \mathbf{r}_2^i & \mathbf{r}_3^i & \mathbf{t}^i\end{bmatrix}\mathbf{x}_w \tag{27}$$

where the matrix $\mathbf{K}$ is the camera calibration matrix, $\mathbf{R}^i$ and $\mathbf{t}^i$ are the rotation and translation that relates the world coordinate system and camera coordinate system, and s is an arbitrary scale factor. Figure 7 shows the relation between a world reference plane and two images taken by a moving camera, showing the homography induced by a plane between these two frames.

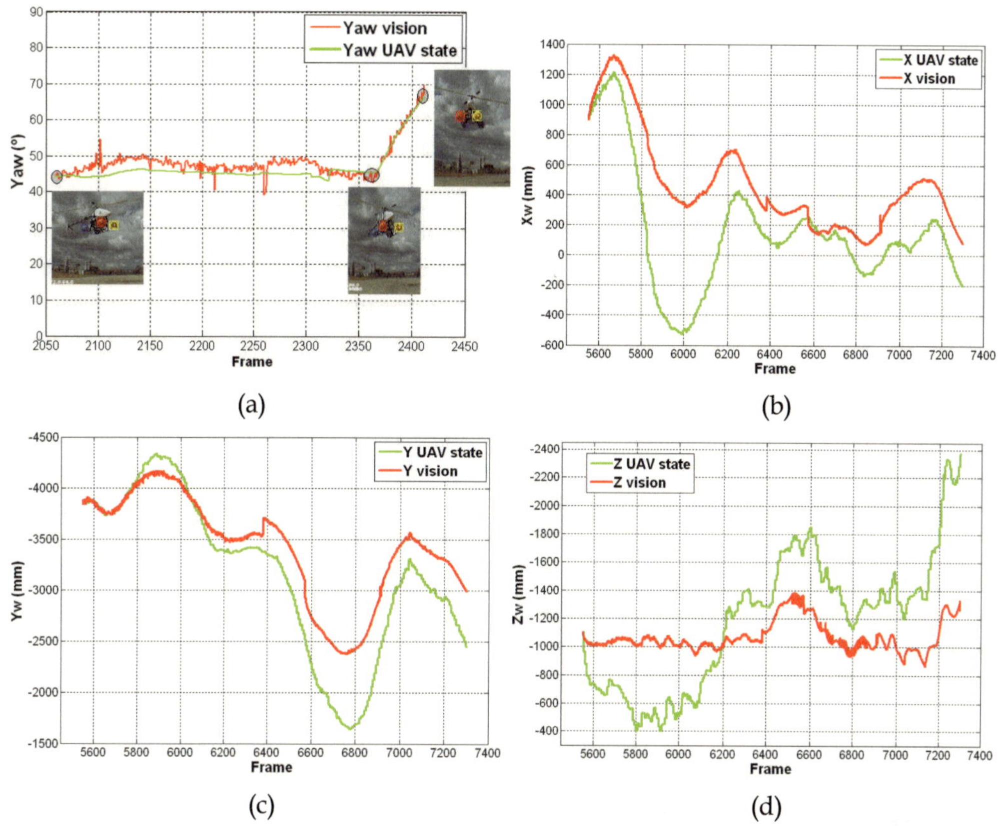

Fig. 6. Vision-based estimation vs. helicopter state estimation. The state values given by the helicopter state estimator after a *Kalman f ilter* (green lines) are compared with a multiple view-based estimation of the helicopter's pose (red lines).

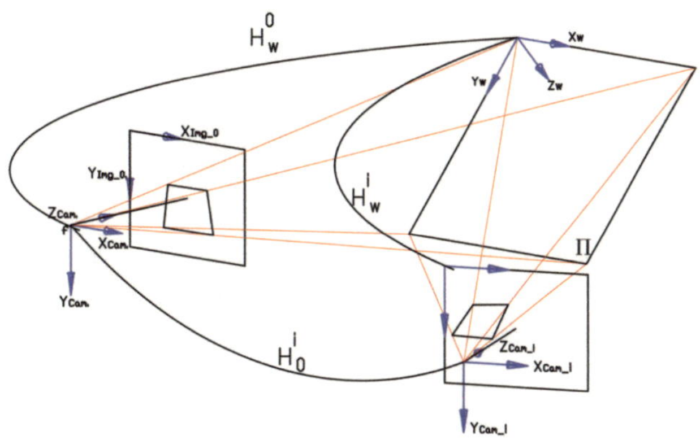

Fig. 7. Projection model on a moving camera and frame-to-frame homography induced by a plane.

If point $\mathbf{x}_w$ is restricted to lie on a plane Π, with a coordinate system selected in such a way that the plane equation of Π is $Z = 0$, the camera projection matrix can be written as equation 28:

$$s\mathbf{x}^i = \mathbf{P}^i\mathbf{x}_\Pi = \mathbf{P}^i \begin{bmatrix} X \\ Y \\ 0 \\ 1 \end{bmatrix} = \langle \mathbf{P}^i \rangle \begin{bmatrix} X \\ Y \\ 1 \end{bmatrix} \tag{28}$$

where $\langle \mathbf{P}^i \rangle$ denotes that this matrix is deprived on its third column or $\langle \mathbf{P}^i \rangle = \mathbf{K}[\, \mathbf{r}_1^i \;\; \mathbf{r}_2^i \;\; \mathbf{t}^i \,]$. The deprived camera projection matrix is a 3×3 projection matrix, which transforms points on the world plane (now in $\mathbb{P}^2$) to the ith image plane (likewise in $\mathbb{P}^2$), that is none other that a planar homography $\mathbf{H}_w^i$ $\mathbf{H}^{iw}$ defined up to scale factor as equation 29 shows.

$$\mathbf{H}_w^i = \mathbf{K}\begin{bmatrix} \mathbf{r}_1^i & \mathbf{r}_2^i & \mathbf{t}^i \end{bmatrix} = \langle \mathbf{P}^i \rangle \tag{29}$$

Equation 29 defines the homography which transforms points on the world plane to the ith image plane. Any point on the world plane $\mathbf{x}_\Pi = [x_\Pi, y_\Pi, 1]^T$ is projected on the image plane as $\mathbf{x} = [x,y,1]^T$. Because the world plane coordinates system is not known for the ith image, $\mathbf{H}_w^i$ can not be directly evaluated. However, if the position of the word plane for a reference image is known, a homography $\mathbf{H}_w^0$, can be defined. Then, the ith image can be related with the reference image to obtain the homography $\mathbf{H}_0^i$. This mapping is obtained using sequential frame-to-frame homographies $\mathbf{H}_{i-1}^i$, calculated for any pair of frames $(i\text{-}1,i)$ and used to relate the ith frame to the first imagen $\mathbf{H}_0^i$ using equation 30:

$$\mathbf{H}_0^i = \mathbf{H}_{i-1}^i \mathbf{H}_{i-2}^{i-1} \cdots \mathbf{H}_0^1 \tag{30}$$

This mapping and the aligning between initial frame to world plane reference is used to obtain the projection between the world plane and the ith image $\mathbf{H}_w^i = \mathbf{H}_0^i\ \mathbf{H}_w^0$. In order to relate the world plane and the ith image, we must know the homography $\mathbf{H}_w^0$. A simple method to obtain it, requires that a user selects four points on the image that correspond to corners of rectangle in the scene, forming the matched points $(0,0) \leftrightarrow (x_1,y_1)$, $(0,\Pi_{Width}) \leftrightarrow (x_2,y_2)$, $(\Pi_{Lenght},0) \leftrightarrow (x_3,y_3)$ and $(\Pi_{Lenght},\Pi_{Width}) \leftrightarrow (x_4,y_4)$. This manual selection generates a world plane defined in a coordinate frame in which the plane equation of Π is $Z = 0$. With these four correspondences between the world plane and the image plane, the minimal solution for homography $\mathbf{H}_w^0 = [\mathbf{h}_1{}_w^0\ \mathbf{h}_2{}_w^0\ \mathbf{h}_3{}_w^0]$ is obtained using the method described on section 2.3.1.

The rotation matrix and the translation vector are computed from the plane to image homography using the method described in (Zhang (2000)). From equation 29 and defining the scale factor $\lambda = 1/s$, we have that:

$$\begin{aligned} &\begin{bmatrix} \mathbf{r}_1 & \mathbf{r}_2 & \mathbf{t} \end{bmatrix} = \lambda \mathbf{K}^{-1}\mathbf{H}_w^i = \lambda \mathbf{K}^{-1}\begin{bmatrix} \mathbf{h}_1 & \mathbf{h}_2 & \mathbf{h}_3 \end{bmatrix} \\ &\text{where} \\ &\mathbf{r}_1 = \lambda \mathbf{K}^{-1}\mathbf{h}_1, \quad \mathbf{r}_2 = \lambda \mathbf{K}^{-1}\mathbf{h}_2, \quad \mathbf{t} = \lambda \mathbf{K}^{-1}\mathbf{h}_3 \end{aligned} \tag{31}$$

The scale factor λ can be calculated using equation 32:

$$\lambda = \frac{1}{\|\mathbf{K}^{-1}\mathbf{h}_1\|} = \frac{1}{\|\mathbf{K}^{-1}\mathbf{h}_2\|} \tag{32}$$

Because the columns of the rotation matrix must be orthonormal, the third vector of the rotation matrix $\mathbf{r}_3$ could be determined by the cross product of $\mathbf{r}_1 \times \mathbf{r}_2$. However, the noise on the homography estimation causes that the resulting matrix $\mathbf{R} = [\mathbf{r}_1\, \mathbf{r}_2\, \mathbf{r}_3]$ does not satisfy the orthonormality condition and we must find a new rotation matrix $\mathbf{R}'$ that best approximates to the given matrix $\mathbf{R}$ according to smallest Frobenius norm for matrices (the root of the sum of squared matrix coefficients) (Sturm (2000), Zhang (2000)). As demonstrated by (Zhang (2000)), this problem can be solved by forming the Rotation Matrix $\mathbf{R} = [\mathbf{r}_1\, \mathbf{r}_2\, \mathbf{r}_2]$ and using singular value decomposition (SVD) to form the new optimal rotation matrix $\mathbf{R}'$ as equation 33 shows:

$$\begin{aligned} \mathbf{R} &= \begin{bmatrix}\mathbf{r}_1 & \mathbf{r}_2 & (\mathbf{r}_1 \times \mathbf{r}_2)\end{bmatrix} = \mathbf{USV}^T \\ \mathbf{S} &= diag(\sigma_1, \sigma_2, \sigma_3) \\ \mathbf{R}' &= \mathbf{UV}^T \end{aligned} \tag{33}$$

Thus, the solution for the camera pose problem is defined by equation 34:

$$\mathbf{x}^i = \mathbf{P}^i\mathbf{X} = \mathbf{K}[\mathbf{R}'|\mathbf{t}]\mathbf{X} \tag{34}$$

4.2 UAV 3D estimation based on planar landmarks

This section shows the use of a pose estimation method based on frame to frame object tracking using robust homographies. The method, makes a matching between consecutive images of a planar reference landmark, using either, homography estimation based on good features to track (Shi & Tomasi (1994)), matched using the pyramidal L-K method, or the ICIA algorithm (Baker & Matthews (2002)) for an object template appearance tracking using a homography warping model. The frame to frame matching is used to estimate a projective transformation between the reference object and the image, using it to obtain the 3D pose of the object with respect to the camera coordinate system.

For these tests a Monocromo CCD Firewire camera with a resolution of 640x480 pixels is used. The camera is calibrated before each test, so the intrinsic parameters are know. The camera is installed in such a way that it is looking downward with relation to the UAV. A know rectangular helipad is used as the reference object to which estimate the UAV 3D position. It is aligned in such a way that its axes are parallel to the local plane North East axes. This helipad was designed in such a way that it produces many distinctive corner for the visual tracking. Figure 8(a), shows the helipad used as reference and figure 8(b), shows the coordinate systems involved in the pose estimation.

The algorithm begins, when a user manually selects four points on the image that correspond to four points on a rectangle in the scene, forming the matched points $(0,0) \leftrightarrow (x_1,y_1)$, $(910mm,0) \leftrightarrow (x_2,y_2)$, $(0,1190mm) \leftrightarrow (x_3,y_3)$ and $(910mm,1190mm) \leftrightarrow (x_4,y_4)$. This manual selection generates a world plane defined in a coordinates frame in which the plane equation of Π is $Z = 0$ (figure 7) and also defining the scale for the 3D results. With these four correspondences between the world plane and the image plane, the minimal solution for homography $\mathbf{H}_w^0$ is obtained.

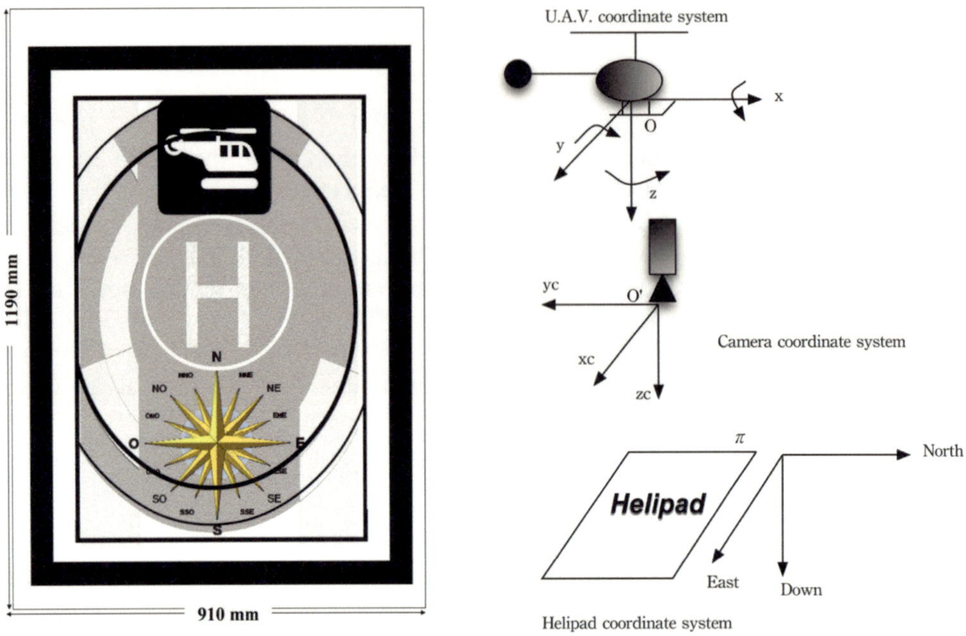

(a) (b)

Fig. 8. 8(a) Helipad used as a plane reference for UAV 3D pose estimation based on homographies. 8(b) Helipad, camera and U.A.V coordinate systems.

Once the alignment between the camera coordinate system and the reference helipad is known ($\mathbf{H}_w^0$) the homographies between consecutive frames are estimated, using either, the Pyramidal L.K. or the ICIA algorithm as is described below:

Optical Flow and RANSAC: good features to track are extracted on the zone corresponding to the projection of the helipad on image I_0. Then a new image I_1 is captured, and for each corner on image I_0, the pyramidal implementation of the Lucas Kanade optical flow method is applied, obtaining for each one either, the corresponding position (velocity vector) on image I_1 (if the corresponding point was found on the second image), or "null" if it was not found. With these points that have been matched or its optical flow was found on image I_1, a Homography $\mathbf{H}_0^1$ is robustly estimated using the algorithm described on section **??**. Homography $\mathbf{H}_0^1$ is used to estimate the alignment between image I_1 and the reference helipad using $\mathbf{H}_w^1 = \mathbf{H}_0^1\,\mathbf{H}_w^0$, which is used to obtain the rotation matrix $\mathbf{R}_w^1$ and the translation vector $\mathbf{t}_w^1$ using the method described on section 4.1. Then, the original frame formed by points $((x_1,y_1), (x_2,y_2), (x_3,y_3)$ and $(x_4,y_4))$ are projected on image I_1 using $\mathbf{x}_{I_1}^i = \mathbf{H}_0^1\,\mathbf{x}_{I_0}^i$, defining the actual position of the helipad on the image I_1. For this position, good features to track are once again estimated and used to calculate a new set of matched points between images I_1 and I_2. These set of matched points are used to calculate $\mathbf{H}_1^2$, and then $\mathbf{H}_0^2$ and $\mathbf{H}_w^2$ from which $\mathbf{R}_w^2$ and $\mathbf{t}_w^2$ is estimated. The process is successively repeated until either, the helipad is lost or the user finishes the process.

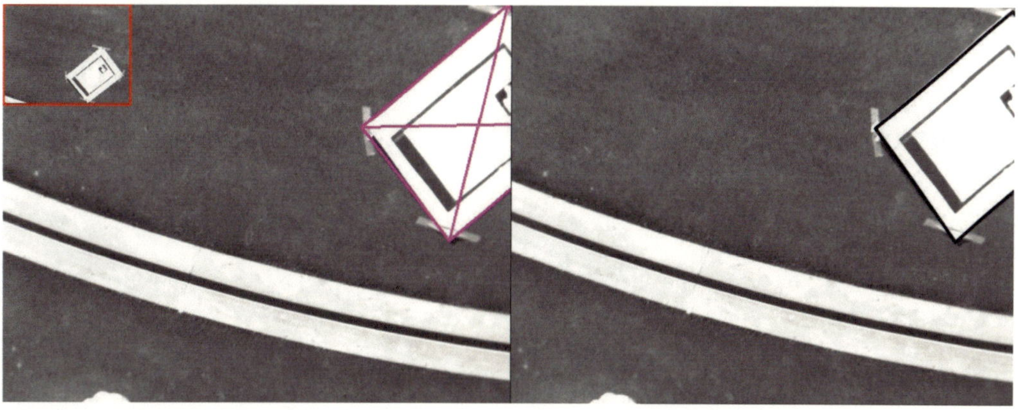

Fig. 9. Homography motion model estimated on a partial occluded image using either, the Lucas-Kanade Algorithm with RANSAC robust function fitting (left) or with the Inverse Compositional Algorithm ICA (right). Superimposed (top left), is the original frame or template under tracking.

ICIA: The zone corresponding to the projection of the helipad on image I_0 is defined as the template to track $T(\mathbf{x})$ on the image sequence. Then for each new image I_k on the sequence, the following equation $\Sigma_{\forall \mathbf{x} \in X}(T(W(\mathbf{x};\, \delta\mu) - I_k(W(\mathbf{x};\mu))^2$ is minimized in order to get the parameters $\mu = (\mu_1, \mu_2, \ldots \mu_n)$ for a Homography motion model (section 2.1), obtaining directly the homography $\mathbf{H}_0^k$ that relates the image I_k with the template $T(\mathbf{x})$ on image I_0. The alignment between frame k and the world plane is obtained using $\mathbf{H}_w^k = \mathbf{H}_w^0\, \mathbf{H}_0^k$ from which $\mathbf{R}_w^k$ and $\mathbf{t}_w^k$ is estimated.

Figure 9 shows the homography estimation using both, the Pyramidal Lucas Kanade tracker and the ICIA algorithm.

The translational vector obtained using the method described on section 4.1, is already scaled based on the dimensions defined for the reference plane during the alignment between the helipad and image I_0, so in our case the resulting vector $\mathbf{t}_w^i$ is in *mm*. The rotation matrix can be decomposed on Tait-Bryan or Cardan Angles. The Tait-Bryan or Cardan angles are formed when the three rotation sequences each occur about a different axis. This is the preferred sequence in flight and vehicle dynamics. Specifically, these angles are formed by the sequence: (1) ψ about z axis (yaw), (2) θ about y_a (pitch), and (3) ϕ about the final x_b axis (roll), where a and b denote the second and third stage in a three-stage sequence or axes. This set of rotation sequences is defined by the rotation matrices as equation 35 shows:

$$\mathbf{R}_{z,\psi} = \begin{bmatrix} \cos\psi & \sin\psi & 0 \\ -\sin\psi & \cos\psi & 0 \\ 0 & 0 & 1 \end{bmatrix}, \quad \mathbf{R}_{y,\theta} = \begin{bmatrix} \cos\theta & 0 & -\sin\theta \\ 0 & 1 & 0 \\ \sin\theta & 0 & \cos\theta \end{bmatrix}, \quad \mathbf{R}_{x,\phi} = \begin{bmatrix} 1 & 0 & 0 \\ 0 & \cos\phi & \sin\phi \\ 0 & -\sin\phi & \cos\phi \end{bmatrix} \quad (35)$$

The final coordinate transformation matrix for Tait-Bryan angles is defined by the composition of the rotations $\mathbf{R}_{x,\phi}\mathbf{R}_{y,\theta}\mathbf{R}_{z,\psi}$ forming the equation 36.

$$\mathbf{R}_{Tait-Bryan} = \begin{bmatrix} \cos\theta\cos\psi & \cos\theta\sin\psi & -\sin\theta \\ \sin\phi\sin\theta\cos\psi - \cos\phi\sin\psi & \sin\phi\sin\theta\sin\psi + \cos\phi\cos\psi & \sin\phi\cos\theta \\ \cos\phi\sin\theta\cos\psi + \sin\phi\sin\psi & \cos\phi\sin\theta\sin\psi - \sin\phi\cos\psi & \cos\phi\cos\theta \end{bmatrix} \quad (36)$$

The angles ψ, θ and ϕ can be obtained from the rotation matrix $\mathbf{R}_w^i$ (remember the rotation sequence order) using the equation 37.

$$\begin{aligned} \mathbf{R}_{Tait-Bryan} = \mathbf{R}_0^i &= \begin{bmatrix} r_{11} & r_{12} & r_{13} \\ r_{21} & r_{22} & r_{23} \\ r_{31} & r_{32} & r_{33} \end{bmatrix} \\ \text{where} \\ \theta &= -\arcsin(r_{13}) \\ \psi &= \arcsin(\frac{r_{12}}{\cos\theta}) \\ \phi &= \arcsin(\frac{r_{23}}{\cos\theta}) \end{aligned} \quad (37)$$

Equation 37 is singular when $\theta = 0$ or $\theta = \pi$.
Figure 10 shows some examples of the 3D pose estimation, based on a reference helipad. This figure shows the original reference image, the current frame, the optical flow between last and current frame, the helipad coordinates in the current frame camera coordinate system and the Tait-Bryan angles obtained from the rotation matrix.
The estimated 3D pose is compared with helicopter position estimated by the Kalman Filter of the controller on the local plane with reference to the takeoff point (Center of the Helipad). Because the local tangent plane to the helicopter is defined in such a way that the X axis is the North position, the Y axis is the East position and Z axis is the Down Position (negative), the measured X and Y values must be rotated according with the helicopter heading or yaw angle, in order to be comparable with the estimated values obtaining from the homographies. Figures 11(a), 11(b) and 12(a) shows the landmark position with respect to the UAV and figure 12(b), shows the estimated *yaw* angle.

5. UAV position control

The 3D pose estimation techniques on sections 3.and 4.are integrated with the UAV control loop using *Position Based Visual Servoing* architectures in *Dynamic Look and Move Systems* (Hutchinson et al. (1996), Chaumette and Hutchinson (2006), Siciliano and Khatib (2008)). In this kind of control, an error between the current and the desired position of the UAV is calculated and used by the low level controller (onboard flight controller) to generate the control commands to move the UAV to the desired position. Depending on the camera configuration in the control system, we will have an *eye-in-hand* or an *eye-to-hand* configuration. In the case of onboard control, it is considered to be an *eye-in-hand*, while in the case of ground control it is an *eye-to-hand* configuration as is shown on figure 13.
When the ground control is used (figure 13(a)), the vision system determines the position of the UAV in the *World Coordinate System*, so that the position $\mathbf{x}_{\mathbf{W}_{\mathbf{setPoint}}}$ and the position information given by the trinocular system $\mathbf{x}_{\mathbf{W}_{\mathbf{uav}}}$, both defined in the *World Coordinate System*, will be compared to generate references to the position controller. These references

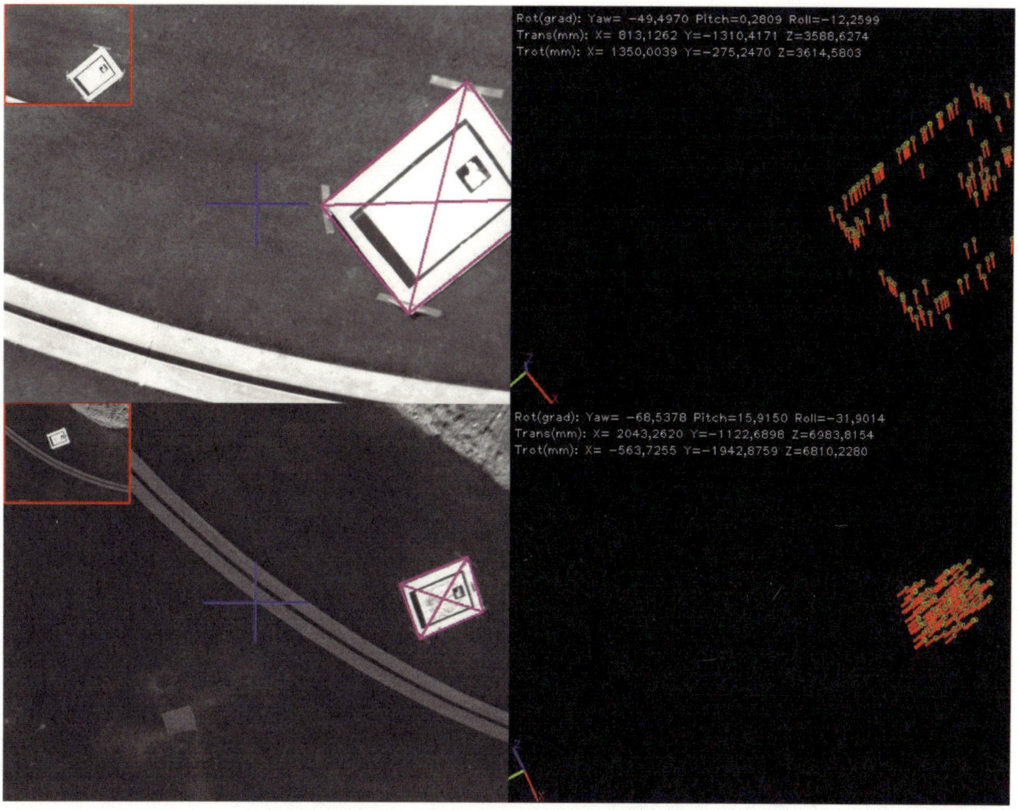

Fig. 10. Two different test for 3D pose estimation based on a helipad tracking using Robust Homography estimation. The reference image is on the small rectangle on the upper left corner. Left it the current frame and Right the Optical Flow between the actual and last frame. Superimposed are the Translation vector and the Tait-Bryan angles.

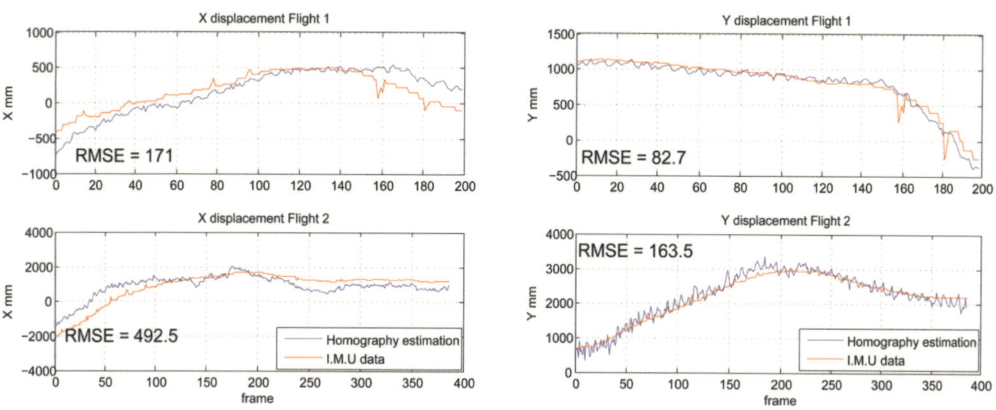

(a) (b)

Fig. 11. Comparison between the homography estimation and IMU data. 11(a) X axis displacement. 11(b) Y axis displacement

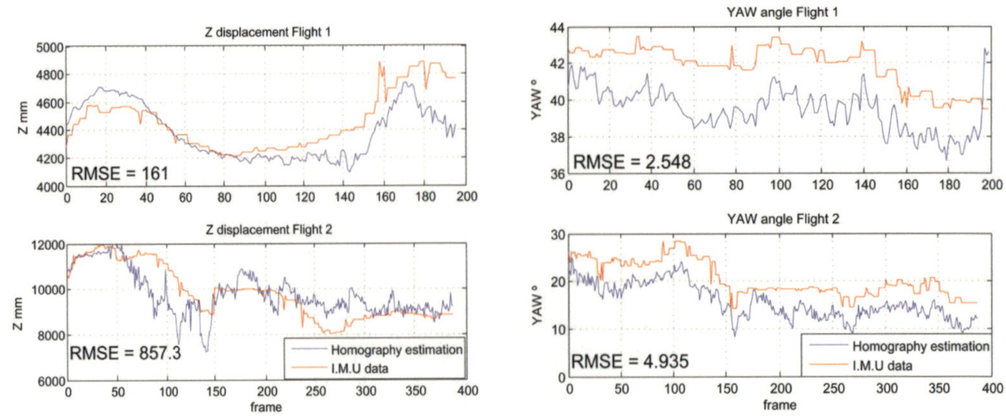

(a) (b)

Fig. 12. Comparison between the homography estimation and IMU data. 12(a) Z axis displacement. 12(b) *Yaw* angle

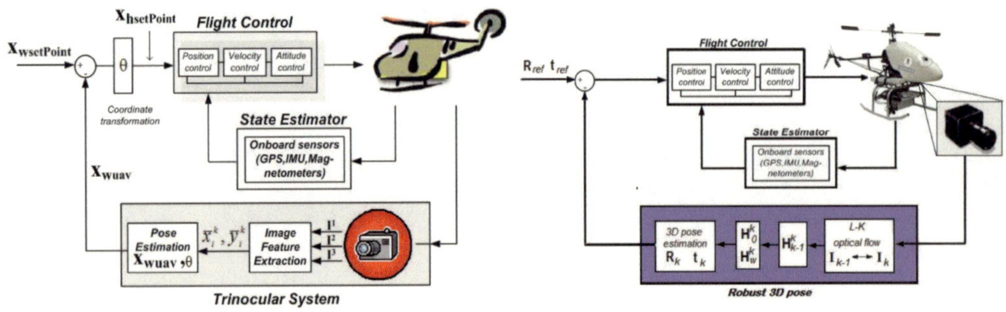

(a) (b)

Fig. 13. UAV visual control system following a *dynamic look-and-move* architecture. 13(a) is an *eye-to-hand* configuration (ground control), while 13(b) is an *eye-in-hand* configuration (onboard control)

are first transformed into commands to the helicopter $\mathbf{x}_{\mathbf{h}_{\text{setPoint}}}$ by taking into account the helicopter's orientation, and then those references are sent to the position controller in order to move the helicopter to the desired position (figure 13(a))

In case of the Onboard (figure 13(b)) control and depending on the control task, a reference point in coordinates relative to the helipad will be defined (e.g. For landing the reference point will be (0,0,0)). Because, the estimated position of the helipad (relative to the camera coordinate system onboard the UAV) is known by the visual system, the reference point can be transformed to coordinates relative to the helicopter coordinate system and will be used to generate the references (X, Y, Z) and (*Heading*) commands, relative to the UAV coordinate system, that will be used by the low-level controller to position the helicopter (e.g. in the landing case the command will be the translation vector obtained by the visual system) (figure 13(b)).

These control architectures have been tested with the COLIBRI III testbed that is shown in figure 14 (COLIBRI (2009), Campoy et al. (2009)). It has a low-level controller based on PID

Fig. 14. COLIBRI III Electric helicopter UAV used in a *dynamic look-and-move* control architecture.

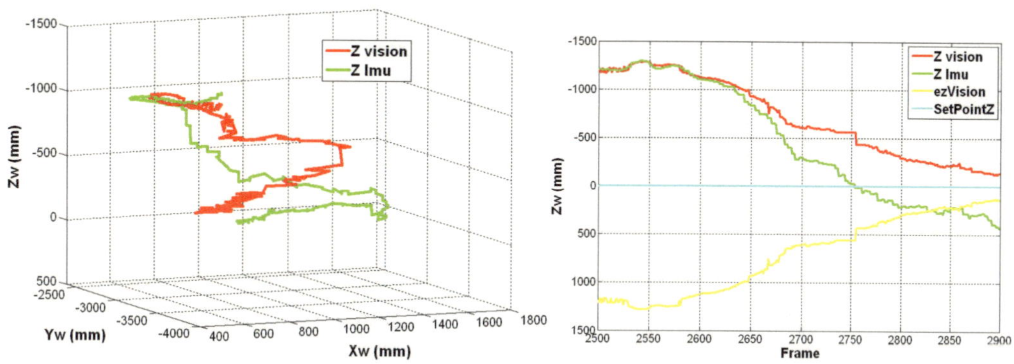

(a) (b)

Fig. 15. UAV control. Vision-based position commands (figure 15(b) yellow line) are sent to the flight controller to develop a vision-based landing task. The vision-based estimation (red line) is compared with the position estimation of the onboard sensors during the task.

control loops to ensure the helicopter' stability, using the state estimation obtained by a Kalman Filter on information given by the GPS, IMU and Magnetometer sensors. In order to enable the UAV to perform onboard image processing, it has a dedicated onboard computer in which the visual systems runs.

The system runs in a client-server architecture using TCP/UDP messages working in a multi-client wireless network, allowing the integration of vision systems and visual tasks with the low level flight control. This architecture allows applications to run both, onboard the autonomous helicopter or with an external processes, through a high level switching layer. The visual control system sends position references to the flight control through this layer using TCP/UDP messages, forming a *dynamic look-and-move* system architecture that is shown in figure 13.

In figure 15, the client server architecture, and the control architectures presented in figure 13 have been used to send position-based commands (figure 15(b) yellow line) to the flight controller in order to develop a vision-based landing task. Those position commands have been generated using the vision-based position estimation (figure 15 red line) obtained with the multi-camera system presented in section 3.. In figure 15(a) the 3D reconstruction of the vision-based position estimation (red line) during the landing task and the position estimation using the onboard sensors (green line) are compared.

6. Fuzzy controllers for visual servoing

This section shows the implementation of a visual control system using a tracker algorithm and three controllers working in parallel. Two of these controllers used to control the camera platform onboard the UAV (one for the pitch axis and the other for the yaw axis) and the third one is used to control the yaw angle of the helicopter (heading). The implementation of the controllers is based on Fuzzy logic, because this controller offers faster setpoint recovery with less overshoot than PID control for both setpoint changes and load changes. At the same time, it offers immunity to process noise when it is near setpoint because the controller develops a nonlinear response analogous to an error-squared PID controller. Also, when the error is larger, the control action is larger than for PID; while when it is smaller, the control action is smaller. However, the nonlinearity is less severe than for an error-squared controller and robustness is not compromised. Also, this controller is ideally suited for large time constants (not dead time) where overshoot and slow recovery are both undesirable. In fact, this controller generally outperforms PID loops in most situations. Another thing in favor is that using Fuzzy controllers it is not necessary to get the model of the helicopter in order to fit the controllers.

The system uses a firewire camera mounted on a pan and tilt platform, that takes images with 320x240 pixels resolution. The visual system is used to track an object of interest, using its position on the image plane (pixels) as the input for the fuzzy system, getting a yaw error (for platform and helicopter) in the range of -160 to 160 pixels, and a range of -120 to 120 pixels error for the platform pitch error.

The fuzzification of the inputs and the outputs are defined by using a triangular and trapezoidal membership functions. The controllers have two inputs, the error between the center of the object and the center of the image (figures 16(a) and 17(a)) and the difference between the last and the actual error (figures 16(b) and 17(b)), derivative of the position or the velocity of the object to track. The platform controllers output represents how many degrees the servo-motor must turn, in the two axis, to gets the center of the object in the center of the image. The output of both variables of the axis of the visual platform have the same output, as is shown in figure 18(a).

The heading controller uses the two same inputs of the yaw controller (figures 16(a) and 16(b)) and the output of the controller represents how many degrees must, the helicopter, turn to line up to the object to track (figure 18(b)).

The process of fuzzification transforms a numerical value to a linguistic value. We defined a linguistic value of each set at the inputs and output of each variables, putting the acronyms in the images of figure 18. The Meaning of these acronyms are shown in the table 1.

The three controllers are working in parallel giving a redundant operation to the yaw axis, but what we want to do with this action is to reduce the error that we have with the yaw-platform controller, where the limitations of the visual algorithm and the movements velocity of the servos hinders us to take a quicker response. The controllers are guided by a 49 rules base. The platform controllers output are defining in such a way that the sector near to the zero response, has more membership functions, as is shown in figure 18(a). This option, give us the possibility to define a very sensible controller when the error is so small (the object is very near to the center of the image), and a very quick respond controller when the object is so far. For the heading controller we defined a trapezoidal part in the middle of the output in order to help the platform controller, just when the object to track is with so far to the center of the image. With these trapezoidal definition we get a more stable behavior of the helicopter, in the situations where the object to track is near to the center, obtaining a 0 value.

Error	VBL	Very Big to the Left
	BL	Big to the Left
	LL	Little to the Left
	C	Center
	LR	Little to the Right
	BR	Big to the Right
	VBR	Very Big to the Right
Derivative Error	VBN	Very Big Negative
	BN	Big Negative
	LN	Little Negative
	Z	Zero
	LP	Little Positive
	BP	Big Positive
	VBP	Very Big Positive
Output: Turn	VBL	Very Big to the Left
	BL	Big to the Left
	L	Left
	LL	Little to the Left
	C	Center
	LR	Little to the Right
	R	Right
	BR	Big to the Right
	VBR	Very Big to the Right

Table 1. Meaning of the acronym of the linguistic value of the fuzzy variables inputs and the output.

DE \ E	**VBL**	**BL**	**LL**	**C**	**LR**	**BR**	**VBR**
VBN	VBL	VBL	VBL	BL	L	LL	Z
BN	VBL	VBL	BL	L	LL	Z	LR
LN	VBL	BL	L	LL	Z	LR	R
Z	BL	L	LL	Z	LR	R	BR
LP	L	LL	Z	LR	R	BR	VBR
BP	LL	Z	LR	R	BR	VBR	VBR
VBP	Z	LR	R	BR	VBR	VBR	VBR

Table 2. Rules base of the Yaw and Pitch controllers. Where *DE* is the derivative error and *E* the error.

For the inference process (in the defuzzification) we used a product classic method, and for the defuzzification part itself, we used the Height Method (equation 38).

$$y = \frac{\sum_{l=1}^{M} \bar{y}^l \prod \left(\mu_{B'}(\bar{y}^l)\right)}{\sum_{l=1}^{M} \prod \left(\mu_{B'}(\bar{y}^l)\right)} \tag{38}$$

In tables 2 and 3 the base of fuzzy rules used by the controllers are shown.

DE \ E	VBL	BL	LL	C	LR	BR	VBR
VBN	BL	BL	BL	BL	L	LL	Z
BN	BL	BL	BL	L	LL	Z	LR
LN	BL	BL	L	LL	Z	LR	R
Z	BL	L	LL	Z	LR	R	BR
LP	L	LL	Z	LR	R	BR	BR
BP	LL	Z	LR	R	BR	BR	BR
VBP	Z	LR	R	BR	BR	BR	BR

Table 3. Rules base of the Heading controller. Where *DE* is the derivative error and *E* the error.

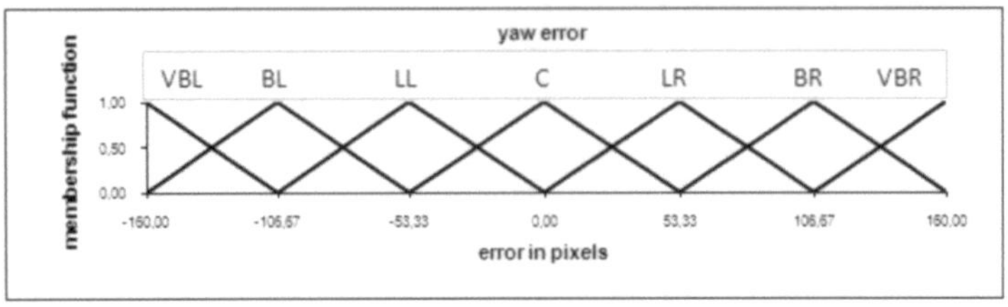

(a) Yaw Error.

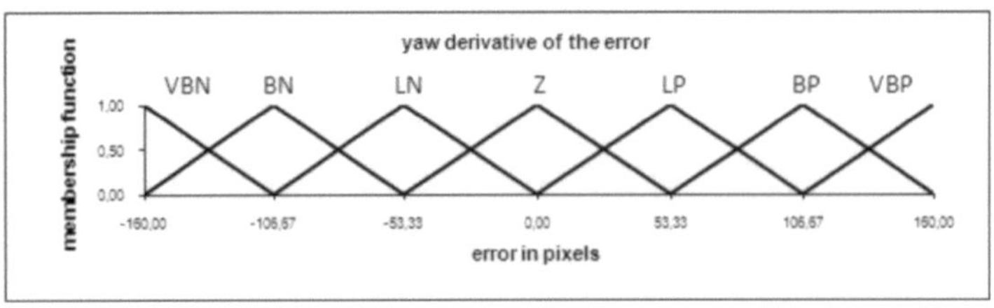

(b) Derivative of the Yaw error.

Fig. 16. Inputs Variables of the Yaw and Heading controllers.

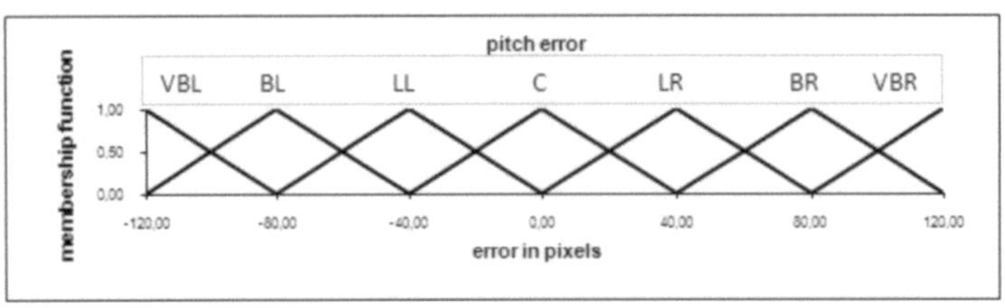

(a) Pitch Error.

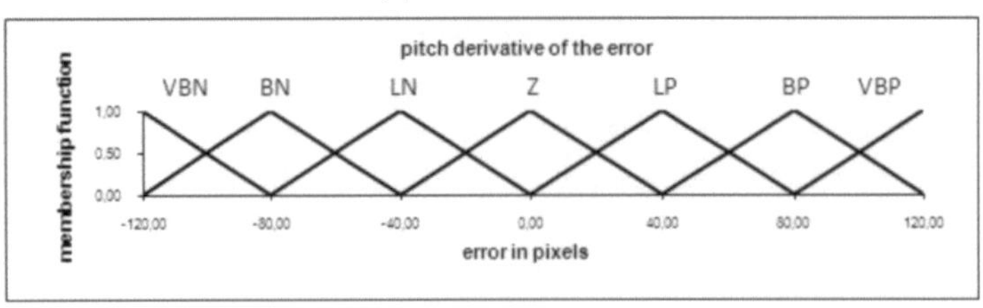

(b) Derivative of the Pitch error Membership functions.

Fig. 17. Inputs Variables of the Pitch controllers.

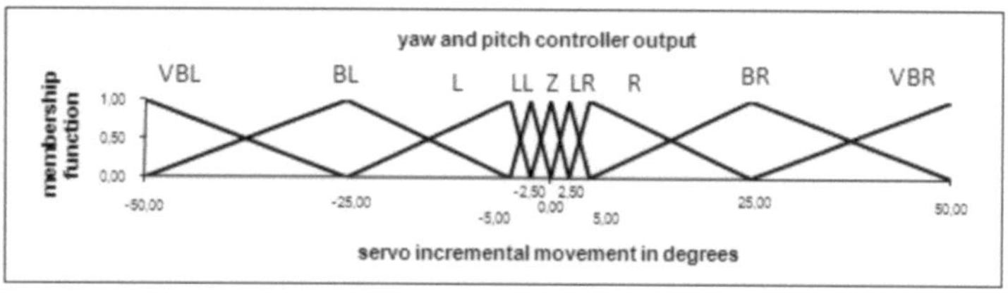

(a) Output of the Yaw and the Pitch Fuzzy Controllers.

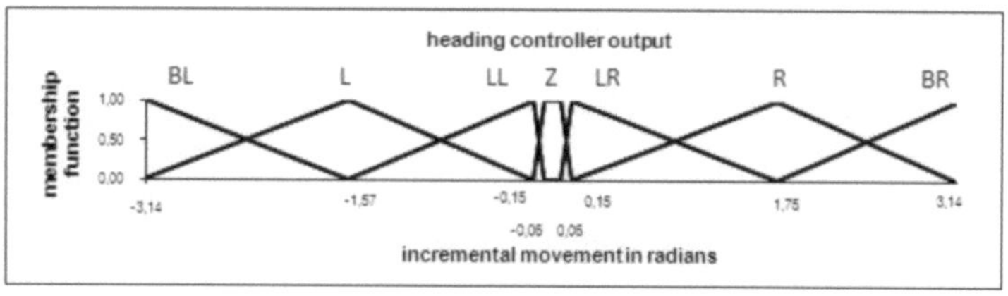

(b) Output of the Heading Fuzzy Controller.

Fig. 18. Variables of the Fuzzy-MOFS controllers.

These controllers are implemented using the software MOFS (Miguel Olivares' Fuzzy Software), with a definition in classes shown in figure 19. Details about this software and the differences between this and others implementations of Fuzzy Logic software can be consulted on Olivares and Madrigal (2007) and Olivares et al. (2008).

In the following paragraphs some results from real tests onboard the UAV, tracking static and moving objects are presented. For these tests, we use the Fuzzy controllers to control the pan and tilt camera platform.

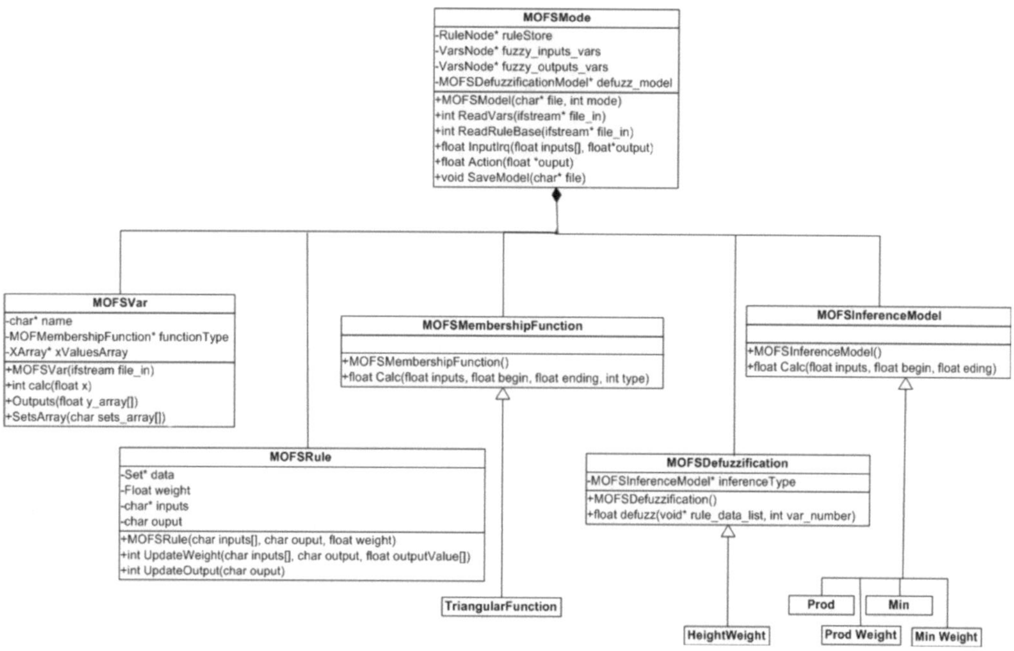

Fig. 19. Software definition.

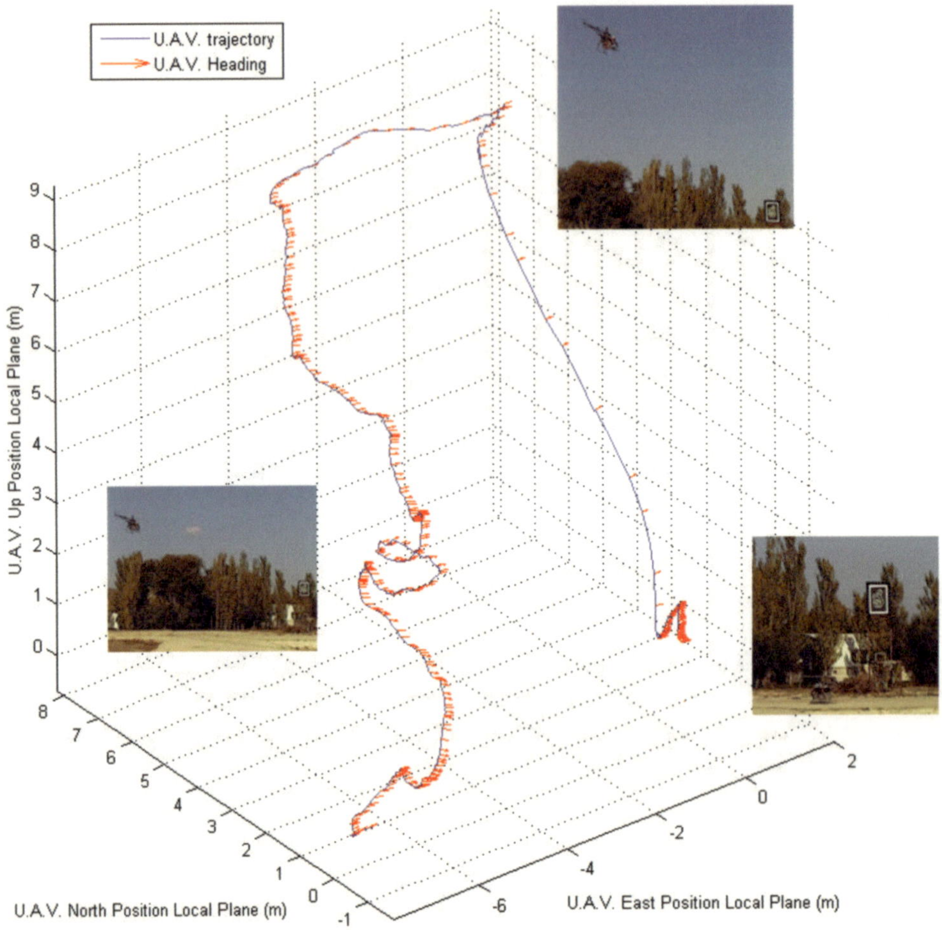

Fig. 20. 3D flight reconstruction from the GPS and the IMU data from the UAV. Where, the 'X' axis represents the NORTH axis of the surface of the tangent of the earth, the 'Y' axis represents the EAST axis of the earth, the 'Z' is the altitude of the helicopter and the red arrows show the pitch angle of the helicopter.

Tracking Static Objects

In this test, we tracked a static object during the full flight of the UAV, from takeoff to landing. This flight was made by sending set-points from the ground station. Figure 20 shows a 3D reconstruction of the flight using the GPS and IMU data on three axes: North (X), East (Y), and Altitude (Z), the first two of which are the axes forming the surface of the local tangent plane. The UAV is positioned over the north axis, looking to the east, where the mark to be tracked is located. The frame rate is 15 frames per second, so those 2500 frames represent a full flight of almost 3 minutes.

Figure 21 shows the UAV's yaw and pitch movements. In figure 23, the output of the two Fuzzy-MOFS controllers in order to compensate the error caused by the changes of the different movements and angle changes of the UAV flight, where we can see the different responses of the controllers, depending the sizes and the types of the perturbations.

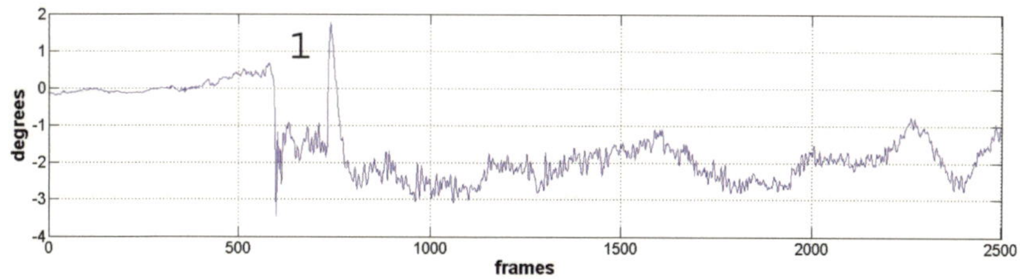

(a) Pitch angle movements.

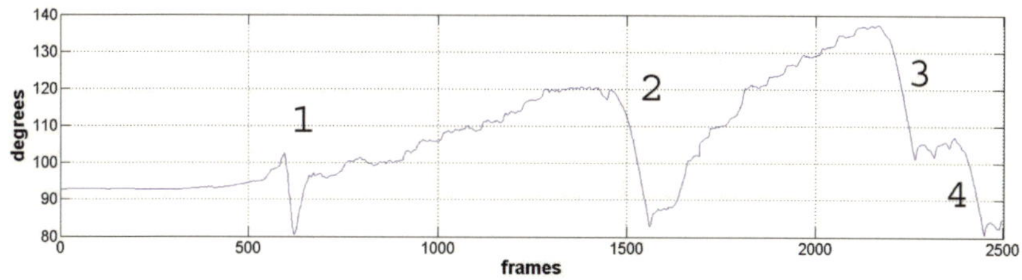

(b) Yaw angle movements.

Fig. 21. Different pitch and yaw movements of the UAV.

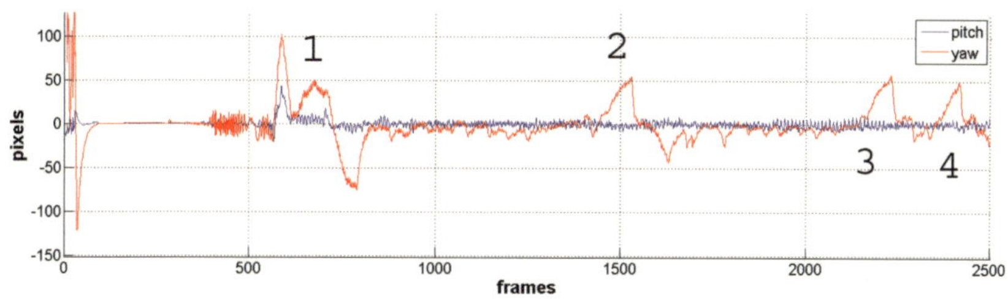

Fig. 22. Error between center of the image and center of the object to track.

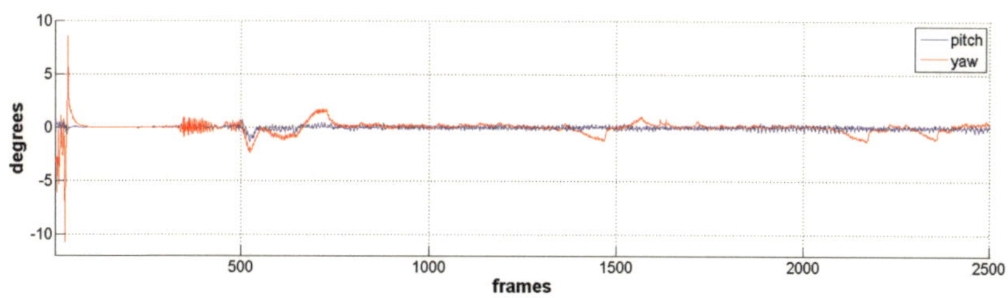

Fig. 23. Output from the Fuzzy Controller.

Tracking Moving Objects

In this part we present the tracking of a van with continuous movements of the helicopter increasing the difficulty of the test. In figure 24 we can see the error in pixels of the two axes of the image. Also, we can see the moments where we deselected the template and re-selected it, in order to increase the difficulty to the controller. These intervals show up as the error remains fixed in one value for a long time. At the same time the pilot move the helicopter in order to increase the difficulty to the controllers, and also, the template was deselected and reselected for made the situation more adverse. In figure 24 it is possible to see the error in pixels of the x and y axis of the image.

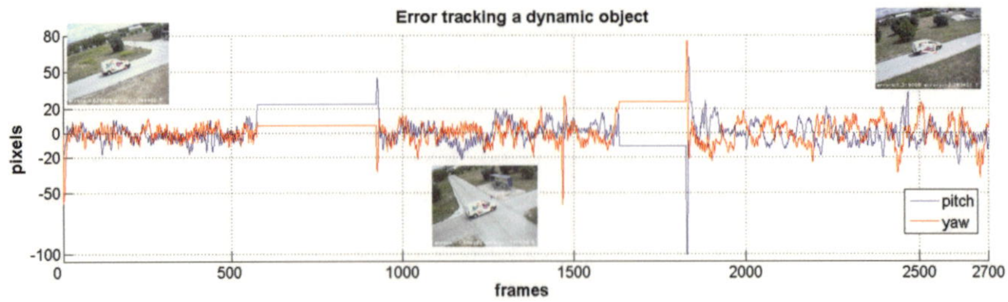

Fig. 24. Error between center of the image and center of the dynamic object (a van) to track.

In figures 25 and 26 we can see the response of the two controllers, showing the large movements sent by the controller to the servos when the mark is re-selected. Notice that in all the figures that show the controller responses, there are no data registered when the mark selection is lost because no motion is tracked. Figure 24 shows the data from the flight log, the black box of the helicopter. We can see that the largest response of the controllers are almost ±10 degrees for the yaw controller and almost 25 degrees for the pitch controller, corresponding to the control correction in a period of fewer than 10 frames.

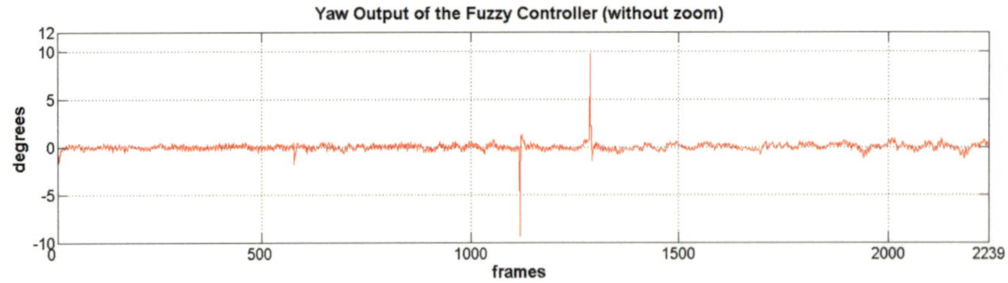

Fig. 25. Response of the Fuzzy control for the Yaw axis of the visual platform tracking a dynamic object (a van).

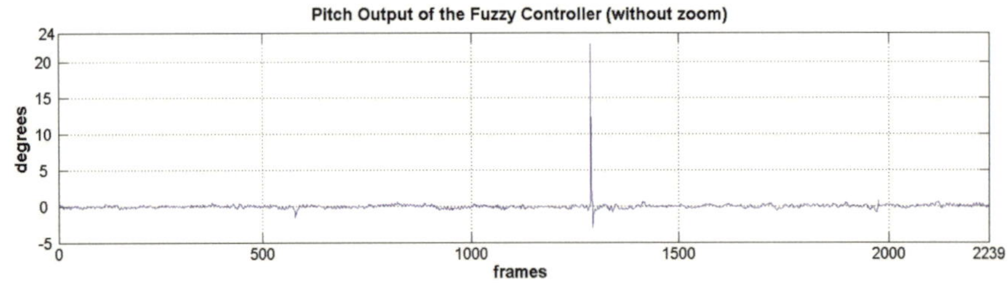

Fig. 26. Response of the Fuzzy control for the Pitch axis of the visual platform tracking a dynamic object (a van).

UAV Heading Control

Finally, we present results of one of the tests where the heading of the helicopter, and the camera platform are controlled using the three controllers explained.

In figure 28 we can see the response of the Fuzzy controller of the visual platform pitch angle, responding very quickly and with good behavior. In addition, figure 29 shows the controller response of the other axis of the platform. We can see a big and rapid movement near 1600 frames, reaching an error of almost 100 pixels. For this change we can see that the response of the controller is very fast, only 10 frames.

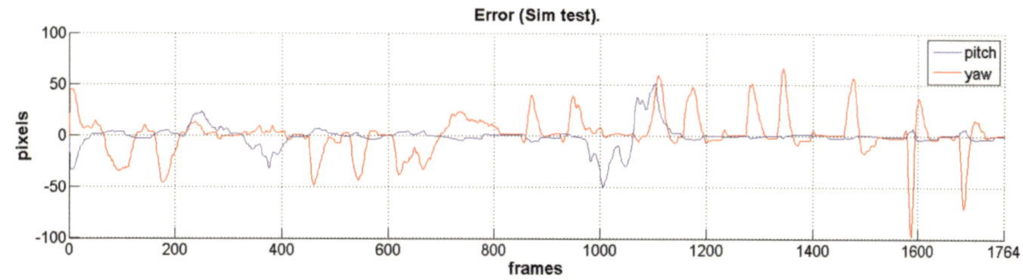

Fig. 27. Error between the static object tracked and the center of the image, running with the UAV simulator.

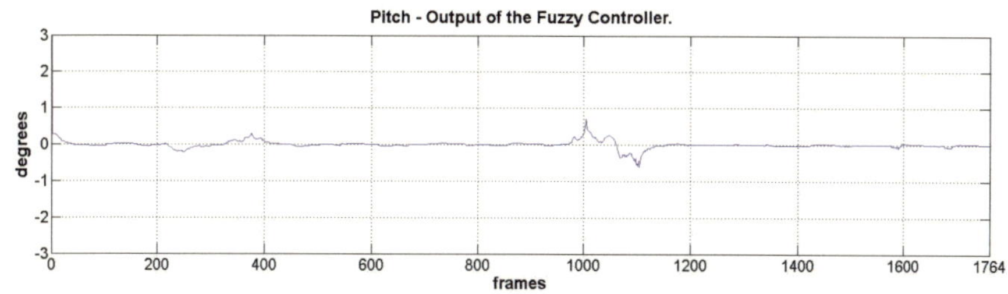

Fig. 28. Response of the Fuzzy control for the Pitch axis of the visual platform tracking a static object with the simulator of the UAV control.

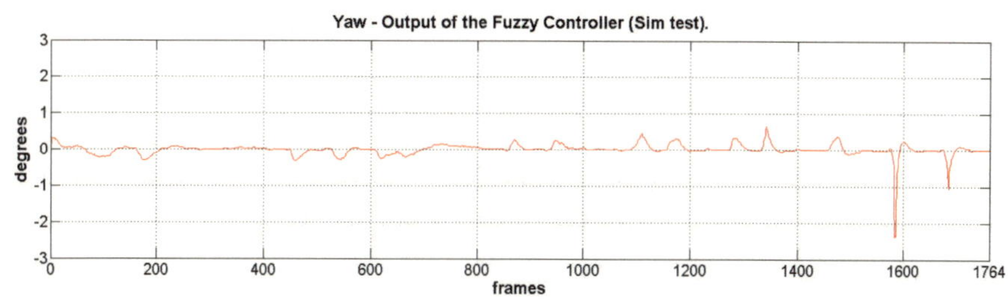

Fig. 29. Response of the Fuzzy control for the Yaw axis of the visual platform tracking a static object with the simulator of the UAV control.

The response of the heading controller is shown in figure 30, where we can see that it only responds to big errors in the yaw angle of the image. Also, we can see, in figure 31, how these signals affect the helicopter's heading, changing the yaw angle in order to collaborate with the yaw controller of the visual platform.

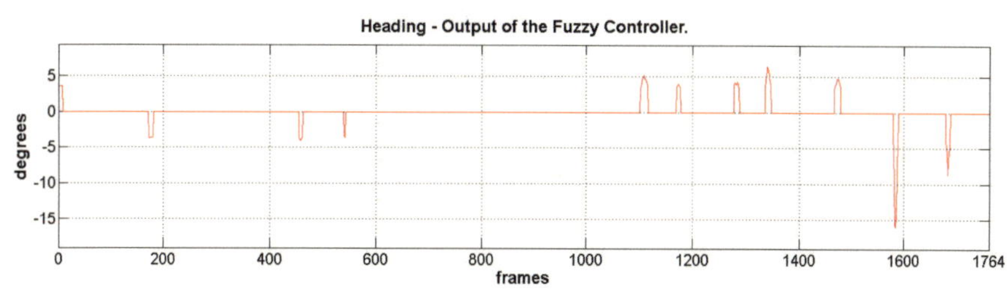

Fig. 30. Response of the Fuzzy control for the heading of the helicopter.

Fig. 31. Heading Response.

7. Conclusion

In this chapter, we have presented some of the techniques used for real time visual servoing on UAVs. These techniques includes visual algorithms for features detection and tracking, pose estimation, visual and pose based control systems, and fuzzy controllers, using them to increase the capabilities of UAVs in situations like object tracking, low altitude tasks such as: positioning and landing.
The methods explained have been integrated in a UAV control architecture, forming both, visual and pose based control systems, that have been tested on real UAV flights, showing the advantages of using visual systems on this kind of robots. Additional examples and videos of the visual systems and process presented in this chapter are available at the Colibri Project web page COLIBRI (2009)

8. Acknowledgments

The work reported in this paper is the product of several research stages at the Computer Vision Group of the Universidad Politécnica de Madrid, sponsored by the Spanish Science and Technology Ministry under grant CICYT DPI 2007-66156. The authors would like to thank Jorge León for supporting the flight trials and the I.A. Institute of the CSIC for collaborating in the flights consecution. The authors also like to thank the Universidad Politécnica de Madrid, the Consejería de Educación de la Comunidad de Madrid and the Fondo Social Europeo (FSE) for some of the authors PhD Scholarships.

9. References

Baker, S. and Matthews, I. (2002). Lucas-kanade 20 years on: A unifying framework: Part 1, *Technical Report CMU-RI-TR-02-16*, Robotics Institute, Carnegie Mellon University, Pittsburgh, PA.

Beis, J. S. and Lowe, D. G. (1997). Shape indexing using approximate nearest-neighbour search in highdimensional spaces, *CVPR '97: Proceedings of the 1997 Conference on Computer Vision and Pattern Recognition (CVPR '97)*, IEEE Computer Society, Washington, DC, USA, p. 1000.

Bouguet Jean Yves (1999). Pyramidal implementation of the lucas-kanade feature tracker, *Technical report*, Intel Corporation. Microprocessor Research Labs, Santa Clara, CA 95052.

Bradski, G. R. (1998). Computer vision face tracking for use in a perceptual user interface, *Intel Technology Journal*.

Buenaposada, J. M., Munoz, E. and Baumela, L. (2003). Tracking a planar patch by additive image registration, *Proc. of International workshop, VLBV 2003*, Vol. 2849 of *LNCS*, pp. 50–57.

Campoy, P., Correa, J. F., Mondrag´on, I., Mart´ınez, C., Olivares, M., Mej´ıas, L. and Artieda, J. (2009). Computer vision onboard UAVs for civilian tasks, *Journal of Intelligent and Robotic Systems.* 54(1-3): 105–135.

Canny, J. (1986). A computational approach to edge detection, *IEEE Trans. Pattern Analysis and Machine Intelligence.* 8(6): 679–698.

Chaumette, F. and Hutchinson, S. (2006). Visual servo control. I. basic approaches, *Robotics & Automation Magazine, IEEE* 13(4): 82–90.

COLIBRI (2009). Universidad Polit´ecnica de Madrid. Computer Vision Group. COLIBRI Project, http://www.disam.upm.es/colibri.

Conticelli, F., Allotta, B. and Khosla, P. (1999). Image-based visual servoing of nonholonomic mobile robots, *Decision and Control, 1999. Proceedings of the 38th IEEE Conference on*, Vol. 4, pp. 3496–3501 vol.4.

Criminisi, A., Reid, I. D. and Zisserman, A. (1999). A plane measuring device, *Image Vision Comput.* 17(8): 625–634.

Duda, R. O. and Hart, P. E. (1972). Use of the hough transformation to detect lines and curves in pictures, *Commun. ACM* 15(1): 11–15.

Feddema, J. and Mitchell, O. (1989). Vision-guided servoing with feature-based trajectory generation [for robots], *Robotics and Automation, IEEE Transactions on* 5(5): 691–700.

Fischer, M. A. and Bolles, R. C. (1981). Random sample concensus: a paradigm for model fitting with applications to image analysis and automated cartography, *Communications of the ACM* 24(6): 381–395.

Harris, C. G. and Stephens, M. (1988). A combined corner and edge detection, *In Proceedings of the 4th Alvey Vision Conference*, pp. 147–151.

Hartley, R. I. and Zisserman, A. (2004). *Multiple View Geometry in Computer Vision*, second edn, Cambridge University Press, ISBN: 0521540518.

Herbert Bay, Tinne Tuytelaars and Luc Van Gool (2006). SURF: Speeded Up Robust Features, *Proceedings of the ninth European Conference on Computer Vision.*

Hutchinson, S., Hager, G. D. and Corke, P. (1996). A tutorial on visual servo control, *IEEE Transaction on Robotics and Automation*, Vol. 12(5), pp. 651–670.

Kragic, S. and Christensen, H. I. (2002). Survey on visual servoing for manipulation, *Tech. Rep ISRN KTH/NA/P–02/01–SE*, Centre for Autonomous Systems,Numerical Analysis and Computer Science, Royal Institute of Technology, Stockholm, Sweden, Fiskartorpsv. 15 A 100 44 Stockholm, Sweden. Available at www.nada.kth.se/ danik/VSpapers/report.pdf.

Lowe, D. G. (2004). Distintive image features from scale-invariant keypoints, *International Journal of Computer Vision* 60(2): 91–110.

Lucas, B. D. and Kanade, T. (1981). An iterative image registration technique with an application to stereo vision, *Proc. of the 7th IJCAI*, Vancouver, Canada, pp. 674–679.

Mariottini, G. L., Oriolo, G. and Prattichizzo, D. (2007). Image-based visual servoing for nonholonomic mobile robots using epipolar geometry, *Robotics, IEEE Transactions on* 23(1): 87–100.

Mart´ınez, C., Campoy, P., Mondragon, I. and Olivares, M. (2009). Trinocular Ground System to Control UAVs, *IEEE/RSJ International Conference on Intelligent Robots and Systems. IROS.*

Masutani, Y., Mikawa, M., Maru, N. and Miyazaki, F. (1994). Visual servoing for non-holonomic mobile robots, *Intelligent Robots and Systems '94. 'Advanced Robotic Systems and the Real World', IROS '94. Proceedings of the IEEE/RSJ/GI International Conference on, Vol. 2*, pp. 1133–1140 vol.2.

Mejias, L. (2006). *Control visual de un vehiculo aereo autonomo usando detecci´on y seguimiento de caracter´ısticas en espacios exteriores.*, PhD thesis, Escuela T´ecnica Superior de Ingenieros Industriales. Universidad Polit´ecnica de Madrid, Spain.

Mej´ıas, L., Mondrag´on, I., Correa, J. F. and Campoy, P. (2007). Colibri: Vision-guided helicopter for surveillance and visual inspection, *Video Proceedings of IEEE International Conference on Robotics and Automation*, Rome, Italy, pp. 2760–2761.

Mejias, L., Roberts, J., Campoy, P., Usher, K. and Corke, P. (2006). Two seconds to touchdown. Visionbased controlled forced landing, *Proceedings of the IEEE/RSJ International Conference on Intelligent Robots and Systems*, Beijing, China, p. to appear.

Olivares, M. and Madrigal, J. (2007). Fuzzy logic user adaptive navigation control system for mobile robots in unknown environments, *Intelligent Signal Processing, 2007. WISP 2007. IEEE International Symposium on* pp. 1–6.

Olivares, M., Campoy, P., , Correa, J., Martinez, C. and Mondragon, I. (2008). Fuzzy control system navigation using priority areas, *Proceedings of the 8th International FLINS Conference*, Madrid,Spain, pp. 987–996.

Rousseeuw, P. J. and Leroy, A. M. (1987). *Robust regression and outlier detection*, JohnWiley & Sons, Inc., New York, NY, USA.

Shi, J. and Tomasi, C. (1994). *Good features to track, 1994 IEEE Conference on Computer Vision and Pattern Recognition* (CVPR'94), pp. 593–600.

Siciliano, B. and Khatib, O. (eds) (2008). *Springer Handbook of Robotics*, Springer, Berlin, Heidelberg.

Simon, G. and Berger, M.-O. (2002). Pose estimation for planar structures, *Computer Graphics and Applications, IEEE* 22(6): 46–53.

Simon, G., Fitzgibbon, A. and Zisserman, A. (2000). Markerless tracking using planar structures in the scene, *Augmented Reality, 2000. (ISAR 2000). Proceedings. IEEE and ACM International Symposium on*, pp. 120–128.

Sobel I., F. G. (1968). A 3x3 isotropic gradient operator for image processing, *presented at a talk at the Stanford Artificial Project.*

Sturm, P. (2000). Algorithms for plane-based pose estimation, *Proceedings of the IEEE Conference on Computer Vision and Pattern Recognition, Hilton Head Island, South Carolina, USA*, pp. 1010–1017.

Swain, M. J. and Ballard, D. H. (1991). Color indexing, *Int. J. Comput. Vision* 7(1): 11–32.

Zhang, Z. (2000). A flexible new technique for camera calibration, *IEEE Transactions on pattern analysis and machine intelligence* 22(11): 1330–1334.

10

Video Watermarking Technique using Visual Sensibility and Motion Vector

Mariko Nakano-Miyatake and Hector Perez-Meana
National Polytechnic Institute of Mexico
Mexico

1. Introduction

Together with the rapid growth of Internet service, copyright violation problems, such as unauthorized duplication and alteration of digital materials, have increased considerably (Langelaar et al., 2001). Therefore copyright protection over the digital materials is a very important issue that requires an urgent solution. The watermarking is considered as a viable technique to solve this problem. Until now, numerous watermarking algorithms have been proposed. Most of them are image watermarking algorithms and relatively few of them are related with video sequences. Although image watermarking algorithms can be used to protect the video signal, generally they are not efficient for this purpose, because image watermarking algorithms does not consider neither temporal redundancy of the video signal nor temporal attacks, which are efficient attacks against video watermarking (Swanson et al., 1998).
Generally, in the watermarking schemes for copyright protection, the embedded watermark signal must be imperceptible and robust against common attacks, such as lossy compression, cropping, noise contamination and filtering (Wolfgang et al., 1999). In addition, video watermarking algorithms must satisfy the following requirements: a blind detection, high speed process and conservation of video file size. The blind detection means that the watermark detection process does not require original video sequence, and the temporal complexity of watermark detection must not affect video decoding time. Also the file size of video sequence must be similar, before and after watermarking. Due to the redundancy of the video sequence, some attacks such as frame dropping and frame averaging can effectively destroy the embedded watermark, without cause any degradation to the video signal. A design of an efficient video watermarking algorithm must consider this type of attacks (Wolfgang et al., 1999).
Basically, video watermarking algorithms can be classified into three categories: watermarking in base band (Wolfgang et al., 1999; Hartung & Girod 1998; Swanson et al., 1998; Kong et al., 2006), watermarking during video coding process (Liu et al., 2004; Zhao et al., 2003; Ueno 2004; Noorkami & Mersereau 2006) and watermarking in coded video sequence (Wang et al., 2004; Biswas et al., 2005; Langelaar & Lagendijk 2002). In the base band technique, the watermarking process is realized in uncompressed video stream, in which almost all image watermarking algorithms can be used, however generally computational complexity for watermark embedding and detection is considerably high for

its practical use. In the algorithm proposed by Wolfgang et al. (1999), Just Noticeable Difference (JND) is used, in the Discrete Cosine Transform (DCT) domain, to determine an adequate watermark embedding energy. Hartung and Girod (1998) proposed an algorithm, in which binary data modulated by pseudo-random sequence is embedded into luminance component of each video frame. Swanson et al. (1998) proposed an algorithm based on the Discrete Wavelet Transform (DWT) through temporal sequences. Kang et al. applied singular value decomposition (SVD) to each frame of video data, and then embedded the watermark signal into the singular values.

The watermarking technique in compressed video data, embed the watermark signal into bit sequence compressed by standard coding, such as MPEG-2 and MPEG-4, etc. Generally this technique has lower computational cost, compared with other methods; however the number of watermark bits must be limited by compression rate. In the algorithm proposed by Wang et al. (2004), the watermark signal is embedded only into the I-frames using JND concept, while Biswas et al. (2005) directly embedded the watermark signal into MPEG compressed video sequence, modifying DCT coefficients. Also in the algorithm proposed by Langelaar and Lagendik (2001), the watermark signal is embedded into the I-frames in the DCT domain.

The watermarking algorithms operating during MPEG coding process are inherently robust against standard compression attacks, without increase the compression rate of the video sequence. Liu et al. (2004) proposed an algorithm, where the watermark signal is embedded into the motion vectors, and using the watermarked motion vectors, MPEG bit sequence is generated. While Zhao et al. (2003) proposed a fast algorithm to estimate motion vectors during the compression process, and also they embed the watermark signal, modifying angle and magnitude of the motion vectors. In the algorithm proposed by Ueno (2004), motion vectors are used to determine an adequate position in DCT coefficients of I-frames for watermark embedding. Noorkami and Mersereau (2006) estimated motion regions, computing spatial distribution of motion through several consecutive frames. Large amount of watermark bits are embedded into dynamic motion regions, while small amount of watermark bits are embedded into statistic regions. In this manner the artifact caused by watermark embedding can be avoided (Noorkami & Mersereau 2006).

In this paper, a video watermarking algorithm is proposed, in which watermark embedding is carried out during MPEG2 coding process. The proposed algorithm uses three criteria based on deficiency of the Human Visual System (HVS) to embed robust watermark, while preserving its imperceptibility. First criterion is based on difference of sensibility of the HVS to basic three color channels (red, green and blue), and second one is based on frequency masking of the HVS proposed by Tong and Venetsanopoulos (1998). Third criterion is based on deficiency of the HVS to trace high speed motion region, which is related directly to the motion vector of each macro-block. The third criterion is only applied to P-frames, while other two criteria are applied both I-frames and P-frames. In the proposed algorithm, B-frames are excluded from the watermark embedding and detection process to reduce computational complexity. In this manner, watermark embedding and detection processes don't cause any delay in coding and decoding processes. Simulation results show the watermark imperceptibility and robustness against common signal processing and some intentional video frame attacks, such as frame dropping, frame averaging and frame swapping. The watermark imperceptibility is measured using the Peak Signal Noise Ratio (PSNR) and a HVS based objective evaluation proposed by Wang and Bovik (2004).

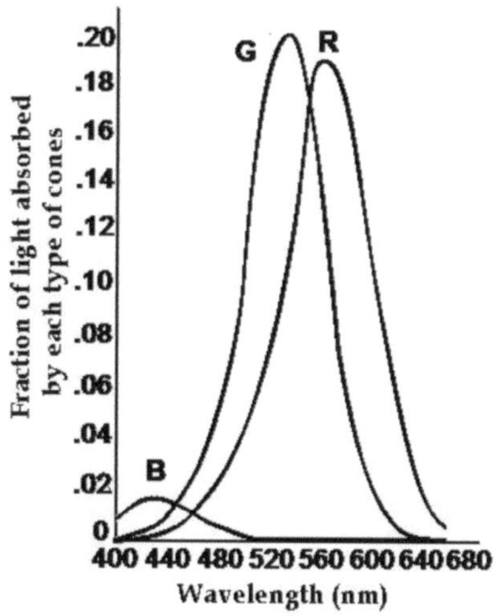

Fig. 1. Sensibility of HVS to different wavelength related to three basic colors.

2. Proposed system

In this section, a detailed description of the proposed video watermarking algorithms is provided.

2.1 HVS based criteria

In the proposed algorithm, the three criteria mentioned in introduction are used to embed imperceptible and robust watermark into a video sequence. These criteria are based on deficiency of sensibility of the HVS to blue channel, regions with details, such as texture region and region with high motion speed.

In the HVS, there are three types of cones that react to the basic three colors: red, green and blue. The number of cones reacted to blue is 30 times smaller than the number of cones reacted to red or green, which means the HVS has deficiency of sensibility to blue color (Sayood, 2000). The figure 1 shows fraction of light absorbed by each type of cone, here R, G, and B represent red, green and blue colors, respectively. The proposed algorithm embeds watermark signal into the blue channel, using its HVS deficiency. Generally color space used for video sequence is YUV or YCrCb, therefore firstly these color spaces are transformed into RBG color space using the transform matrix given by (1) and (2). (Plataniotis & Venesanopoulos 2000).

$$\begin{bmatrix} R \\ G \\ B \end{bmatrix} = \begin{bmatrix} 0.299 & 0.587 & 0.114 \\ 0.596 & -0.275 & -0.321 \\ 0.212 & -0.523 & 0.311 \end{bmatrix}^{-1} \begin{bmatrix} Y \\ C_r \\ C_b \end{bmatrix} \tag{1}$$

$$\begin{bmatrix} R \\ G \\ B \end{bmatrix} = \begin{bmatrix} 0.299 & 0.587 & 0.114 \\ -0.147 & -0.289 & 0.436 \\ 0.615 & -0.5149 & -0.100 \end{bmatrix}^{-1} \begin{bmatrix} Y \\ U \\ V \end{bmatrix} \tag{2}$$

The second criterion is based on the difference of the HVS's sensibility to spatial features, such as texture, edge and plain regions. Each I-frame and P-frame is divided into blocks of size 8x8, and then 2D-DCT is applied to each block. Classification of each block is carried out using algorithm proposed by Tong and Venetsnopoulos (1998), which is described briefly as follows:

1. Each DCT block is divided into 4 areas denoted by DC, L, E and H, as shown by figure 2.
2. The sum of absolute value of coefficients belonging to DC, L, E and H are denoted as S_{DC}, S_L, S_E and S_H, respectively.
3. Using the following conditions, each block of DCT is classified as "edge block", "Texture block" or "plain block".

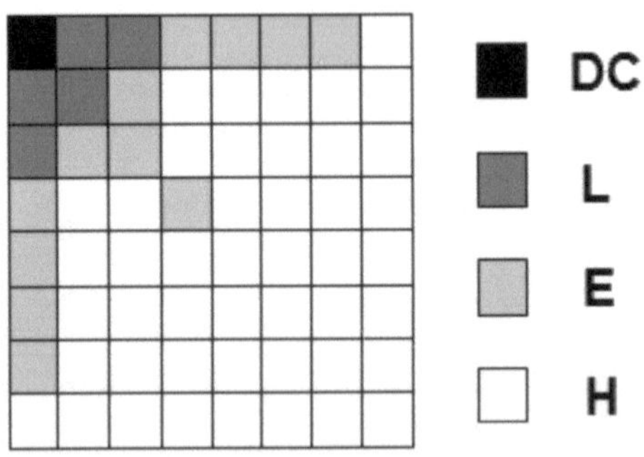

Fig. 2. Four regions of a DCT block

Conditions for "edge block"

If either of two conditions A or B is satisfied, then the DCT block is classified as "edge block".

Condition-A

$$\text{If } (S_E + S_H \le \mu_1) \wedge \left(\frac{S_L}{S_E} \ge \alpha_1\right) \wedge \left(\frac{S_L + S_E}{S_H} \ge \beta_1\right)$$
$$\vee$$
$$\text{If } (S_E + S_H \le \mu_1) \wedge \left(\frac{S_L}{S_E} \ge \beta_1\right) \wedge \left(\frac{S_L + S_E}{S_H} \ge \alpha_1\right)$$
$$\vee$$
$$\text{If } (S_E + S_H \le \mu_1) \wedge \left(\frac{S_L + S_E}{S_H} \ge \gamma\right).$$

Condition-B

$$\text{If } (S_E + S_H > \mu_1) \wedge \left(\frac{S_L}{S_E} \ge \alpha_2\right) \wedge \left(\frac{S_L + S_E}{S_H} \ge \beta_2\right)$$
$$\vee$$
$$\text{If } (S_E + S_H > \mu_1) \wedge \left(\frac{S_L}{S_E} \ge \beta_2\right) \wedge \left(\frac{S_L + S_E}{S_H} \ge \alpha_2\right)$$
$$\vee$$
$$\text{If } (S_E + S_H > \mu_1) \wedge \left(\frac{S_L + S_E}{S_H} \ge \gamma\right)$$

where the operations $\wedge$ and $\vee$ mean logical multiplication and logical addition, and six parameters used in the conditions are $\mu_1 = 900$, $\alpha_1 = 2.3, \beta_1 = 1.6, \alpha_2 = 1.4, \beta_2 = 1.1$ y $\gamma = 4$.

Condition for "texture block"

If the condition-A is not satisfied and $S_E + S_H > \kappa$, or if the condition-B is not satisfied then the block is classified as "texture block", where $\kappa = 290$.

Condition for "plain block"

If $S_E + S_H \le \mu_2$ is satisfied or the condition-A is not satisfied and $S_E + S_H \le \kappa$, then the block is classified as "plain block", where $\mu_2 = 125$.

(a) (b)

Fig. 3. An example of the block classification using second criterion

The figure 3 shows an example of block classification using the above algorithm (Tong and Venetsanopoulos, 1998). Here black blocks, gray blocks and white blocks indicate "plain blocks", "texture blocks" and "edge blocks", respectively.

The last criterion is based on deficiency of the HVS to trace regions with high speed motion. Actually the MPEG coding uses this deficiency to reduce temporal redundancy of video sequence. The macro-blocks, whose motion vector has large magnitude, can be classified as regions with high speed motion. The macro-blocks classified as high motion speed regions are adequate to embed a high energy watermark signal without causing any visual distortion. The magnitude of the motion vector is computed by (3).

$$Mmv_i = \sqrt{mvh_i^2 + mvv_i^2}, \quad i = 1..MB \tag{3}$$

where mvh_i, mvv_i are horizontal and vertical components of the motion vector of i-th macro-block, and MB is total number of macro-blocks. To determine macro-blocks with high speed motion, a threshold value Th_mv is introduced, which value is computed by (4)

$$Th\_mv = \frac{1}{MB}\sum_{i=1}^{MB} Mmv_i \tag{4}$$

Using this value, macro-block is classified as follows.

If $Mmv_i < Th\_mv$ then i-th macro-block doesn't have motion (static region).
If $Mmv_i \ge Th\_mv$ then i-th macro-block has motion (dynamic region).

The macro-blocks, whose magnitude of motion vector is smaller than the threshold, are considered as static blocks and the motion vectors of the static blocks are ignored. The figure

4(a) and (b) show two consecutive frames, all motion vectors before classification are depicted in fig 4(c) and fig. 4(d) shows only the motion vectors classified as high speed motion.

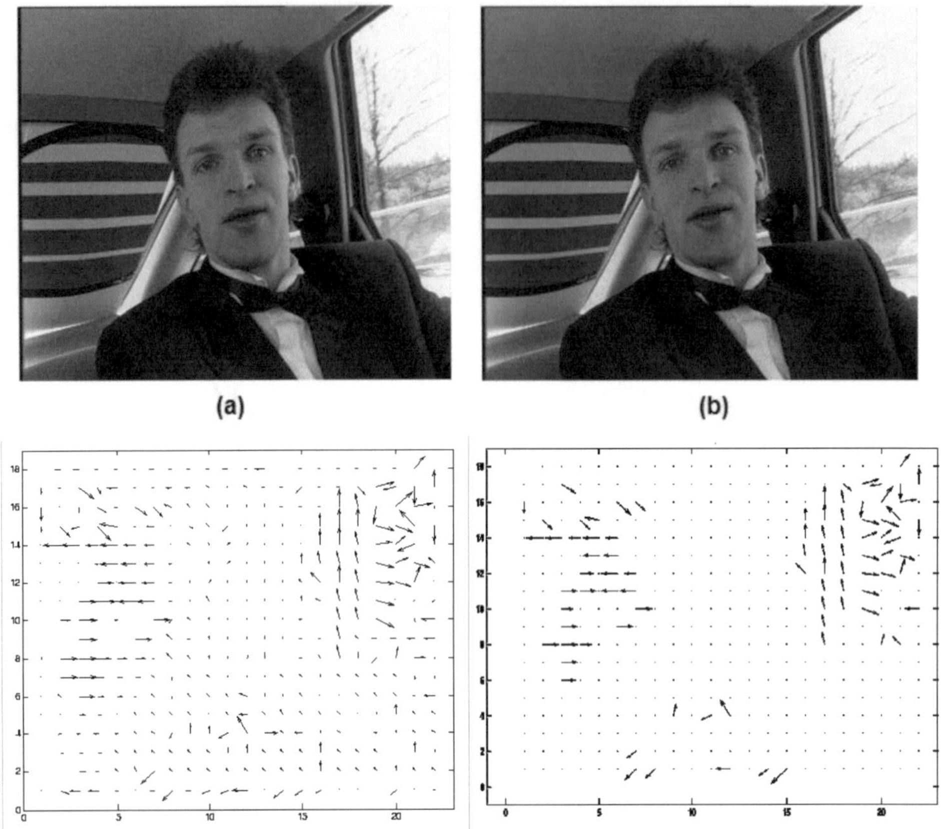

Fig. 4. (a) and (b) are two consecutive frames, (c) all motion vectors got from two consecutive frames and (d) motion vectors with high speed motion.

Energy	$B\in$ Plain	$B\in$ Texture	$B\in$ *Edge*
$B\in C_I$	0.4	0.6	0.9
$B\in C_P$ $B\notin Mb_{motion}$	0.4	0.6	0.9
$B\in C_P$ $B\in Mb_{motion}$	0.8	1.2	1.8

Table 1. Watermark embedding energy

Combining latter two criteria, the spatial feature of 8x8 blocks and the motion feature of macro-blocks (16x16), the watermark embedding energy for I-frames and P-frames is determined experimentally using 10 video sequences. The embedding energies of different type of blocks are shown by the table 1; in which, *B* means blocks of size 8x8 of I-frames and P-frames, and Mb_{motion} means macro-blocks with high speed motion. Each macro-block contains 4 blocks *B*, and C_I, C_P mean I-frames and P-frames, respectively. Figure 5 shows an example of block classification together with the watermark embedding energy assigned to each block.

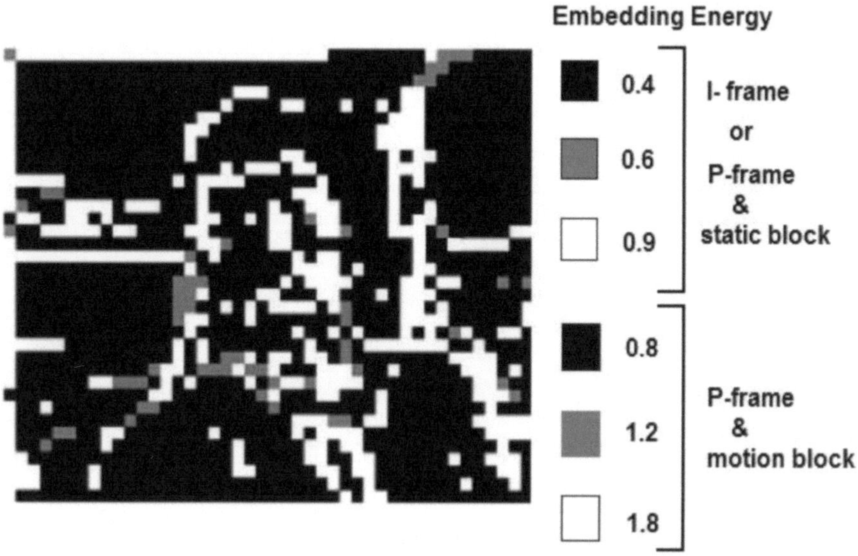

Fig. 5. An example of watermark embedding energy assignation

2.2 Watermark embedding process

The watermark embedding process of the proposed algorithm consists of two parts: first part and second part. In the first part, an adequate watermark embedding energy for each block is calculated, and in the second part, the watermark signal is embedded into each block using the adequate embedding energy computed in the first part. The video sequence is decomposed by RGB color space, and only the blue channel is used for watermark embedding. The blue channel is divided into blocks of 8x8 and then each block is transformed by 2D-DCT. Each block is classified into three categories: plain block, texture block and edge block. For P-frames, the macro-blocks are generated by combining four neighbor blocks of 8x8, and then each macro-block is classified between static block and block with high speed motion. Using table 1, the watermark embedding energy is assigned to each block (8x8). The watermark signal is a pseudo-random sequence generated by the secret user's key. Watermark embedding is performed by (5).

$$
\begin{aligned}
& DCT_k(i,j) = DCT_k(i,j) + \alpha_k \left| DCT_k(i,j) \right| W_k \\
& (i,j) = (1,2) \ \text{ for } C_I \\
& (i,j) \in \{(1,2),(1,3),(2,1),(2,2),(3,1)\} \ \text{ for } C_P
\end{aligned}
\tag{5}
$$

where $DCT_k(i,j)$ is *(i,j)-th* DCT coefficient of *k-th* block and α_k is the embedding energy assigned to *k-th* block. For I-frames, a watermark bit is embedded only into a AC coefficient with lowest frequency, while for P-frames; five watermark bits are embedded into five lowest AC coefficients. The figure 6 shows the watermark embedding process.

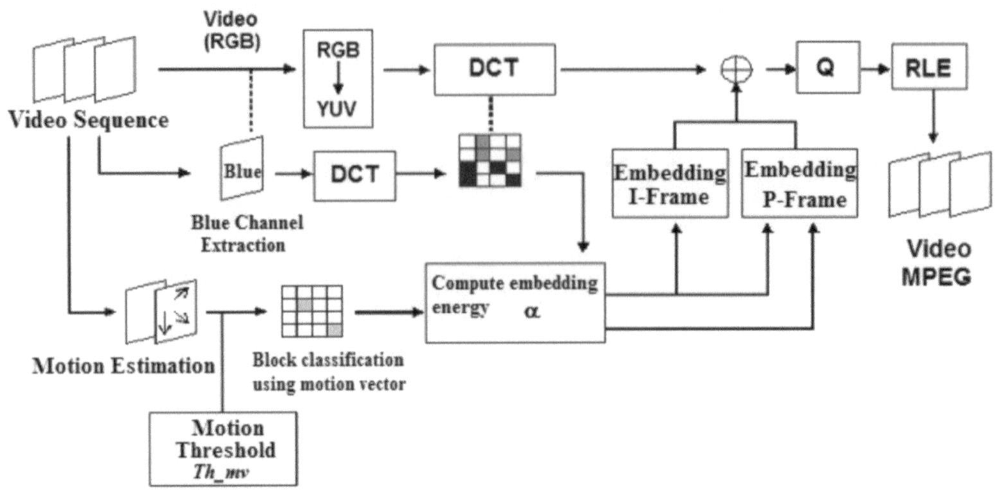

Fig. 6. Watermark embedding process

2.3 Watermark detection process

In the watermark detection process, only the blue channel of the watermarked and possibly distorted video sequences is used, which is divided into blocks of 8x8 pixels. 2D-DCT is applied to each block of I-frames and P-frames. The lowest AC coefficient of each block of I-frame and the five lowest coefficients of each block of P-frame are extracted. The extracted coefficients are concatenated through all blocks of I and P frames to generate an extracted watermark sequence Y. Finally to determine if owner's watermark is presented in the video sequence or not, the cross-correlation between the extracted watermarked coefficients *Y* and the owner's watermark sequence *W*, is calculated as shown by (6).

$$C = \frac{1}{L}\sum_{i=1}^{L} W_i Y_i \tag{6}$$

where L is watermark length.

If the cross-correlation value C is bigger than a predetermined threshold value Th_w, it is considered that the owner's watermark signal is presented in the video sequence; otherwise the video sequence was not watermarked or watermarked by another watermark sequence. Here the threshold value plays a very important role and this value is determined considering two probabilities: probability of detection error and probability of false alarm error. In the proposed algorithm, adaptive threshold value is used, which is given by (7). This threshold value guarantees that false alarm error probability is smaller than 10^{-6}. (Piva et al., 1997).

$$Th_w = \frac{1}{3L}\sum_{i=1}^{L} |Y_i| \alpha_i \tag{7}$$

3. Experimental results

To evaluate the proposed algorithm, several video sequences with format YUV-CIF are used. The figure 7 shows some video sequences used in the evaluation. The proposed algorithm is evaluated from embedded watermark imperceptibility and robustness points of view.

Fig. 7. Some video sequences used for evaluation of the proposed algorithm

3.1 Imperceptibility

Embedded watermark imperceptibility of the proposed algorithm is evaluated using PSNR and the universal quality index (UQI) proposed by Wang et al. (2004). The UQI is an objective quality assessment related with perceptual distortion, which was originally developed to assess a visual quality for images as given by (8). In order to assess perceptual quality of the video sequence, UQI value of each macro-block is compensated according to the motion speed of the macro-block.

$$UQI = \frac{4\sigma_{xy}\bar{x}\cdot\bar{y}}{\left(\sigma_x^2+\sigma_y^2\right)\left[(\bar{x})^2+(\bar{y})^2\right]} \tag{8}$$

where $\bar{x}, \bar{y}$ are mean values of original image x and processed image y, respectively, σ_x^2 and σ_y^2 are variances of x and y. respectively, and σ_{xy} is their covariance. The value range of UQI is [0, 1.0], when the video sequence under analysis is identical with the original one, the UQI value is equal to 1.0; and as the visual distortion increase, the UQI value decrease.
The figure 8 shows the original I-frame and P-frame, and their watermarked versions, respectively, together with the PSNR values. From this figure, we can considered that the embedded watermark is imperceptible, because the PSNR value is bigger than 40dB for I-frame, and it is approximately 40dB for P-frame, and UQI of the watermarked video sequence is equal to 0.96, which indicates that the embedded watermark is imperceptible by the HVS.

	Original frame	Watermarked frame	PSNR (dB)
I-frame			42.91
P-frame			39.83

Fig. 8. Watermark imperceptibility of I-frame and P-frame

3.2 Watermark robustness

To evaluate the watermark robustness of the proposed algorithm, the watermarked video sequences are attacked using common signal and image processing tasks, such as coding rate change, impulsive and Gaussian noise contamination, frame dropping, frame swapping, frame averaging and cropping. The figure 9 shows the watermark robustness against the coding rate change, applying quantized matrix with different quality factor during MPEG coding process. Fig 9(a) shows one frame of the watermarked video without any attack, fig. 9 (c) shows one frame of the watermarked and compressed video with quality factor equal to 30. Figure 9(b) and fig. 9(d) show the detector responses. From these figures, it is concluded that the embedded watermark is robust to high rate compression; actually the watermark signal survived after compression with quality factor of 30. If the watermarked video sequence is compressed using a lower quality factor than 30, the embedded watermark signal can be lost, however in this situation the distortion caused by compression is not acceptable and the attacked video sequence no longer has commercial value. In all robustness evaluations, the watermarked videos are analyzed using 1000 possible watermark signals generated by 1000 different keys; and the embedded watermark generated by the owner corresponds to the key equal to 450. In all figures, the horizontal line represents the threshold value calculated by (7).

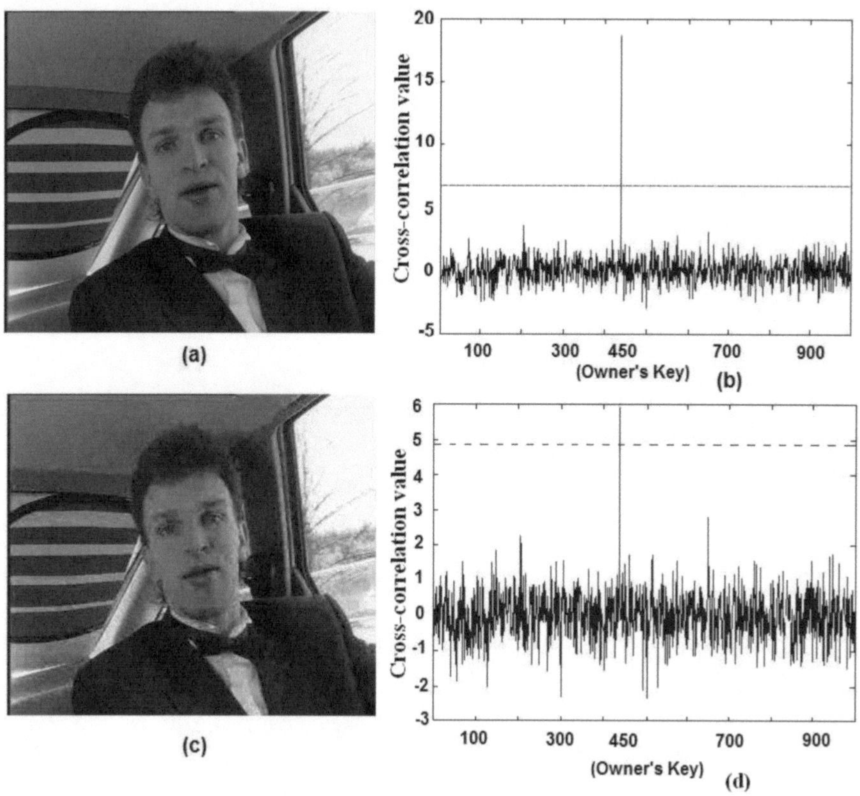

Fig. 9. (a) Watermarked frame, (c) watermarked and compressed frame, (b) and (d) detector responses for (a) and (c), respectively

The figure 10 shows the watermark robustness against impulsive noise contamination. The fig. 10(a) and fig. 10(c) show the watermarked frames that received impulsive noise contamination with a density of 3% and 10%, respectively; here fig. 10(b) and fig 10(d) are the detector responses of fig. 10(a) and fig. 10(c). Fig. 11 shows the watermark robustness against Gaussian noise contamination, here fig. 11(a) and fig. 11(c) show a watermarked and contaminated frame by Gaussian noise with variance 0.01 and 0.05, respectively; and fig. 11(b) and fig. 11(d) are detector responses of both cases, respectively. From these figures, we can conclude that the embedded watermark is sufficiently robust to impulsive and Gaussian noise contamination.

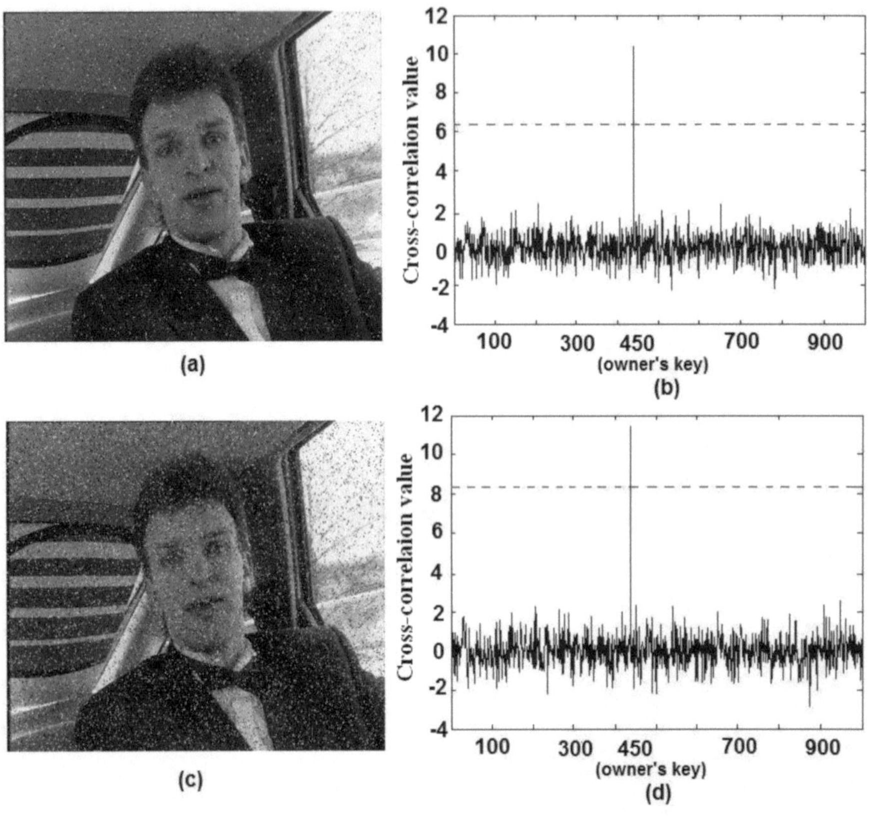

Fig. 10. Watermark Robustness against impulsive noise contamination.

Also in the proposed algorithm, the embedded watermark is sufficiently robust against cropping attack. The figure 12 shows the watermarked cropped frames and watermark detection performance from the cropped video sequence. Here fig. 12(a) and fig. 12(c) show a video sequence in which 40% and 75% of all frames of watermarked video sequence are cropped, and fig. 12(b) and fig. 12(d) are detector responses of both cases, respectively.

Frame dropping, frame swapping and frame averaging are intentional attacks for watermarked video sequences (Zhyang et al., 2004). These frame attacks take advantage of the temporal redundancy of video sequences and try to destroy efficiently the embedded watermark signal, without causing any visual degradation in the video sequence. Frame dropping, frame swapping and frame averaging attacks are described by (9), (10) and (11), respectively.

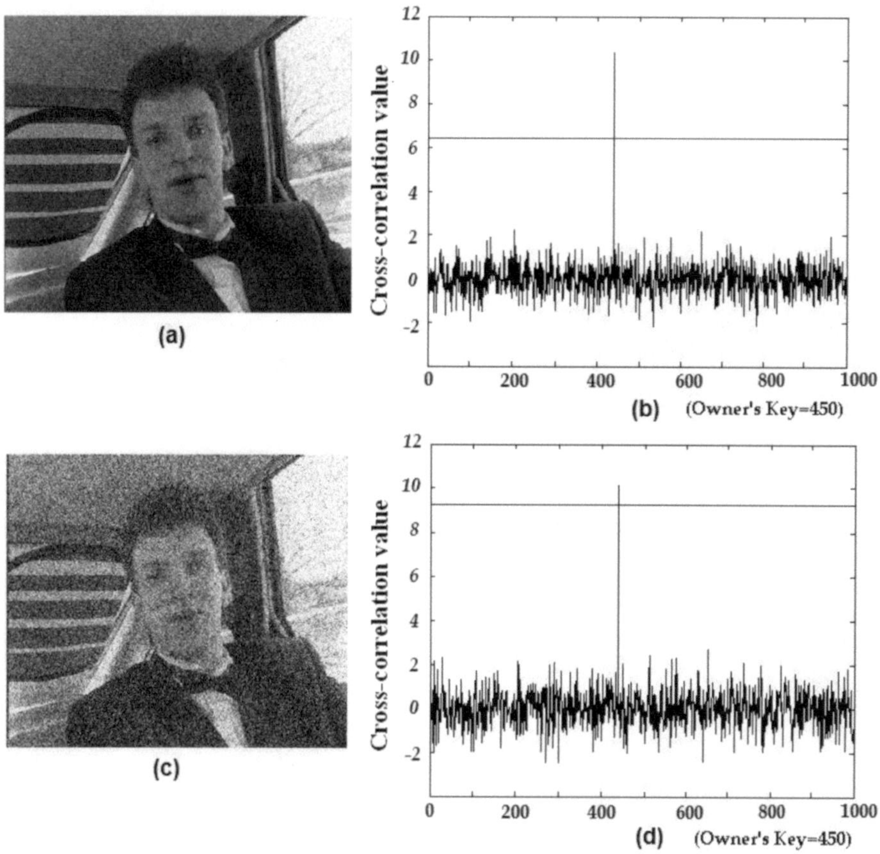

Fig. 11. Watermark Robustness against Gaussian noise contamination

$$V_{attacked} = V_{watermarked} - \{F_{r1}, F_{r2}, ...F_{rn}\} \quad (9)$$

where $V_{attacked}$ and $V_{watermarked}$ are attacked watermarked sequences by frame dropping and the watermarked sequence without attack, respectively, and $[r1, r2, ...rn]$ are the numbers of frames selected randomly.

$$\tilde{F}_k \Leftrightarrow F_{k+1} \;\; \tilde{F}_{k+1} \Leftrightarrow F_k \quad (10)$$

$$\tilde{F}_k = \frac{1}{3}[F_{k-1} + F_k + F_{k+1}] \quad (11)$$

where $F_k, \tilde{F}_k$ are *k-th* frames of watermarked and attacked videos, respectively.

Because the proposed algorithm embeds watermark signals through temporal video sequences, the embedded watermark signal is inherently robust to frame attacks. Figure 13 shows the watermark robustness against frame averaging attack. Fig. 13(a) shows a result of averaging of one I-frame and two P-frames, and fig. 13(b) shows a result of averaging three P-frames; and fig. 13(b) and fig. 13(d) are the detector responses of both cases.

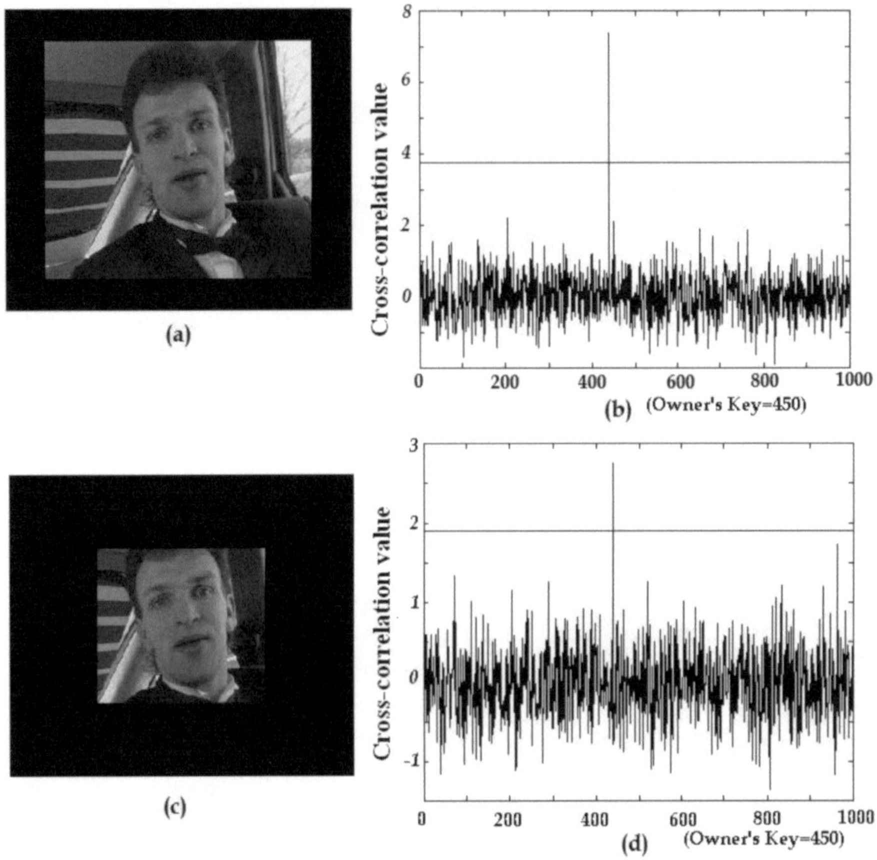

Fig. 12. Watermark robustness against cropping. (a) 40% of video data is cropped and (b) 75% of video data is cropped and (c) and (d) Detector responses of both cases.

3.3 Computational cost for watermark embedding and detection

In the video watermarking techniques, time consuming caused by watermark embedding and detection processes must be minimized. In our experiment, to evaluate the consumed time for watermarking operation, consider the processing time of MPEG coding with/without the proposed watermarking process. As shown by table 2, the processing time with the watermarking operation increases approximately 40% for I-frames and 10% for P-frames, compared with the processing time without watermarking process. Considering that the number of P-frames is approximately 4 times larger than that of the I-frames and that the B-frames are excluded for watermarking process, the overall time required for watermarking operation is smaller than 10% of the total time required by the MPEG compression. This result means that the proposed watermarking algorithm is suitable for an actual implementation.

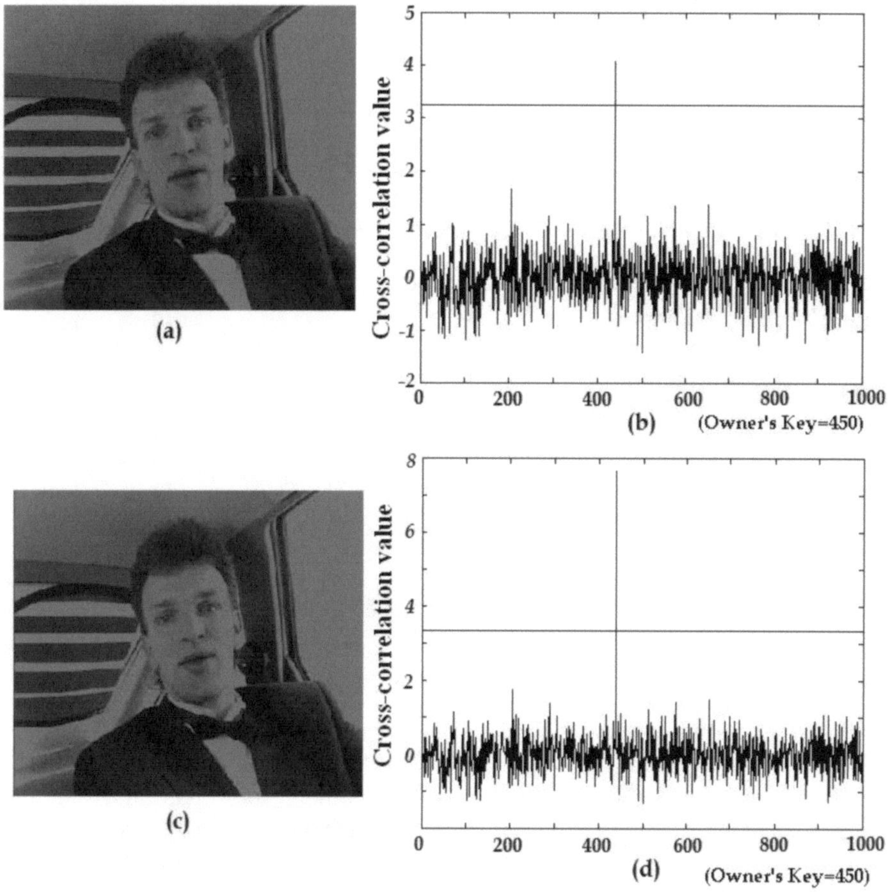

Fig. 13. (a) Frame averaging with I-frame, P-frames (b) detector response of (a), (c) Frame averaging with P-frames and (d) detector response of (c).

Frame Type	**Coding Time without watermarking**	**Coding time With watermarking**
I	0.83 sec /frame	1.2 sec /frame
P	0.3 sec / frame	0.33 sec /frame

Table 2. Computational cost of watermarking process

3.4 Comparison with other algorithms

The proposed algorithm is compared with other previously proposed algorithms with similar objective. Selected algorithms are as follows: A) Base band algorithm using 3D Wavelet Transform (Zhuang et al., 2004), B) The algorithm based on motion vector modification during MPEG coding (Zhao et al., 2003) and C) The watermarking algorithm based on the DCT domain, which performs during MPEG coding process (Zhang et al.,

2001). The table 3 shows the watermark robustness comparison, here symbol 'O' means that the embedded watermark is robust against the indicated attack, while the symbol 'X' means that the embedded watermark cannot be detected after the attack is applied. As shown in

Attacks	A	B	C	Proposed
MPEG	X	O	O	O
Frame average	O	O	X	O
Frame dropping	O	O	O	O
Frame swapping	O	O	O	O
Cropping	X	X	O	O
Impulsive noise	O	X	O	O
Gaussian noise	O	X	O	O

Table 3. Comparison among three algorithms reported in literature and the proposed one.

4. Conclusions

The proposed algorithm performs watermark embedding and detection during MPEG-2 coding process, in which firstly an adequate embedding energy is computed using the HVS based three criteria: sensibility of different color channels (red, green and blue channel), sensibility of spatial region with different features, such as plain, texture and edge regions, and finally sensibility of regions with different motion speed. Due to the lower sensibility of the HVS to blue channel, the watermark embedding is carried out only in blue channel. And using the latter two criteria, an adequate watermark embedding energy is assigned to each block of 8×8 DCT coefficients of I-frames and P-frames. In the proposed algorithm, B-frames are not used for watermarking in order to reduce watermark embedding and detection time. The proposed algorithm was evaluated from watermark imperceptibility and robustness points of view. The watermark imperceptibility was evaluated using the PSNR and Universal Quality Index (UQI). Both values show the watermark imperceptibility of the proposed algorithm, especially perceptual distortion evaluated using UQI shows that the watermark is imperceptible by the HVS. To evaluate the watermark robustness against some common attacks including video frame attacks such as: frame dropping, frame swapping and frame averaging. The simulation results show the watermark high robustness to above mentioned attacks. Also the proposed algorithm is compared with other algorithms with similar objective; the comparison results show that the proposed watermarking algorithm is more robust against a wider range of attacks than other watermarking algorithms.

The additional processing time to the MPEG-2 standard coding caused by watermarking is also measured. Since in the proposed algorithm, watermarking is carried out only in the I-frames and P-frames, the overall additional time is less than 10% of the MPEG-2 standard coding. Therefore we can conclude that the proposed watermarking algorithm is suitable for a real implementation.

5. References

Biswas S., S. R. Das and E. M. Petriu, An Adaptive Compressed MPEG-2 Video Watermarking Scheme, IEEE Trans. on Instrumentation and Measurement: 54(5), 1853-1861 (2005).

Hartung, F. and B. Girod, Watermarking of uncompressed and compressed video, Signal Processing: 66, 283-301 (1998)

Kong W., B. Yang, D. Wu and X. Niu, SVD Based Blind Video Watermarking Algorithm, IEEE Int. Conf. on Innovative Computing, Information and Control: (2006)

Langelaar, G. C. and R. L. Lagendijk, Optimal differential energy watermarking of DCT encoded images and video, IEEE Trans. on Image Processing: 10(1), 148-158 (2001).

Liu, Z., H. Liang, X. Niu and Y. Yang, A Robust Video Watermarking in Motion Vectors, IEEE Int. Conf. Signal Processing: 2358-2361 (2004).

Noorkami M. and R. M. Mersereau, Improving Perceptual Quality in Video Watermarking Using Motion Estimation, IEEE Int. Conf. on Image Processing (ICIP): 2, 520-523 (2006).

Piva, A., M. Barni, F. Bartolini and V. Cappellini, DCT-Based Watermark Recovering without Resorting to the Uncorrupted Original Image, IEEE Int. Conf. on Image Processing (ICIP): 1, 520-523 (1997).

Plataniotis, N. and A. N. Venetsanopoulos, Color Image Processing and Application, First Edition, Springer-Verlag, (2000).

Sayood K., Introduction to Data Compression, 2nd Edition, Morgan Kaufmann Publishers (2000).

Swanson, M.D., B. Zhu and A. H. Tewfik, Multi-resolution scene-based video watermarking using perceptual models, IEEE J. Select areas Communication: 16, 540-550 (1998).

Tong, H. Y. and A. N. Venetsanopoulos, A Perceptual model for JPEG applications based on block classification, texture masking and luminance masking, Int. Conf. on Image Processing (ICIP), 3, 428-432 (1998).

Ueno, Y. and A Digital Video watermarking method by association with the Motion Estimation, IEEE Int. Conf. on Signal Processing (ICSP), 2576-2579 (2004).

Wang, J., A. R. Steele and J. Liu, Efficient Integration of Watermarking with MPEG Compression, IEEE Int. Conf. on Multimedia and Expo (ICME), 911-914 (2004)

Wang, Z., Lu L. and Bovik A. C., Video quality assessment based on structural distortion measurement, Elsevier Signal Processing: Image Communication: 19, 121-132 (2004).

Wolfgang, R. B, C. I. Podilchuk and E. J. Delp, Perceptual watermarks for digital images and video, Proceeding IEEE: 87(7): 1108-1126 (1999).

Zhao, Z., N. Yu and X. Li, A Novel Video Watermarking Scheme in Compressed Domain Based of Fast Motion Estimation, IEEE Int. Conf. on Communication Technology (ICCT), 1878-1882 (2003)

Zhang, J., J. Li, and L. Zhang, Video Watermark Technique in Motion Vector, XIV Brazilian Symposium on Computer Graphics and Image Processing (2001).

Zhuang, H., Y. Li and C. Wu, A Blind Spatial-temporal Algorithm based on 3D Wavelet for Video Watermarking, IEEE int. Conf. on Multimedia and Expo (ICME), 3, 1727-1730 (2004).